HADRONIC PHYSICS

To learn more about the AIP Conference Proceedings, including the
Conference Proceedings Series, please visit the webpage
http://proceedings.aip.org/proceedings

HADRONIC PHYSICS

Joint Meeting Heidelberg-Liège-Paris-Rostock

HLPR 2004

Spa, Belgium   16 – 18 December 2004

EDITORS
Jean-Philippe Lansberg
Jean-René Cudell
Joseph Cugnon
University of Liège
Liège, Belgium

SPONSORING ORGANIZATIONS
University of Liège, Département de Physique de la Faculté des Sciences
Fonds National de la Recherche Scientifique de la Communauté Française de Belgique
Ministère de l'Enseignement supérieur, de l'Enseignement de promotion sociale et de la Recherche scientifique

Melville, New York, 2005
AIP CONFERENCE PROCEEDINGS ■ VOLUME 775

Editors:

Jean-Philippe Lansberg
Jean-René Cudell
Joseph Cugnon

University of Liège
Physics Department
allée du 6 Août, 17 Bât. B5
B-4000 Liège 1
Belgium

E-mail: jph.lansberg@ulg.ac.be
         jr.cudell@ulg.ac.be
         cugnon@plasma.theo.phys.ulg.ac.be

Authorization to photocopy items for internal or personal use, beyond the free copying permitted under the 1978 U.S. Copyright Law (see statement below), is granted by the American Institute of Physics for users registered with the Copyright Clearance Center (CCC) Transactional Reporting Service, provided that the base fee of $22.50 per copy is paid directly to CCC, 222 Rosewood Drive, Danvers, MA 01923, USA. For those organizations that have been granted a photocopy license by CCC, a separate system of payment has been arranged. The fee code for users of the Transactional Reporting Services is: 0-7354-0260-4/05/$22.50.

© 2005 American Institute of Physics

Permission is granted to quote from the AIP Conference Proceedings with the customary acknowledgment of the source. Republication of an article or portions thereof (e.g., extensive excerpts, figures, tables, etc.) in original form or in translation, as well as other types of reuse (e.g., in course packs) require formal permission from AIP and may be subject to fees. As a courtesy, the author of the original proceedings article should be informed of any request for republication/reuse. Permission may be obtained online using Rightslink. Locate the article online at http://proceedings.aip.org, then simply click on the Rightslink icon/"Permission for Reuse" link found in the article abstract. You may also address requests to: AIP Office of Rights and Permissions, Suite 1NO1, 2 Huntington Quadrangle, Melville, NY 11747-4502, USA; Fax: 516-576-2450; Tel.: 516-576-2268; E-mail: rights@aip.org.

L.C. Catalog Card No. 2005927493
ISBN 0-7354-0260-4
ISSN 0094-243X
Printed in the United States of America

CONTENTS

Preface vii
Committees ix

PENTAQUARKS AND SPECTROSCOPY

Mixing and Decays of the Antidecuplet in Context of Approximate
SU(3) Symmetry 3
    V. Guzey
Ambiguities in the Calculation of Leptonic Decays of Excited
Heavy Quarkonium 11
    J. P. Lansberg
Highly Excited Baryons in Large-N_c QCD 22
    N. Matagne and F. Stancu
Dynamics of Pentaquarks in Constituent Quark Models:
Recent Developments 32
    F. Stancu
Review of Experimental Aspects of Pentaquark Physics 41
    I. I. Strakovsky, R. A. Arndt, Y. I. Azimov, M. V. Polyakov, and
    R. L. Workman

DIAGONAL AND OFF-DIAGONAL STRUCTURE

Probing the Partonic Structure of Exotic Particles in Hard
Electroproduction 51
    I. V. Anikin, B. Pire, L. Szymanowski, O. V. Teryaev, and S. Wallon
A Toy Model for Generalised Parton Distributions 61
    J. R. Cudell, F. Bissey, J. Cugnon, M. Jaminon, J. P. Lansberg, and
    P. Stassart
Deep Inelastic Lepton Nucleus Scattering and Hadronization at
HERMES Energies 71
    D. Grünewald
Gluon Transversity in the Hard Exclusive Reactions 81
    N. Kivel
New Insight on Global QCD Fits Using Regge Theory 88
    G. Soyez

DIFFRACTION

Fluctuating Pulled Fronts & Pomerons 99
    E. Iancu
Odderon, Redone 111
    K. Itakura

Hard Diffractive Production of Vector Mesons 120
    R. Kirschner

HIGH TEMPERATURE AND DENSITY

Heavy-Quarkonium Interaction and the Temperature Dependence of
the QCD Static Potential 133
    F. Arleo
Light Front Approach to Hot and Dense Quark Matter 139
    M. Beyer, S. Mattiello, S. Strauß, T. Frederico, and H. J. Weber
Hadronic Resonances above the QCD Phase Transition 151
    D. Blaschke
RG Flow of the Polyakov-Loop Potential—First Status Report 162
    J. Braun, H. Gies, and H.-J. Pirner
Restoration of $U_A(1)$ Symmetry and Meson Spectrum in Hot
or Dense Matter 173
    P. Costa, M. C. Ruivo, C. A. de Sousa, and Y. L. Kalinovsky
Cooling Evolution of Hybrid Stars 182
    H. Grigorian
Volume Dependence of the Pion Mass from Renormalization
Group Flows 193
    B. Klein, J. Braun, and H. J. Pirner
Statistical Physics on the Light-Front 202
    J. Raufeisen
Light Front NJL Model at Finite Temperature 212
    S. Strauß, M. Beyer, and S. Mattiello
Random Matrix Models for Chiral and Diquark Condensation 220
    B. Vanderheyden and A. D. Jackson

FUNDAMENTAL QUESTIONS

The Future of Nuclear Energy 235
    J. Cugnon
How to Become a Nobel Laureate 245
    J. Hüfner
QCD—NLC 252
    H. J. Pirner
Confinement of Quarks in Dual Superconductor Models 262
    G. Ripka

List of Participants 269
Author Index 273

Preface

The *Three days of Hadronic Physics* conference was held from the 16th to the 18th of December 2004 in Spa, a small city not far from Liège, renowned for its spring waters. It was organised by the Fundamental Theoretical Physics Group of the University of Liège. This conference is the eighth of a series, which emanated from weekly seminars on the quark-gluon plasma, organised in Orsay by Dominique Vautherin during the academic year 1987-1988, and which was animated by a few physicists from Orsay and Saclay and by visiting scientists from the Universities of Heidelberg and Liège. At the end of the academic year, the participants decided to continue this scientific collaboration in some way. They came to the idea of setting more or less regular meetings of physicists from the theory groups of the Universities of Orsay, Liège and Heidelberg and from the theory division of the Centre d'Etudes Nucléaires de Saclay. Orsay and Saclay being (almost) at the same geographical place, these meetings were named "triangular". A few years ago, the organisers have decided to open progressively the audience to other institutions and the adjective "triangular" is no longer valid, though it has been kept up to now.

Similarly the topics discussed at the successive conferences, first closely related to the quark-gluon plasma and heavy-ion collisions, have considerably widened, as can be seen from these proceedings. They cover many aspects of Quantum Chromodynamics and of hadronic interactions in general. Yet, they are closely connected with the main research interests of the respective laboratories. The organisers of the meetings have tried to preserve the informal and friendly atmosphere of the initial seminars, while keeping high standard discussions.

In Spa, more than thirty speakers addressed a whole variety of subjects pertaining to low-energy QCD dynamics (chiral symmetry breaking, hadron spectroscopy, including pentaquarks and exotics), to QCD at high temperature and high density, to parton distributions and diffractive collisions, and to QCD on the light cone. Cross-links between these topics appeared in the discussions, witnessing the progress toward a more coherent view of hadronic interactions. Inclusion of side topics is also a tradition of these conferences. This time, the organisers set a presentation on the sociological aspects of the Noble prizes in physics and a discussion evening on nuclear energy.

On behalf of the Organising Committee, we want to express our gratitude to the members of the Scientific Advisory Committee for their valuable suggestions in the preparation of the conference program. Our thanks also go to the speakers for the quality of their presentations, to all the participants, who contributed to make this conference very lively, and to most of the contributors for their efforts to complete their manuscripts on time.

We are especially grateful to the *Fonds National de la Recherche Scientifique de la Communauté Française de Belgique* and the *Ministère de l'Enseignement supérieur, de l'Enseignement de promotion sociale et de la Recherche scientifique* for their financial support. We also benefitted from a funding from the *Patrimoine* and from the *Département de Physique de la Faculté des Sciences* of the University of Liège.

We are thankful to the Delta Lloyd bank and to the personnel of the Sol-Cress organisation for their help in the material organisation of the conference. We would also like to express our gratefulness to L. Szymanowski and V.L. Yudichev for the photographs appearing in these proceedings.

We hope that this meeting has propagated the spirit of the participants of the original seminar in Orsay, who wanted to discuss up-to-date topics in hadronic physics in a pleasant atmosphere, favouring scientific and human contacts among physicists of different countries.

The Editors
March 2005

Scientific Advisory Committee

Jean-Paul BLAIZOT, ECT* Trento
David BLASCHKE, Universität Rostock
Joseph CUGNON, Université de Liège
Yura L. KALINOVSKY, JINR Dubna
Hans-Jürgen PIRNER, Universität Heidelberg
Maxim POLYAKOV, Université de Liège
Cristina VOLPE, Université Paris-Sud, Orsay

Organising Committee

Jean-René CUDELL, Université de Liège
Joseph CUGNON, Université de Liège
Jean-Marc GERARD, Université Catholique de Louvain
Jean-Philippe LANSBERG, Université de Liège

PENTAQUARKS AND SPECTROSCOPY

Mixing and decays of the antidecuplet in context of approximate SU(3) symmetry

V. Guzey

Institut für Theoretische Physik II, Ruhr-Universität Bochum, D-44780 Bochum, Germany
E-mail: vadimg@tp2.rub.de

Abstract. We consider mixing of the antidecuplet with three $J^P = 1/2^+$ octets (the ground-state octet, the octet containing $N(1440)$, $\Lambda(1600)$, $\Sigma(1660)$ and $\Xi(1690)$ and the octet containing $N(1710)$, $\Lambda(1800)$, $\Sigma(1880)$ and $\Xi(1950)$) in the framework of approximate flavor SU(3) symmetry and in the limit of small mixing angles. We give general expressions for the partial decay widths of all members of the antidecuplet as functions of the two mixing angles and Γ_{Θ^+} and $\Sigma_{\pi N}$. Identifying $N_{\overline{10}}$ with the $N(1670)$ observed by the GRAAL experiment, we show that the considered mixing scenario can accommodate all present experimental and phenomenological information on the Θ^+ and $N_{\overline{10}}$ decays: Θ^+ could be as narrow as 1 MeV; the $N_{\overline{10}} \to N\eta$ decay is sizable, while the $N_{\overline{10}} \to N\pi$ decay is suppressed and the $N_{\overline{10}} \to \Lambda K$ decay is possibly suppressed. Constraining the mixing angles by the $N_{\overline{10}}$ decays, we make definite predictions for the $\Sigma_{\overline{10}}$ decays. We point out that $\Sigma_{\overline{10}}$ with mass near 1770 MeV could be searched for in the available data on $K_S p$ invariant mass spectrum, which already revealed the Θ^+ peak. It is important to experimentally verify the decay properties of $\Sigma(1770)$ because its mass and $J^P = 1/2^+$ make it an attractive candidate for $\Sigma_{\overline{10}}$.

Keywords: Exotic baryon spectroscopy
PACS: 11.30.Hv, 14.20.Jn

1. INTRODUCTION

Approximate flavor SU(3) symmetry of strong interactions allows to group all hadrons into certain multiplets. Only singlets, octets (8) and decuplets (10) have been realized in Nature. The discoveries of the Θ^+ and Ξ^{--}, if confirmed, mean the existence of a new physical multiplet: the antidecuplet ($\overline{10}$) [1].

Approximate SU(3) symmetry works surprisingly well: Mass splittings and partial decay widths of all baryon multiplets (singlets, octets, decuplets) are described and predicted with good accuracy [2, 3, 4, 5].

Approximate SU(3) symmetry suits best for establishing in a model-independent way the overall structure of a given SU(3) multiplet: Mass splittings, necessity of mixing with other multiplets due to SU(3) breaking, correlations between partial decay widths. This is exactly what one needs for the antidecuplet: a reliable overall picture of $\overline{10}$ and its mixing with other multiplets and a way to sort out the present experimental information on the $\overline{10}$ decays.

Since SU(3) symmetry is broken, states from different multiplets with the same spin and parity can mix. Because of the small width of Θ^+ [1], even small mixing with non-exotic multiplets dramatically affects predictions for the $\overline{10}$ decays. At the same time, small mixing with $\overline{10}$ affects very little non-exotic multiplets. This means that one can use the results of SU(3) analysis of the non-exotic multiplets (three octets in our case) in the SU(3) analysis of $\overline{10}$ decays.

After the SU(3) picture of $\overline{10}$ is established using the scarce experimental info on $\overline{10}$ decays, one can make model-independent predictions for unmeasured decays and assess available models of the $\overline{10}$ mixing.

2. ANTIDECUPLET MIXING WITH THREE OCTETS

We consider the scenario that $\overline{10}$ mixes with three $J^P = 1/2^+$ octets: the ground-state octet, the octet containing $N(1440)$ $\Lambda(1600)$, $\Sigma(1660)$ and $\Xi(1690)$, and the octet containing $N(1710)$, $\Lambda(1800)$, $\Sigma(1880)$ and $\Xi(1950)$. The mixing involves the $N_{\overline{10}}$ and $\Sigma_{\overline{10}}$ and the corresponding N and Σ octet states:

$$\begin{pmatrix} |N_1^{\text{phys}}\rangle \\ |N_2^{\text{phys}}\rangle \\ |N_3^{\text{phys}}\rangle \\ |N_{\overline{10}}^{\text{phys}}\rangle \end{pmatrix} = \begin{pmatrix} 1 & 0 & 0 & \sin\theta_1 \\ 0 & 1 & 0 & \sin\theta_2 \\ 0 & 0 & 1 & \sin\theta_3 \\ -\sin\theta_1 & -\sin\theta_2 & -\sin\theta_3 & 1 \end{pmatrix} \begin{pmatrix} |N_1\rangle \\ |N_2\rangle \\ |N_3\rangle \\ |N_{\overline{10}}\rangle \end{pmatrix}. \quad (1)$$

We assume that the θ_i mixing angles are small, $\theta_i = \mathcal{O}(\epsilon)$, where ϵ is a small parameter of SU(3) breaking. We systematically neglect $\mathcal{O}(\epsilon^2)$ terms.

It is important to note that the $|N_1\rangle$, $|N_2\rangle$ and $|N_3\rangle$ states can mix among themselves, i.e. they can belong to several different octets. Using the χ^2 fit to the measured decays, we find that $|N_2\rangle$ and $|N_3\rangle$ states are slightly mixed and that the $|N_1\rangle$ state is unmixed [6]. After this mixing is taken into account, it is sufficient to consider only the mixing of each individual $|N_i^{\text{phys}}\rangle$ with $|N_{\overline{10}}^{\text{phys}}\rangle$.

The mixing angles θ_i and θ_i^Σ are related [7],

$$\sin\theta_i \left(N_i^{\text{phys}} - N_{\overline{10}}^{\text{phys}} \right) = \sin\theta_i^\Sigma \left(\Sigma_i^{\text{phys}} - \Sigma_{\overline{10}}^{\text{phys}} \right), \quad (2)$$

which becomes $\theta_i = \theta_i^\Sigma$ ignoring the $\mathcal{O}(\epsilon^2)$ terms.

The Gell-Mann–Okubo mass formulas, which describe the mass splitting inside SU(3) multiplets, are not sensitive to small mixing

$$N_i^{\text{phys}} \equiv \langle N_i^{\text{phys}} | \hat{M} | N_i^{\text{phys}} \rangle = N_i + \sin^2\theta_i N_{\overline{10}} = N_i + \mathcal{O}(\epsilon^2). \quad (3)$$

Hence, it is not legitimate to estimate the mixing angles from the Gell-Mann–Okubo mass formula. Instead, one has to consider decays which contain $\mathcal{O}(\epsilon)$ terms.

In this work, we assume that SU(3) symmetry is violated by non-equal masses inside a given multiplet and mixing among multiplets and that SU(3) is exact in decay vertices. This enables us to express the antidecuplet partial decay widths in terms of a finite number of universal SU(3) coupling constants.

In our analysis, the total width of the Θ^+, Γ_{Θ^+}, and the pion-nucleon sigma term, $\Sigma_{\pi N}$, are external parameters, which are varied in the following intervals: $1 \leq \Gamma_{\Theta^+} \leq 5$ MeV; $45 \leq \Sigma_{\pi N} \leq 75$ MeV. $\Sigma_{\pi N}$ determines the θ_1 mixing angle with the ground state octet [8]; Γ_{Θ^+} determines the $G_{\overline{10}}$ and $H_{\overline{10}}$ ($H_{\overline{10}} = 2G_{\overline{10}} - 18$) coupling constants [1, 9] though the $\Theta^+ NK$ coupling constant

$$g_{\Theta^+ NK} = \frac{1}{\sqrt{5}}\left(G_{\overline{10}} + \sin\theta_1 H_{\overline{10}} \frac{\sqrt{5}}{4}\right). \tag{4}$$

Therefore, at given Γ_{Θ^+} and $\Sigma_{\pi N}$, the $\overline{10}$ decays can be parameterized in terms of the θ_2 and θ_3 mixing angles, which are treated as free parameters.

3. DECAYS OF $N_{\overline{10}}$

General expressions for the $N_{\overline{10}}$ coupling constants have the following form

$$g_{N_{\overline{10}} N\pi} = \frac{1}{2\sqrt{5}}\left(G_{\overline{10}} + \sin\theta_1\left(H_{\overline{10}}\frac{\sqrt{5}}{4} - G_8\frac{7}{\sqrt{5}}\right) - \sum_{i=2,3}\sin\theta_i\, g_{N_i N\pi}\right),$$

$$g_{N_{\overline{10}} N\eta} = \frac{1}{2\sqrt{5}}\left(-G_{\overline{10}} + \sin\theta_1\left(H_{\overline{10}}\frac{\sqrt{5}}{4} - G_8\frac{1}{\sqrt{5}}\right) + \sum_{i=2,3}\sin\theta_i\, g_{N_i N\eta}\right),$$

$$g_{N_{\overline{10}} \Lambda K} = \frac{1}{2\sqrt{5}}\left(G_{\overline{10}} + \sin\theta_1 G_8 \frac{4}{\sqrt{5}} + \sum_{i=2,3}\sin\theta_i\, g_{N_i \Lambda K}\right),$$

$$g_{N_{\overline{10}} \Delta\pi} = \frac{2}{\sqrt{5}}\left(\sin\theta_1 G_8 + \sum_{i=2,3}\sin\theta_i\, g_{N_i \Delta\pi}\right). \tag{5}$$

The $g_{N_i BP}$ coupling constants are determined by a χ^2 fit to the measured decays of the octets [6]. Numerically $g_{N_i BP} > G_{\overline{10}}$, which justifies our conjecture that the mixing with the antidecuplet can be neglected in the decays of the involved non-exotic octets.

Note that the $N_{\overline{10}} \to \Delta\pi$ decay is possible only due to the mixing.

There exists an important correlation between the coupling constants in Eq. (5), which we extensively used in our analysis: Because of the minus sign in front of $G_{\overline{10}}$ in the expression for $g_{N_{\overline{10}} N\eta}$, mixing with the octets can decrease $g_{N_{\overline{10}} N\pi}$ and simultaneously increase $g_{N_{\overline{10}} N\eta}$.

The partial $B_1 \to B_2 + P$ decay widths are found from

$$\Gamma(B_1 \to B_2 + P) = 3|g_{B_1 B_2 P}|^2 \frac{|\vec{p}|^3}{2\pi(M_1 + M_2)^2} \frac{M_2}{M_1}, \tag{6}$$

where B_1 denotes the initial baryon with mass M_1; B_2 denotes the final baryon with mass M_2; P denotes the pseudoscalar meson; p is the momentum of the particles in the final state in the center of mass system.

While approximate SU(3) symmetry allows to express the $N_{\overline{10}}$ partial decay widths in terms of a small number of free parameters (SU(3) coupling constants and mixing angles), in order to make quantitative predictions one has to constrain

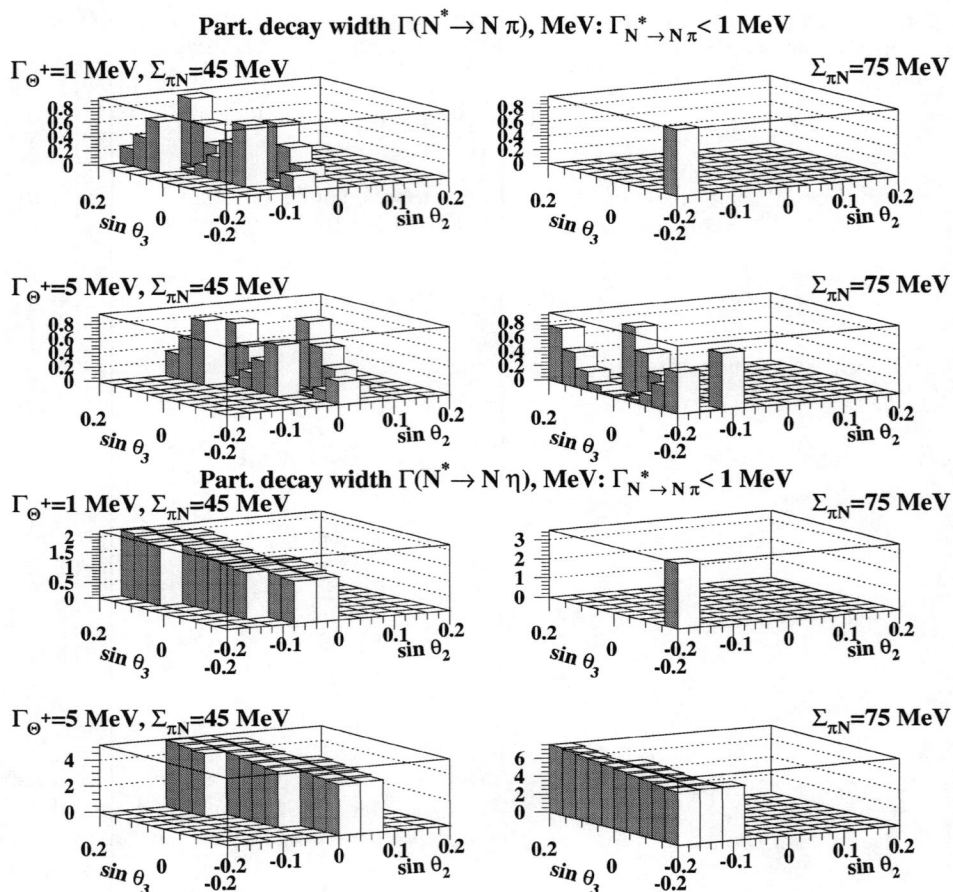

FIGURE 1. $\Gamma_{N_{\overline{10}} \to N\pi}$ and $\Gamma_{N_{\overline{10}} \to N\eta}$ as functions of θ_2 and θ_3 at given Γ_{Θ^+} and $\Sigma_{\pi N}$. The decay widths are shown only where $\Gamma_{N_{\overline{10}} \to N\pi} \leq 1$ MeV.

the free parameters. This can be achieved by using the available experimental and phenomenological information on the decays of those N^* states, which can serve as candidates for $N_{\overline{10}}$.

First, the PWA analysis of [9] gives two candidate N^* states with masses 1680 MeV and 1730 MeV. Both states should have $\Gamma_{N_{\overline{10}} \to N\pi} \leq 0.5$ MeV.

Second, the GRAAL experiment [10] observes a narrow nucleon resonance near 1670 MeV in the reaction $\gamma n \to n\eta$, while no such resonant structure is visible in the simultaneously measured $\gamma p \to p\eta$ reaction. This can be interpreted as that $\Gamma_{N_{\overline{10}} \to N\eta}$ should not be too small. The much stronger photocoupling to the neutron than to the proton is a benchmark property of the $N_{\overline{10}}$ state, as predicted in the chiral quark soliton model [11].

Third, the STAR experiment observes a narrow peak at 1734 MeV and only a weak indication of a narrow peak at 1693 MeV in the ΛK_S invariant mass [12]. A possible interpretation of this result is that the $\Gamma_{N_{\overline{10}} \to \Lambda K}$ partial decay width is possibly suppressed.

Identifying $N_{\overline{10}}$ with the GRAAL's $N(1670)$, we find that the above discussed regularities in the $N_{\overline{10}}$ decays can be realized by a suitable choice of θ_i. In particular, we impose the $\Gamma_{N_{\overline{10}} \to N\pi} \leq 1$ MeV cut and find unsuppressed $\Gamma_{N_{\overline{10}} \to N\eta}$ and somewhat suppressed $\Gamma_{N_{\overline{10}} \to \Lambda K}$. Note that the $N_{\overline{10}} \to \Lambda K$ decay is anyway suppressed by the phase space factor. An example of such an analysis for the $\Gamma_{N_{\overline{10}} \to N\pi}$ and $\Gamma_{N_{\overline{10}} \to N\eta}$ is illustrated in Fig. 1.

In summary, identifying $N_{\overline{10}}$ with the $N(1670)$ observed by the GRAAL experiment [10], the following picture of the $N_{\overline{10}}$ decays emerges: the $N_{\overline{10}} \to N\eta$ decay is sizable, while the $N_{\overline{10}} \to N\pi$ decay is suppressed and the $N_{\overline{10}} \to \Lambda K$ decay is possibly suppressed.

4. DECAYS OF $\Sigma_{\overline{10}}$

The general expressions for the (selected) $\Sigma_{\overline{10}}$ coupling constants read

$$g_{\Sigma_{\overline{10}} \Lambda \pi} = \frac{1}{2\sqrt{5}} \left(G_{\overline{10}} - \sin\theta_1^\Sigma G_8 \frac{3}{\sqrt{5}} - \sum_{i=2,3} \sin\theta_i^\Sigma g_{\Sigma_i \Lambda \pi} \right),$$

$$g_{\Sigma_{\overline{10}} \Sigma \pi} = \frac{1}{\sqrt{30}} \left(G_{\overline{10}} + \sin\theta_1 \left(H_{\overline{10}} \frac{\sqrt{5}}{2} - G_8 \sqrt{5} \right) - \sum_{i=2,3} \sin\theta_i^\Sigma g_{\Sigma_i \Sigma \pi} \right),$$

$$g_{\Sigma_{\overline{10}} N \overline{K}} = \frac{1}{\sqrt{30}} \left(-G_{\overline{10}} + \sin\theta_1 H_{\overline{10}} \frac{\sqrt{5}}{2} + \sin\theta_1^\Sigma G_8 \frac{4}{\sqrt{20}} + \sum_{i=2,3} \sin\theta_i^\Sigma g_{\Sigma_i N \overline{K}} \right),$$

$$g_{\Sigma_{\overline{10}} \Sigma_{10} \pi} = \frac{\sqrt{30}}{15} \left(G_8 \sin\theta_1 + \sum_{i=2,3} \sin\theta_i^\Sigma g_{\Sigma_i \Sigma_{10} \pi} \right). \quad (7)$$

Unlike the $N_{\overline{10}}$ case, there are no distinct correlations among the partial decay widths of $\Sigma_{\overline{10}}$ when $\theta_{2,3}$ are free. However, imposing the $\Gamma_{N_{\overline{10}} \to N\pi} \leq 1$ MeV cut, certain correlations for most of the $\theta_{2,3}$ range appear, such as e.g. $\Gamma_{\Sigma_{\overline{10}} N \overline{K}} > \Gamma_{\Sigma_{\overline{10}} \Lambda \pi}, \Gamma_{\Sigma_{\overline{10}} \Sigma \pi}$. An example of such an analysis is shown in Fig. 2.

$\Sigma_{\overline{10}}$ is the least known member of $\overline{10}$. In our numerical analysis, we use $m_{\Sigma_{\overline{10}}} = 1765$ MeV ($\Sigma_{\overline{10}}$ is equally spaced between the $N(1670)$ and $\Xi^{--}(1862)$).

Among the experiments reporting the Θ^+ signal, there were four experiments, where the Θ^+ was observed as a peak in the pK_S invariant mass. Since $\Sigma_{\overline{10}}$ decays in the same final state ($N\overline{K}$), the four experiments give direct information on the $\Sigma_{\overline{10}} \to N\overline{K}$ decay!. The analysis of [13] reveals a number of peaks in the

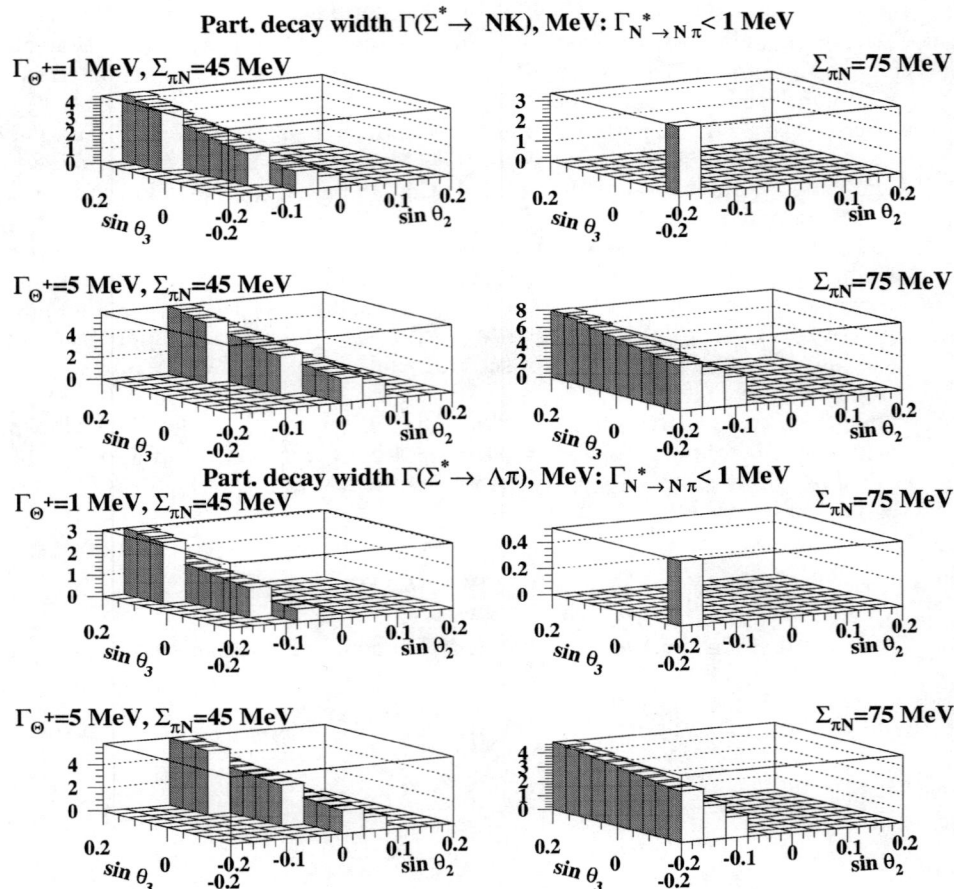

FIGURE 2. $\Gamma_{\Sigma_{\overline{10}} \to N\overline{K}}$ and $\Gamma_{\Sigma_{\overline{10}} \to \Lambda\pi}$ as functions of θ_2 and θ_3 at given Γ_{Θ^+} and $\Sigma_{\pi N}$. The decay widths are shown only where $\Gamma_{N_{\overline{10}} \to N\pi} \leq 1$ MeV.

$1650 < M_{pK_S} < 1850$ MeV mass range. The analysis of [14] contains at least two prominent peaks in the 1700-1800 MeV mass range, before the cuts to enhance the Θ^+ signal are imposed. The HERMES [15] and ZEUS [16] pK_S invariant mass spectra extend only up to 1.7 MeV. Therefore, a possibly narrow $\Sigma_{\overline{10}}(1765)$ should be searched for in the available data!

It is curious that the Particle Data Group baryon listing contains a candidate

$\Sigma_{\overline{10}}$ state, $\Sigma(1770)$, with the required $J^P = 1/2^+$ but only one-star rating [17]. For $\Sigma(1770)$, the $N\overline{K}$ branching ratio is larger than the branching ratios in the $\Lambda\pi$ and $\Sigma\pi$ final states, which can be realized in our picture of the $\Sigma_{\overline{10}}$ decays, see Fig. 2. In addition, our comprehensive SU(3) analysis of baryon multiplets [6] fails to convincingly assign $\Sigma(1770)$ to any non-exotic multiplet, and, hence, the antidecuplet seems like a very attractive place for $\Sigma(1770)$.

5. DISCUSSION AND CONCLUSIONS

In order to estimate the theoretical uncertainty of our predictions, we add an additional mixing of the antidecuplet with a 27-plet [8]. We arrive at two scenarios. In the first case, the qualitative picture of the $N_{\overline{10}}$ decays does not change. The correlations between the $\Sigma_{\overline{10}}$ partial decay widths do change, which makes it impossible to identify $\Sigma_{\overline{10}}$ with $\Sigma(1770)$.

In the second case, the picture of $N_{\overline{10}}$ decays becomes only marginally compatible with experiment. The pattern of the $\Sigma_{\overline{10}}$ decays is similar to $\Sigma(1770)$. In this scenario, both $N_{\overline{10}}$ and $\Sigma_{\overline{10}}$ states are very narrow.

We find that the scenario of large (ideal) mixing [18] is inconsistent with the experimental data on the $N(1440)$ and $N(1710)$ decays [17]. Identifying $N_{\overline{10}}$ with $N(1710)$, which is nearly ideally mixed with the Roper $N(1440)$, the χ^2 fit fails to simultaneously accommodate $\Gamma_{\Theta^+} \leq 10$ MeV and a large $\Gamma_{N(1440) \to N\pi}$. An acceptably low χ^2 can only be found by assuming very small mixing, $|\theta_N| \approx 4^0$.

In conclusion, we considered mixing of the antidecuplet with three $J^P = 1/2^+$ octets in the framework of approximate flavor SU(3) symmetry and in the limit of small mixing angles. We presented general expressions for the $\overline{10}$ partial decay widths as functions of the three mixing angles and Γ_{Θ^+}.

Identifying $N_{\overline{10}}$ with the $N(1670)$ observed by GRAAL [10], we arrive at the following picture of $\overline{10}$ decays: Θ^+ could be as narrow as 1 MeV; the $N_{\overline{10}} \to N\eta$ decay is sizable, while the $N_{\overline{10}} \to N\pi$ decay is suppressed and the $N_{\overline{10}} \to \Lambda K$ decay is possibly suppressed.

Constraining the mixing angles by the $N_{\overline{10}}$ decays, we make predictions for the $\Sigma_{\overline{10}}$ decays. We point out that $\Sigma_{\overline{10}}(1765)$ could be searched for in the available data on $K_S p$ invariant mass spectrum, which already revealed the Θ^+ peak.

Also, it is important to experimentally verify the decay properties of $\Sigma(1770)$ because its mass and $J^P = 1/2^+$ make it an attractive candidate for $\Sigma_{\overline{10}}$.

REFERENCES

1. D. Diakonov, V. Petrov and M. Polyakov, Z. Phys. A **359**, 305 (1997).
2. M. Gell-Mann and Y. Ne'eman, *The Eightfold Way*, W.A. Benjamin, Inc., Amsterdam and New York, 1964.
3. J.J.J. Kokkedee, *The quark model*, W.A. Benjamin, Inc., Amsterdam and New York, 1969.
4. F.E. Close, *An introduction to quarks and partons*, Academic Press, London, 1979.
5. N.P. Samios, M. Goldberg, and B.T. Meadows, Rev. Mod. Phys. **46**, 49 (1974).
6. V. Guzey and M. Polyakov, in preparation.
7. D. Diakonov and V. Petrov, Phys. Rev. D **69**, 094011 (2004) [hep-ph/310212].

8. J. Ellis, M. Karliner, M. Praszalowicz, JHEP 0405 (2004) 002 [hep-ph/0401127]; M. Praszalowicz, Acta Phys. Polon. B **35** (2004) 1625 [hep-ph/0402038].
9. R.A. Arndt, Ya.I. Azimov, M.V. Polyakov, I.I. Strakovsky, and R.L. Workman, Phys. Rev. C **69**, 035208 (2004) [nucl-th/0312126].
10. V. Kuznetsov for the GRAAL Collab., hep-ex/0409032.
11. M.V. Polyakov and A. Rathke, Eur. Phys. J. A **18**, 691 (2003) [hep-ph/0303138].
12. S. Kabana for the STAR Collab., hep-ex/0406032.
13. A.E. Asratyan, A.G. Dolgolenko and M.A. Kubantsev, Yad. Fiz. **67**, 704 (2004) [Phys. At. Nucl. **67**, 682 (2004)] [hep-ex/0309042].
14. SVD Collab., A. Aleev et al., hep-ex/0401024, submitted to Yad. Fiz.
15. HERMES Collab., A. Airapetian et al., Phys. Lett. B **585**, 213 (2004) [hep-ex/0312044].
16. ZEUS Collab., S. Chekanov et al., Phys. Lett. B **591**, 7 (2004) [hep-ex/0403051].
17. The Review of Particle Physics 2004, Particle Data Group, S. Eidelman et al. Phys. Lett. B. **592** (2004) 1.
18. R. Jaffe and F. Wilczek, Phys. Rev. Lett. **91**, 232003 (2003) [hep-ph/0307341].

Ambiguities in the calculation of leptonic decays of excited heavy quarkonium

J.P. Lansberg

*Physique théorique fondamentale, Département de Physique, Université de Liège,
allée du 6 Août 17, bât. B5, B-4000 Liège 1, Belgium
E-mail: JPH.Lansberg@ulg.ac.be*

Abstract. We point out that the determination of the leptonic decay width of radially-excited quarkonia is strongly dependent on the position of the node typical of these excitations. We suggest that this feature could be related with the longstanding $\rho - \pi$ puzzle.

Keywords: Decays of quarkonia, Excited states, rho-pi puzzle
PACS: 14.40.Gx, 13.20.Gd, 13.25.Gv, 11.10.St

1. INTRODUCTION

To study processes involving heavy quarkonia, such as decay and production mechanisms, in field theory, all the information needed can be parameterised by vertex functions, which describe the coupling of the bound state to its constituents and contain the information about the size of the bound state, the amplitude of probability for given quark configurations and the normalisation of the bound-state wave functions.

In a parallel work [1, 2], we have have considered production processes of J/ψ, ψ' and Υ at hadron colliders. We relied on a phenomenological approach with vertex functions, and after restoration of gauge invariance, we have reached an interesting agreement with production cross sections and polarisation measurements by CDF [3, 4, 5, 6] and PHENIX [7].

In the case of J/ψ and $\Upsilon(1S)$, we have used the leptonic decay width to fix the normalisation of these vertex functions. This procedure has appeared as robust and reliable for 1S states [1] and is explained in the following. However in the case of radially excited states, an important feature has emerged from calculations: the decay width, and thus the normalisation, are strongly dependent on the position of the node appearing in the vertex function. This ambiguity in the determination of the decay width might also exist for hadronic decays, such as $\psi' \to \rho\pi$, and is perhaps closely related to the $\rho - \pi$ puzzle.

2. OUR PHENOMENOLOGICAL APPROACH

Our approach to build the quarkonium vertex functions does not rely on the Bethe-Salpeter equation [8] (BSE), usually used to constrain the properties of vertex functions; for example, to relate the mass of the bound state to the mass of its constituent quarks. Indeed, besides problems with gauge invariance, all predictions coming from BSE are realised within Euclidean space (see *e.g.* [9, 10, 11]) and we have shown that the continuation to Minkowski space can be problematic [1].

We have therefore chosen to describe the quarkonia by a phenomenological vertex function and decided to remain in Minkowski space, at variance with other phenomenological models (see *e.g.* [12, 13]). The price to pay for not using BSE is an additional uncertainty due to the functional dependence of the vertex. Fortunately, it can be cancelled if one fixes the normalisation from the study of the leptonic decay width, which is our main concern here.

2.1. Choice for the vertex function

It has been shown elsewhere (see *e.g.* [10]) that other Dirac structures than γ^μ for vector bound state are suppressed by one order of magnitude in the case of light mesons, and that this suppression is increased for the ϕ meson. Therefore their effect is expected to be even more negligible in heavy quarkonia.

As a consequence, the following phenomenological Ansatz for the vector meson vertex, inspired by spin-projection operators, is likely to be sufficient for our purposes. It reads:

$$V_\mu(p,P) = \Gamma(p,P)\gamma_\mu, \qquad (1)$$

with P the total momentum of the bound state, $P = p_1 - p_2$, and p the relative one, $p = (p_1 + p_2)/2$ as drawn on Fig. 1. This Ansatz amounts to multiplying the *point* vertex (corresponding to a structureless particle) by a function, $\Gamma(p,P)$.

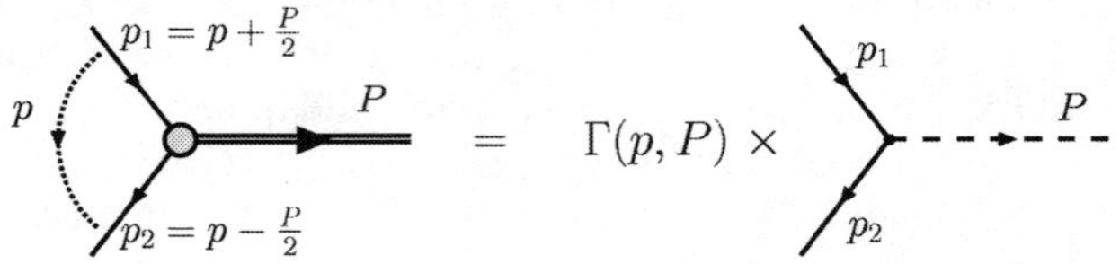

FIGURE 1. Phenomenological vertex obtained by multiplying a *point vertex*, representing a structureless particle, by a vertex function (or form factors).

The function Γ, which is called the 3-point vertex function, can be chosen in different ways. Two simple Ansätze are commonly used in the context of BSE. We consider both. They correspond to two extreme choices at large distances: a dipolar form which decreases gently with its argument, and a gaussian form:

$$\Gamma_0(p,P) = \frac{N}{(1-\frac{p^2}{\Lambda^2})^2} \text{ and } \Gamma_0(p,P) = N e^{\frac{p^2}{\Lambda^2}}, \qquad (2)$$

both with a free size parameter Λ. As $\Gamma(p,P)$ in principle depends on p and P, we may shift the variable and use $p^2 - \frac{(p.P)^2}{M^2}$ instead of p^2 which has the advantage of reducing to $-|\vec{p}|^2$ in the rest frame of the bound state. We refer to it as the *vertex function with shifted argument*. In the latter cases, we have –in that frame–,

$$\Gamma(p,P) = \frac{N}{(1+\frac{|\vec{p}|^2}{\Lambda^2})^2} \text{ and } \Gamma(p,P) = Ne^{\frac{-|\vec{p}|^2}{\Lambda^2}}. \tag{3}$$

2.2. Excited states

As is well-known, the number of nodes in the wave function, in whatever space, increases with the principal quantum number n. This simple feature can be used to differentiate between $1S$ and $2S$ states.

We thus have simply to determine the position of the node of the wave function in momentum space. To what concerns the vertex function, working in the meson rest frame, the node comes through a prefactor, $1 - \frac{|\vec{p}|}{a_{node}}$, which multiplies the vertex function for the $1S$ state. Explicitly, $\Gamma_{2S}(p,P)$, for a node a_{node}, reads

$$N'\left(1 - \frac{|\vec{p}|}{a_{node}}\right)\frac{1}{(1+\frac{|\vec{p}|^2}{\Lambda^2})^2} \text{ and } N'\left(1 - \frac{|\vec{p}|}{a_{node}}\right)e^{\frac{-|\vec{p}|^2}{\Lambda^2}}. \tag{4}$$

In order to determine the node position in momentum space, we can use two methods. The first is to fix a_{node} from its known value in position space, e.g. from potential studies, and to Fourier-transform the vertex function. In the case of a gaussian form, this can be carried out analytically.

The second method is to impose the following relation[1] between the $1S$ and $2S$ vertex functions:

$$\int |\vec{p}|^2 d|\vec{p}| e^{\frac{-|\vec{p}|^2}{\Lambda^2}} \left(1 - \frac{|\vec{p}|}{a_{node}}\right) e^{\frac{-|\vec{p}|^2}{\Lambda^2}} = 0. \tag{5}$$

The two methods give compatible results.

3. NORMALISING: THE LEPTONIC DECAY WIDTH

The width in terms of the decay amplitude $\mathcal{M}$, is given by

$$\Gamma_{\ell\ell} = \frac{1}{2M}\frac{1}{(4\pi^2)}\int |\bar{\mathcal{M}}|^2 d_2(PS), \tag{6}$$

where $d_2(PS)$ is the two-particle phase space [14].

The amplitude is obtained as usual through Feynman rules, for which we use our vertex function at the meson-quark-antiquark vertex. At leading order, the square of the amplitude is obtained from the cut-diagram drawn in Fig. 2.

---

[1] inspired by the orthogonality between the $1S$ and $2S$ wave functions.

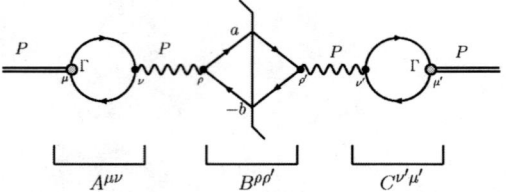

FIGURE 2. Feynman diagram for $^3S_1 \to \ell\bar\ell$.

In terms of the sub-amplitudes $A^{\mu\nu}$, $B^{\mu\nu}$ and $C^{\mu\nu}$ defined in Fig. 2, we have[2]:

$$\int |\bar{\mathcal{M}}|^2 \, d_2(PS) = \frac{1}{3} \Delta_{\mu\mu'} A^{\mu\nu} \left(\frac{-ig_{\nu\rho}}{M^2}\right) B^{\rho\rho'} \left(\frac{-ig_{\rho'\nu'}}{M^2}\right) C^{\nu'\mu'}, \quad (7)$$

where the factor $\Delta_{\mu\nu} = (g_{\mu\nu} - \frac{P_\mu P_{\nu'}}{M^2}) = \sum_i \varepsilon_{i,\mu} \varepsilon^\star_{i,\mu'}$ results from the sum over polarisations of the meson and the factor $\frac{1}{3}$ accounts for the averaging on these initial polarisations.

3.1. Sub-amplitude calculation

To what concerns the sub-amplitude $B^{\rho\rho'}$, from the Feynman rules, and after integration on the two-particle phase space, we have

$$B^{\rho\rho'} = (ie)^2 \left[\pi M^2 g^{\rho\rho'} - 8\frac{\pi M^2}{24}\left(g^{\rho\rho'} + 2\frac{P^\rho P^{\rho'}}{M^2}\right)\right] = (ie)^2 \frac{2\pi}{3} M^2 \underbrace{\left[g^{\rho\rho'} - \frac{P^\rho P^{\rho'}}{M^2}\right]}_{\Delta^{\rho\rho'}}. \quad (8)$$

For $A^{\mu\nu}$, from the Feynman rules and using the vertex functions discussed above, we have (see Fig. 3)

$$iA^{\mu\nu} = -3e_Q \int \frac{d^4k}{(2\pi)^4} \text{Tr}\left((i\Gamma(k,P)\gamma^\mu) \frac{i(\slashed{k} - \frac{1}{2}\slashed{P} + m)}{(k-\frac{P}{2})^2 - m^2 + i\varepsilon}(ie\gamma^\nu)\frac{i(\slashed{k} + \frac{1}{2}\slashed{P} + m)}{(k+\frac{P}{2})^2 - m^2 + i\varepsilon}\right)$$

$$= -3e_Q \int \frac{d^4k}{(2\pi)^4} \Gamma(k,P) \frac{g^{\mu\nu}(M^2 + 4m^2 - 4k^2) + 8k^\mu k^\nu - 2P^\mu P^\nu}{((k-\frac{P}{2})^2 - m^2 + i\varepsilon)((k+\frac{P}{2})^2 - m^2 + i\varepsilon)}, \quad (9)$$

e_Q is the heavy quark charge, -1 comes for the fermionic loop, 3 is the colour factor.

---

[2] In the Feynman gauge. The calculation can be shown to be gauge-invariant, though.

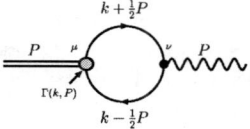

FIGURE 3. Feynman diagram for $^3S_1 \to \gamma$.

We can easily motivate our choice of the argument of the vertex function, beside its simple relation with that of wave functions. Its main virtue is to regularise the integration in all spatial directions. In the meson rest frame, the k^0 integral can be done by standard residue techniques, and the remaining $\vec{k}$ integral is guaranteed to converge. If one uses k^2 instead of $|\vec{k}|^2$ as an argument, it is possible to show that along the light cone, one obtains logarithmic divergences, which would presumably need to be renormalised. Besides, the k^0 integral then also becomes dependent on the singularity structure of the vertex function. These two reasons make us prefer the shifted vertex in our calculation.

A notable simplification can be obtained by guessing the tensorial form of $A^{\mu\nu}$. Indeed, current conservation (gauge invariance) for the photon can be expressed as:

$$A^{\mu\nu}P_\nu = 0. \tag{10}$$

It is equivalent to

$$iA^{\mu\nu} = iF(g^{\mu\nu} - \frac{P^\mu P^\nu}{M^2}) = iF\Delta^{\mu\nu}, \tag{11}$$

the coefficient F being $\frac{A^\mu_\mu}{3}$ since $A^\mu_{\ \mu} = A^{\mu\nu}g_{\nu\mu} = 3F$. This assumption can be easily verified in the bound-state rest frame $P = (M,0,0,0)$ for which Eq. (10) reduces to $A^{\mu 0}M = 0 \Rightarrow A^{\mu 0} = 0$. We shall check this at the end of this calculation.

It is therefore sufficient to compute the following quantity, where we set $|\vec{k}|^2 \equiv K^2$ and define $\Gamma(-K^2) = \Gamma(k,P)$,

$$iA^\mu_{\ \mu} = -3e_Q \int_0^\infty 4\pi K^2 dK \Gamma(-K^2) \times$$
$$\int_{-\infty}^\infty \frac{dk_0}{(2\pi)^4} \frac{4(-2(k^2 - \frac{M^2}{4}) + 4m^2)}{((k-\frac{P}{2})^2 - m^2 + i\varepsilon)((k+\frac{P}{2})^2 - m^2 + i\varepsilon)}. \tag{12}$$

Let us first integrate on k_0 by residues. Defining $E = \sqrt{K^2 + m^2}$, we determine the position of the pole on k_0 (still in the bound-state rest frame) as

$$(k \pm \frac{P}{2})^2 - m^2 + i\varepsilon = (k_0 \pm \frac{M}{2})^2 - E^2 + i\varepsilon. \tag{13}$$

To calculate the integral on k_0, we choose the contour as drawn in Fig. 4. Two poles $-E - \frac{M}{2}$ and $-E + \frac{M}{2}$ are located in the upper half-plane and the two others in the lower half-plane. We shall therefore have two residues to consider. The contribution of the contour $\mathcal{C}_R$ vanishes as R tends to ∞.

$$iA^\mu_{\ \mu} = \frac{-3e_Q}{\pi^3} \int_0^\infty K^2 dK \Gamma(-K^2) \times$$

$$2i\pi \left[\frac{-2(-E-\frac{M}{2})^2 + \frac{M^2}{2} + 2E^2 + 2m^2}{(-2E)(-M)(-2(E+\frac{M}{2}))} + \frac{-2(-E+\frac{M}{2})^2 + \frac{M^2}{2} + 2E^2 + 2m^2}{M(-2(E-\frac{M}{2}))(-2E)} \right] \quad (14)$$

$$= \frac{-3e_Q}{\pi^3} \int_0^\infty K^2 dK \Gamma(-K^2) \frac{2i\pi}{4ME} \underbrace{\left[-\frac{2EM + 2m^2}{E+\frac{M}{2}} + \frac{2EM + 2m^2}{E-\frac{M}{2}} \right]}_{\frac{M(4E^2+2m^2)}{E^2 - \frac{M^2}{4}}}.$$

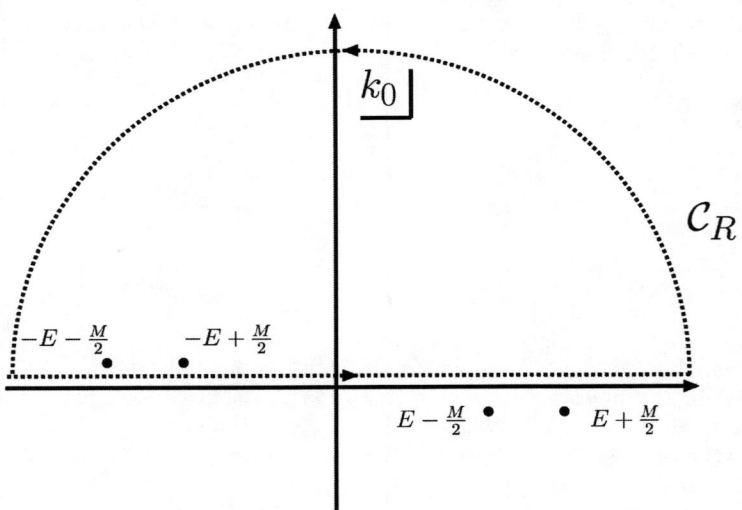

FIGURE 4. Illustration of the contour chosen to integrate on k_0.

The procedure is correct only if the poles of the upper-plane are on the left side and those of the lower-plane are on the right[3].

This crossing would first occur between the points $-E + \frac{M}{2}$ and $E - \frac{M}{2}$. Solving for E, we have:

$$-E + \frac{M}{2} = E - \frac{M}{2} \Rightarrow \frac{M^2}{4} - m^2 = K^2 > 0. \quad (15)$$

---

[3] If they cross, the integral acquires a discontinuity from the pinch and becomes complex.

The crossing is therefore impossible when $M < 2m$. Therefore, to get a coherent description of all charmonia (resp. bottomonia) below the open charm (resp. beauty) threshold, we shall set the quark mass high enough to avoid crossing for all of these. This sets m_c to 1.87 GeV (m_D) and m_b to 5.28 GeV (m_B). Considering the variety of results obtained from potential models, this seems to be a sensible choice.

Putting it all together, we get

$$A^\mu{}_\mu = \frac{-3e_Q}{\pi^2} \int_0^\infty dK \frac{K^2 \Gamma(-K^2)}{\sqrt{K^2+m^2}} \frac{(2K^2+3m^2)}{(K^2+m^2-\frac{M^2}{4})}. \tag{16}$$

One is left with the integration on K for which we define the integral I depending on the vertex function whose normalisation is pulled off,

$$I(\Lambda, M, m) \equiv \int_0^\infty \frac{dK K^2}{\sqrt{K^2+m^2}} \frac{\Gamma(-K^2)}{N} \frac{(2K^2+3m^2)}{(K^2+m^2-\frac{M^2}{4})}, \tag{17}$$

I is a function of Λ through the vertex function $\Gamma(-K^2)$ and is not in general computable analytically. In the following we shall leave it as is and express $A^{\mu\nu}$ as:

$$A^\mu{}_\mu = \frac{-3e_Q}{\pi^2} N I(\Lambda, M, m) \Rightarrow A^{\mu\nu} = \frac{-e_Q}{\pi^2} N I(\Lambda, M, m) \Delta^{\mu\nu}. \tag{18}$$

Finally, to what concerns the sub-amplitude C, we simply have

$$C^{\nu'\mu'} = (A^{\nu'\mu'})^\dagger = A^{\nu'\mu'}. \tag{19}$$

3.2. Results

Now that all quantities in Eq. (7) are determined, we can combine them. Hence, we obtain[4]:

$$\int |\bar{\mathcal{M}}|^2 d_2(PS) = \frac{\Delta_{\mu\mu'}}{3} \frac{-1}{M^4} \left(\frac{e_Q}{\pi^2} N I(\Lambda, M, m)\right)^2 \left((ie)^2 \frac{2\pi}{3} M^2\right) \Delta^{\mu\nu} g_{\nu\rho} \Delta^{\rho\rho'} g_{\rho'\nu'} \Delta^{\nu'\mu'} \tag{20}$$

The leptonic decay width eventually reads from Eq. (6):

$$\Gamma_{\ell\ell} = N^2 \frac{e^2}{12\pi M^3} \left(\frac{e_Q}{\pi^2} I(\Lambda, M, m)\right)^2. \tag{21}$$

3.2.1. Numerical results

The value of N is in practice obtained by replacing $\Gamma_{\ell\ell}$ by its measured value, e_Q by $\frac{2e}{3}$ for c quark and $\frac{-e}{3}$ for b quark and, finally, by introducing the value obtained

---

[4] Recall that the projector $\Delta_{\mu\nu}$ satisfies $\Delta_{\mu\nu}\Delta^{\mu\nu} = 3$ and $\Delta_{\mu\nu}\Delta^{\mu\nu'} = \Delta_\nu{}^{\nu'}$.

for I for the chosen value of Λ. We sketch in Fig. 5 and Fig. 6 some plots of N for the J/ψ and for the $\Upsilon(1S)$ to show its dependence on Λ and m_Q and the value it actually takes.

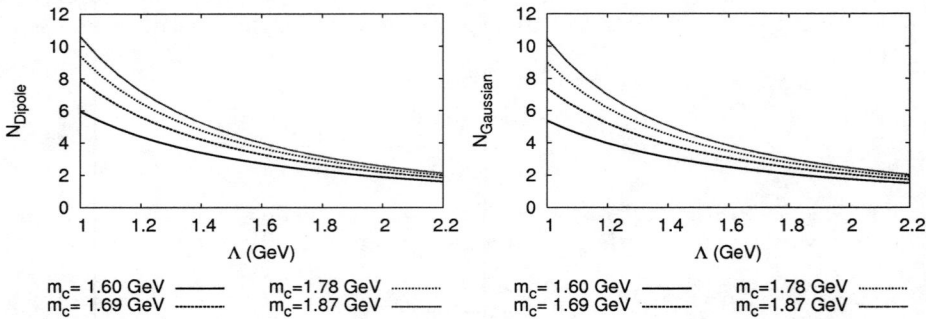

FIGURE 5. Normalisation for a dipolar (resp. gaussian) form for J/ψ as a function of Λ: right (resp. left).

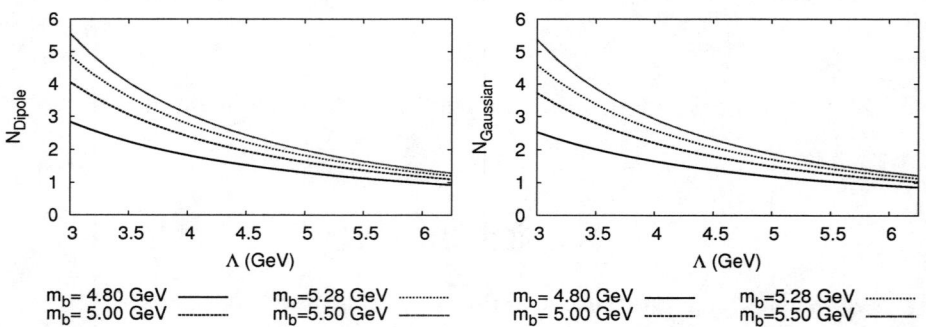

FIGURE 6. Normalisation for a dipolar (resp. gaussian) form for $\Upsilon(1S)$ as a function of Λ: right (resp. left).

TABLE 1. Set of values for m_c, Λ and N obtained for the J/ψ within BSE approach [15].

m_c (GeV)	Vertex functions			
	Gaussian		Dipole	
	N	Λ (GeV)	N	Λ (GeV)
1.6	1.59	2.34	1.33	2.87
1.7	2.28	2.08	1.87	2.54
1.87	3.31	1.97	2.76	2.31

We have chosen the range of size of the mesons, or Λ, following BSE studies [15] and other phenomenological models [12, 13], where the commonly accepted values for the J/ψ are from around 1 to 2.4 GeV, and for the $\Upsilon(1S)$ from around 3.0 to 6.5 GeV. For illustration, we put in Tab. 1 several values of Λ obtained with BSE [15], accompanied with the vertex-function normalisation.

3.2.2. Result for ψ' and implication for the ρ-π puzzle

To what concerns the ψ', we give the results for $\Lambda=1.8$ GeV, $m_c = 1.87$ GeV and a gaussian vertex function (see Fig. 7). The normalisation N' diverges for $a_{node} \simeq 1.35$ GeV. This comes from the cancellation of the leptonic decay width for this value of a_{node}.

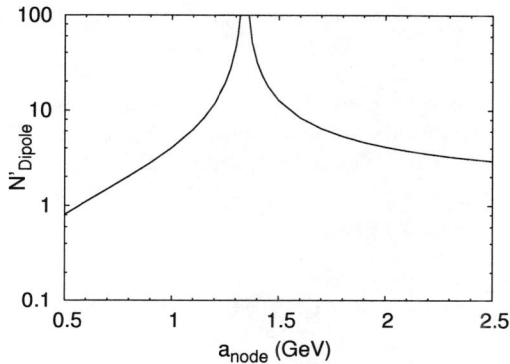

FIGURE 7. N' for the ψ' as a function of a_{node}.

This cancellation can be traced back to the integrand of I (see Eq. (17)), which we note $\frac{dI}{dK}$ (see Fig. 8). The cancellation of the positive contribution by the negative one therefore occurs in this case for $a_{node} = 1.35$ GeV.

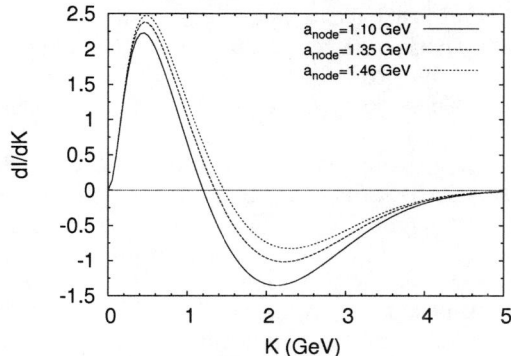

FIGURE 8. Evolution of the integrand in I as a function of K and for three values of a_{node} in the ψ' case.

The constraint for the node position Eq. (5) gives $a_{node} = 1.46$ GeV for these values of Λ and m_c. For this value of a_{node}, the normalisation N' is 16.36. This was the value that we have retained for the numerical applications of [1].

In the case of the hadronic decay $\psi' \to \rho\pi$, we expect it to proceed via an off-shell ω (see Fig. 9). Consequently, the integral I, denoted $I_{\rho\pi}$, will be different due to the finite-size effects of the ω, parametrised by the vertex function $\Gamma_\omega(p_{rel})$.

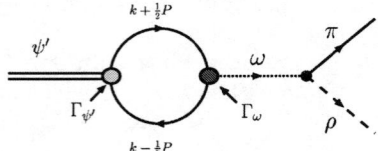

FIGURE 9. Decay of a ψ' into a ρ - π pair via an off-shell ω.

If the form of $I_{\rho\pi}$ is such that the point where it vanishes is closer to the natural value of a_{node}, we expect a severe suppression of the decay amplitude compared to the nodeless, *i.e.* J/ψ, case. This suppression could in turn explains the $\rho-\pi$ puzzle, namely that the measured ratio $\frac{\Gamma(\psi'\to\rho\pi)}{\Gamma(J/\psi\to\rho\pi)}$ is not of the order of 15 % (expected from the leptonic decays) but rather smaller than 1%.

4. CONCLUSION

We have explained our approach to describe on a phenomenological basis the internal dynamics of heavy quarkonia in the context of a Feynman-diagram calculation. We have also provided a way to represent the distinct features of $2S$ states, like the ψ'. We have seen that a simple but robust constraint on the vertex functions for $1S$-states could be achieved by imposing that their normalisation reproduces the leptonic decay width, through a simple leading order calculation.

An interesting simplification can be also obtained by shifting the argument of the vertex function, namely the relative momentum of the quark inside the quarkonium, into a quantity that reduce to the tri-dimensional relative momentum in the meson rest frame. This enable us to work out analytically, for whatever vertex function, the integration on k_0.

Furthermore, we have shown that the leptonic decay width was very dependent on the node position, which, incidentally, is not a completely constrained parameter. This induces the same effects on the normalisation (see Fig. 7). This has consequences on production processes where our approach to describe internal dynamics of heavy quarkonia can be applied (see [1, 2]). It should be interesting to see whether this happens for other excited states. Finally, we suggest that this feature typical of radially excited states could be the awaited explanation for the longtstanding $\rho-\pi$ puzzle.

ACKNOWLEDGMENTS

This work is supported by a IISN (Interuniversity Institute of Nuclear Science, Belgium) Research Fellowship and has been done in close collaboration with J.R. Cudell and Yu.L. Kalinovsky. We would also like to gratefully thank Yu.L. Kalinovsky and P. Costa for sharing their results from the BSE approach.

REFERENCES

1. J.P Lansberg, *Quarkonium Production at High-Energy Hadron Colliders*, Ph.D. Thesis, ULg, Liège, Belgium, 2005.
2. J.R. Cudell, Yu.L. Kalinovsky, J.P. Lansberg (in preparation).
3. F. Abe *et al.* [CDF Collaboration], Phys. Rev. Lett. **79** (1997) 572.
4. F. Abe *et al.* [CDF Collaboration], Phys. Rev. Lett. **79** (1997) 578.
5. T. Affolder *et al.* [CDF Collaboration], Phys. Rev. Lett. **85** (2000) 2886 [arXiv:hep-ex/0004027].
6. D. Acosta *et al.* [CDF Collaboration], Phys. Rev. Lett. **88** (2002) 161802.
7. S. S. Adler *et al.* [PHENIX Collaboration], Phys. Rev. Lett. **92** (2004) 051802 [arXiv:hep-ex/0307019].
8. E. E. Salpeter and H. A. Bethe, Phys. Rev. **84** (1951) 1232.
9. P. Maris and C. D. Roberts, Int. J. Mod. Phys. E **12** (2003) 297 [arXiv:nucl-th/0301049].
10. C. J. Burden, L. Qian, C. D. Roberts, P. C. Tandy and M. J. Thomson, Phys. Rev. C **55** (1997) 2649 [arXiv:nucl-th/9605027].
11. M. A. Ivanov, Y. L. Kalinovsky and C. D. Roberts, Phys. Rev. D **60** (1999) 034018 [arXiv:nucl-th/9812063].
12. M. A. Ivanov, J. G. Korner and P. Santorelli, Phys. Rev. D **63** (2001) 074010 [arXiv:hep-ph/0007169].
13. M. A. Ivanov, J. G. Korner and P. Santorelli, Phys. Rev. D **70** (2004) 014005 [arXiv:hep-ph/0311300].
14. V.D. Barger and R.J.N. Philips, *Collider Physics*, Addison-Wesley, Menlo Park, 1987.
15. P. Costa (private communication).

Highly Excited Baryons in Large-N_c QCD

N. Matagne and Fl. Stancu

Physique théorique fondamentale, Département de Physique, Université de Liège, allée du 6 Août 17, bât. B5, B-4000 Liège 1, Belgium
E-mail: nmatagne@ulg.ac.be, fstancu@ulg.ac.be

Abstract. We use the $1/N_c$ expansion of QCD to analyse the spectrum of positive parity resonances with strangeness $S = 0, -1, -2$ and -3 in the 2–3 GeV mass region, supposed to belong to the [**56**, 4^+] multiplet. The mass operator is similar to that of [**56**, 2^+], previously studied in the literature. The analysis of the latter is revisited. In the [**56**, 4^+] multiplet we find that the spin-spin term brings the dominant contribution and that the spin-orbit term is entirely negligible in the hyperfine interaction, in agreement with the constituent quark model practice, where this interaction is usually neglected. More data are strongly desirable, especially in the strange sector in order to fully exploit the power of this approach. We discuss possibilities of extending the calculations to other excited baryons belonging to the $N = 2$ or the $N = 4$ band.

Keywords: baryon spectroscopy, large N_c QCD
PACS: 12.39.-x,11.15.Pg,11.30.Hv

1. INTRODUCTION

The $1/N_c$ expansion of QCD suggested by 't Hooft [1] about 30 years ago and analysed in greater detail by Witten [2] has lead to a powerful algebraic method to study baryon spectroscopy. This method is based on the result that the SU(6) spin-flavor symmetry is exact in the large N_c limit [3]. It has been applied with a great success to the ground state baryons which correspond to the SU(6) **56** multiplet [4, 5, 6, 7, 8, 9] as well as to some excited states, as for example the [**70**, 1^-] negative parity states [10, 11, 12, 13, 14, 15, 16, 17, 18, 19].

The applicability of the large N_c QCD to the description of excited baryon states is a problem of active investigation. Here we analyse the applicability of the method to the [**56**, 4^+] multiplet and develop considerations for the treatment of multiplet as [**70**, 0^+] or [**70**, 2^+].

2. THE MASS OPERATOR

The study of the [**56**, 4^+] multiplet is similar to that of [**56**, 2^+] as analysed in Ref. [18], where the mass spectrum is calculated in the $1/N_c$ expansion up to and including $\mathcal{O}(1/N_c)$ effects. The SU(2) isospin symmetry is supposed to be exact. The SU(3) symmetry breaking is implemented to $\mathcal{O}(\varepsilon)$, where $\varepsilon \sim 0.3$ gives a measure of this breaking by the strange quark mass. As the **56** is a symmetric representation of SU(6), it is not necessary to distinguish between excited and core quarks for the construction of a basis of mass operators, as explained in Ref. [18]. Then the mass operator of the SU(3) multiplets has the following structure

$$M = \sum_i c_i O_i + \sum_i b_i \bar{B}_i \qquad (1)$$

given in terms of the linearly independent operators O_i and $\bar{B}_i$ defined in Table 1. Here O_i ($i = 1, 2, 3$) are rotational invariants and SU(3) flavor singlets [10], $\bar{B}_1$ is the strangeness quark number operator with negative sign, and the operators $\bar{B}_i$ ($i = 2, 3$) are also rotational invariants but contain the SU(6) spin-flavor generators G_{i8} as well. The operators $\bar{B}_i$ ($i = 1, 2, 3$) provide SU(3) breaking and are defined to have vanishing matrix elements for nonstrange baryons. The relation (1) contains the effective coefficients c_i and b_i as parameters. They represent reduced matrix elements that encode the QCD dynamics. The values of the corresponding coefficients which we obtained from fitting the experimentally known masses are given in Table 1 both for the $[56, 4^+]$ and the presently revisited $[56, 2^+]$.

3. THE $[56, 2^+]$ MULTIPLET REVISITED

As mentioned above, the study of the $[56, 4^+]$ multiplet is similar to that of $[56, 2^+]$. Here we first revisit the $[56, 2^+]$ multiplet for two purposes: 1) to get a consistency test of our procedure of calculating matrix elements of the operators in Table 1 and 2) to analyse a new assignement of the $\Delta_{5/2^+}$ resonances.

The matrix elements of O_1, O_2, O_3 and $\bar{B}_1$ are trivial to calculate for both multiplets under study. For the $[56, 2^+]$ one can find them in Table 2 of Ref. [18] and for the $[56, 4^+]$ they are given in the next section.

To calculate the diagonal and off-diagonal matrix elements of $\bar{B}_2$ we use the definition

$$G_{i8} = G^{i8} = \frac{1}{2\sqrt{3}} \left(S^i - 3 S_s^i \right), \qquad (2)$$

where S^i and S_s^i are components of the total spin and of the total strange-quark spin respectively [8]. Using (2) we can rewrite $\bar{B}_2$ from Table 1 as

$$\bar{B}_2 = -\frac{\sqrt{3}}{2N_c} \vec{l} \cdot \vec{S}_s \qquad (3)$$

with the decomposition

$$\vec{l} \cdot \vec{S}_s = l_0 S_{s0} + \frac{1}{2} \left(l_+ S_{s-} + l_- S_{s+} \right). \qquad (4)$$

We calculated the matrix elements from the wave functions used in constituent quark model studies, where the center of mass coordinate has been removed and only the internal Jacobi coordinates appear (see, for example, Ref. [20]). The expressions we found for the matrix elements of $\bar{B}_2$ were identical with those of

Ref. [18], based on Hartree wave functions, exact in the $N_c \to \infty$ limit only. This proves that in the Hartree approach no center of mass corrections are necessary for the [**56**, 2^+] multiplet. We expect the same conclusion to stand for any [**56**, ℓ^+]. For mixed representations the situation is more intricate [12], see section 5.

For $\bar{B}_3$, we used the following relation [14]

$$S_i G_{i8} = \frac{1}{4\sqrt{2}} \left[3I(I+1) - S(S+1) - \frac{3}{4} N_s(N_s+2) \right] \tag{5}$$

in agreement with [8]. Here I is the isospin, S is the total spin and N_s the number of strange quarks. As for the matrix elements of $\bar{B}_2$, we found identical results to those of Ref. [18]. Note that only $\bar{B}_2$ has non-vanishing off-diagonal matrix elements. Their role is very important in the state mixing in particular in the octet-decuplet mixing. We found that the diagonal matrix elements of $O_2, O_3, \bar{B}_2$ and $\bar{B}_3$ of strange baryons satisfy the following relation

$$\frac{\bar{B}_2}{\bar{B}_3} = \frac{O_2}{O_3}, \tag{6}$$

for any state, irrespective of the value of J in both the octet and the decuplet. Such a relation also holds for the multiplet [**56**, 4^+] studied in the next section and might possibly be a feature of all [**56**, ℓ^+] multiplets. It can be used as a check of the analytic expressions in Table 3. In spite of the relation (6) which holds for the diagonal matrix elements, the operators O_i and $\bar{B}_i$ are linearly independent, as it can be easily proved.

The other issue for the [**56**, 2^+] multiplet is that the analysis performed in Ref. [18] is based on the standard identification of resonances due to the pioneering work of Isgur and Karl [21]. In that work the spectrum of positive parity resonances was calculated from a Hamiltonian containing a harmonic oscillator confinement and a hyperfine interaction of one-gluon exchange type. The mixing angles in the $\Delta_{5/2^+}$ sector turned out to be

State	Mass (MeV)	Mixing angles	
$^4\Delta[\mathbf{56}, 2^+]\frac{5}{2}^+$	1940	0.94	0.38
$^4\Delta[\mathbf{70}, 2^+]\frac{5}{2}^+$	1975	-0.38	0.94

which shows that the lowest resonance at 1940 MeV is dominantly a [**56**, 2^+] state. As a consequence, the lowest observed F_{35} $\Delta(1905)$ resonance was interpreted as a member of the [**56**, 2^+] multiplet.

In a more realistic description, based on a linear confinement [22], the structure of the $\Delta_{5/2^+}$ sector appeared to be different. The result was

State	Mass (MeV)	Mixing angles	
$^4\Delta[\mathbf{56},2^+]\frac{5}{2}^+$	1962	0.408	0.913
$^4\Delta[\mathbf{70},2^+]\frac{5}{2}^+$	1985	0.913	−0.408

which means that in this case the higher resonance, of mass 1985 MeV, is dominantly $[\mathbf{56},2^+]$. Accordingly, here we interpret the higher experimentally observed resonance F_{35} $\Delta(2000)$ as belonging to the $[\mathbf{56},2^+]$ multiplet instead of the lower one. Thus we take as experimental input the mass 1976 ± 237 MeV, determined from the full listings of the PDG [23] in the same manner as for the one- and two-star resonances of the $[\mathbf{56},4^+]$ multiplet (see below). The fit for $[\mathbf{56},2^+]$ multiplet based on this assignement is shown in Table 1. The χ^2_{dof} obtained is 0.58, as compared to $\chi^2_{\text{dof}} = 0.7$ of Ref. [18]. The contribution of the spin-orbit operator O_2 is slightly smaller here than in Ref. [18]. Although $\Delta(2000)$ is a two-star resonance only, the incentive of making the above choice was that the calculated pion decay widths of the $\Delta_{5/2^+}$ sector were better reproduced [24] with the mixing angles of the model [22] than with those of the standard model of Ref. [21]. It is well known that decay widths are useful to test mixing angles. Moreover, it would be more natural that the resonances $\Delta_{1/2}$ and $\Delta_{5/2}$ would have different masses, contrary to the assumption of Ref. [18] where these masses were taken to be identical.

TABLE 1. Operators of Eq. (1) and coefficients resulting from the fit with $\chi^2_{\text{dof}} \simeq 0.58$ for $[\mathbf{56},2^+]$ and $\chi^2_{\text{dof}} \simeq 0.26$ for $[\mathbf{56},4^+]$.

Operator	Fitted coef. (MeV)	
	$[\mathbf{56},2^+]$	$[\mathbf{56},4^+]$
$O_1 = N_c \mathbf{1}$	$c_1 = 540 \pm 3$	$c_1 = 736 \pm 30$
$O_2 = \frac{1}{N_c} l_i S_i$	$c_2 = 14 \pm 9$	$c_2 = 4 \pm 40$
$O_3 = \frac{1}{N_c} S_i S_i$	$c_3 = 247 \pm 10$	$c_3 = 135 \pm 90$
$\bar{B}_1 = -\mathcal{S}$	$b_1 = 213 \pm 15$	$b_1 = 110 \pm 67$
$\bar{B}_2 = \frac{1}{N_c} l_i G_{i8} - \frac{1}{2\sqrt{3}} O_2$	$b_2 = 83 \pm 40$	
$\bar{B}_3 = \frac{1}{N_c} S_i G_{i8} - \frac{1}{2\sqrt{3}} O_3$	$b_3 = 266 \pm 65$	

4. THE $[\mathbf{56}, 4^+]$ MULTIPLET

Tables 2 and 3 give all matrix elements needed for the octets and decuplets belonging to the $[\mathbf{56}, 4^+]$ multiplet. They are calculated following the prescription of the previous section. This means that the matrix elements of O_1, O_2, O_3 and $\bar{B}_1$ are straightforward and for $\bar{B}_3$ we use the formula (5). The matrix elements of $\bar{B}_2$ are calculated from the wave functions given explicitly in Ref. [25], firstly derived and employed in constituent quark model calculations [24]. One can see that the relation (6) holds for this multiplet as well.

As mentioned above, only the operator $\bar{B}_2$ has non-vanishing off-diagonal matrix elements, so $\bar{B}_2$ is the only one which induces mixing between the octet and decuplet states of $[\mathbf{56}, 4^+]$ with the same quantum numbers, as a consequence of the SU(3) flavor breaking. Thus this mixing affects the octet and the decuplet Σ and Ξ states. As there are four off-diagonal matrix elements (Table 3), there are also four mixing angles, namely, θ_J^Σ and θ_J^Ξ, each with $J = 7/2$ and $9/2$. In terms of these mixing angles, the physical Σ_J and Σ'_J states are defined by the following basis states

$$|\Sigma_J\rangle = |\Sigma_J^{(8)}\rangle \cos\theta_J^\Sigma + |\Sigma_J^{(10)}\rangle \sin\theta_J^\Sigma, \quad (7)$$

$$|\Sigma'_J\rangle = -|\Sigma_J^{(8)}\rangle \sin\theta_J^\Sigma + |\Sigma_J^{(10)}\rangle \cos\theta_J^\Sigma, \quad (8)$$

and similar relations hold for Ξ. The masses of the physical states become

$$M(\Sigma_J) = M(\Sigma_J^{(8)}) + b_2\langle\Sigma_J^{(8)}|\bar{B}_2|\Sigma_J^{(10)}\rangle \tan\theta_J^\Sigma, \quad (9)$$

$$M(\Sigma'_J) = M(\Sigma_J^{(10)}) - b_2\langle\Sigma_J^{(8)}|\bar{B}_2|\Sigma_J^{(10)}\rangle \tan\theta_J^\Sigma, \quad (10)$$

where $M(\Sigma_J^{(8)})$ and $M(\Sigma_J^{(10)})$ are the diagonal matrix of the mass operator (1), here equal to $c_1 O_1 + c_2 O_2 + c_3 O_3 + b_1 \bar{B}_1 + b_2 \bar{B}_2 + b_3 \bar{B}_3$, for Σ states and similarly for Ξ states (see Table 4). If replaced in the mass operator (1), the relations (9) and (10) and their counterparts for Ξ, introduce four new parameters which should be included in the fit. Actually the procedure of Ref. [18] was simplified to fit the coefficients c_i and b_i directly to the physical masses and then to calculate the mixing angle from

$$\theta_J = \frac{1}{2}\arcsin\left(2\frac{b_2\langle\Sigma_J^{(8)}|\bar{B}_2|\Sigma_J^{(10)}\rangle}{M(\Sigma_J) - M(\Sigma'_J)}\right). \quad (11)$$

for Σ_J states and analogously for Ξ states.

Due to the scarcity of data in the 2–3 GeV mass region, even such a simplified procedure is not possible at present in the $[\mathbf{56}, 4^+]$ multiplet.

The fit of the masses derived from Eq. (1) and the available empirical values used in the fit, together with the corresponding resonance status in the Particle Data Group [23] are listed in Table 4. The values of the coefficients c_i and b_1 obtained from the fit are presented in Table 1, as already mentioned. For the four and three-star resonances we used the empirical masses given in the summary table. For the others, namely the one-star resonance $\Delta(2390)$ and the two-star resonance $\Delta(2300)$ we adopted the following procedure. We considered as "experimental"

TABLE 2. Matrix elements of SU(3) singlet operators.

	O_1	O_2	O_3
$^2 8_{7/2}$	N_c	$-\frac{5}{2N_c}$	$\frac{3}{4N_c}$
$^2 8_{9/2}$	N_c	$\frac{2}{N_c}$	$\frac{3}{4N_c}$
$^4 10_{5/2}$	N_c	$-\frac{15}{2N_c}$	$\frac{15}{4N_c}$
$^4 10_{7/2}$	N_c	$-\frac{4}{N_c}$	$\frac{15}{4N_c}$
$^4 10_{9/2}$	N_c	$\frac{1}{2N_c}$	$\frac{15}{4N_c}$
$^4 10_{11/2}$	N_c	$\frac{6}{N_c}$	$\frac{15}{4N_c}$

mass the average of all masses quoted in the full listings. The experimental error to the mass was defined as the quadrature of two uncorrelated errors, one being the average error obtained from the same references in the full listings and the other was the difference between the average mass relative to the farthest off observed mass. The masses and errors thus obtained are indicated in the before last column of Table 4.

Due to the lack of experimental data in the strange sector it was not possible to include all the operators $\bar{B}_i$ in the fit in order to obtain some reliable predictions. As the breaking of SU(3) is dominated by $\bar{B}_1$ we included only this operator in Eq. (1) and neglected the contribution of the operators $\bar{B}_2$ and $\bar{B}_3$. At a later stage, when more data will hopefully be available, all analytical work performed here could be used to improve the fit. That is why Table 1 contains results for c_i ($i = $ 1, 2 and 3) and b_1 only. The χ^2_{dof} of the fit is 0.26, where the number of degrees of freedom (dof) is equal to one (five data and four coefficients).

The first column of Table 4 contains the 56 states (each state having a $2I+1$ multiplicity from assuming an exact SU(2) isospin symmetry[1]). The columns two to five show the partial contribution of each operator included in the fit, multiplied by the corresponding coefficient c_i or b_1. The column six gives the total mass according to Eq. (1). The errors shown in the predictions result from the errors on the coefficients c_i and b_1 given in Table 1. As there are only five experimental data available, nineteen of these masses are predictions. The breaking of SU(3) flavor due to the operator $\bar{B}_1$ is 110 MeV as compared to 200 MeV produced in the $[\mathbf{56}, 2^+]$ multiplet.

---

[1] Note that the notation Σ_J, Σ'_J is consistent with the relations (9), (10) inasmuch as the contribution of $\bar{B}_2$ is neglected (same remark for Ξ_J, Ξ'_J and corresponding relations).

TABLE 3. Matrix elements of SU(3) breaking operators. Here, $a_J = 5/2, -2$ for $J = 7/2, 9/2$, respectively and $b_J = 5/2, 4/3, -1/6, -2$ for $J = 5/2, 7/2, 9/2, 11/2$, respectively.

	$\bar{B}_1$	$\bar{B}_2$	$\bar{B}_3$
N_J	0	0	0
Λ_J	1	$\frac{\sqrt{3}\,a_J}{2N_c}$	$-\frac{3\sqrt{3}}{8N_c}$
Σ_J	1	$-\frac{\sqrt{3}\,a_J}{6N_c}$	$\frac{\sqrt{3}}{8N_c}$
Ξ_J	2	$\frac{2\sqrt{3}\,a_J}{3N_c}$	$-\frac{\sqrt{3}}{2N_c}$
Δ_J	0	0	0
Σ_J	1	$\frac{\sqrt{3}\,b_J}{2N_c}$	$-\frac{5\sqrt{3}}{8N_c}$
Ξ_J	2	$\frac{\sqrt{3}\,b_J}{N_c}$	$-\frac{5\sqrt{3}}{4N_c}$
Ω_J	3	$\frac{3\sqrt{3}\,b_J}{2N_c}$	$-\frac{15\sqrt{3}}{8N_c}$
$\Sigma^8_{7/2} - \Sigma^{10}_{7/2}$	0	$-\frac{\sqrt{35}}{2\sqrt{3}N_c}$	0
$\Sigma^8_{9/2} - \Sigma^{10}_{9/2}$	0	$-\frac{\sqrt{11}}{\sqrt{3}N_c}$	0
$\Xi^8_{7/2} - \Xi^{10}_{7/2}$	0	$-\frac{\sqrt{35}}{2\sqrt{3}N_c}$	0
$\Xi^8_{9/2} - \Xi^{10}_{9/2}$	0	$-\frac{\sqrt{11}}{\sqrt{3}N_c}$	0

Our results can be summarized as follows:

- The main part of the mass is provided by the spin-flavor singlet operator O_1, which is $\mathcal{O}(N_c)$.
- The spin-orbit contribution given by $c_2 O_2$ is small. This fact reinforces the practice used in constituent quark models where the spin-orbit contribution is usually neglected in order to obtain a good fit. It is also consistent with the intuitive picture of Ref. [26] where the spin-orbit interaction vanishes at high excitation energy.
- The breaking of the SU(6) symmetry keeping the flavor symmetry exact is mainly due to the spin-spin operator O_3. This hyperfine interaction produces a splitting between octet and decuplet states of approximately 130 MeV which is smaller than that obtained in the $[\mathbf{56}, 2^+]$ case [18], which gives 240 MeV.
- The contribution of $\bar{B}_1$ per unit of strangeness, 110 MeV, is also smaller here than in the $[\mathbf{56}, 2^+]$ multiplet [18], where it takes a value of about 200 MeV. That may be quite natural, as one expects a shrinking of the spectrum with the excitation energy.
- As it was not possible to include the contribution of $\bar{B}_2$ and $\bar{B}_3$ in our fit, a degeneracy appears between Λ and Σ.

TABLE 4. Masses (in MeV) of the $[\mathbf{56},4^+]$ multiplet predicted by the $1/N_c$ expansion as compared with the empirically known masses. The partial contribution of each operator is indicated for all masses. Those partial contributions in blank are equal to the one above in the same column.

	$1/N_c$ expansion results					Empirical	Name, status
	Partial contribution (MeV)				Total (MeV)	(MeV)	
	$c_1 O_1$	$c_2 O_2$	$c_3 O_3$	$b_1 \bar{B}_1$			
$N_{7/2}$	2209	-3	34	0	2240 ± 97		
$\Lambda_{7/2}$				110	2350 ± 118		
$\Sigma_{7/2}$				110	2350 ± 118		
$\Xi_{7/2}$				220	2460 ± 166		
$N_{9/2}$	2209	2	34	0	2245 ± 95	2245 ± 65	$N(2220)$****
$\Lambda_{9/2}$				110	2355 ± 116	2355 ± 15	$\Lambda(2350)$***
$\Sigma_{9/2}$				110	2355 ± 116		
$\Xi_{9/2}$				220	2465 ± 164		
$\Delta_{5/2}$	2209	-9	168	0	2368 ± 175		
$\Sigma_{5/2}$				110	2478 ± 187		
$\Xi_{5/2}$				220	2588 ± 220		
$\Omega_{5/2}$				330	2698 ± 266		
$\Delta_{7/2}$	2209	-5	168	0	2372 ± 153	2387 ± 88	$\Delta(2390)$*
$\Sigma'_{7/2}$				110	2482 ± 167		
$\Xi'_{7/2}$				220	2592 ± 203		
$\Omega_{7/2}$				330	2702 ± 252		
$\Delta_{9/2}$	2209	1	168	0	2378 ± 144	2318 ± 132	$\Delta(2300)$**
$\Sigma'_{9/2}$				110	2488 ± 159		
$\Xi'_{9/2}$				220	2598 ± 197		
$\Omega_{9/2}$				330	2708 ± 247		
$\Delta_{11/2}$	2209	7	168	0	2385 ± 164	2400 ± 100	$\Delta(2420)$****
$\Sigma_{11/2}$				110	2495 ± 177		
$\Xi_{11/2}$				220	2605 ± 212		
$\Omega_{11/2}$				330	2715 ± 260		

In conclusion we have studied the spectrum of highly excited resonances in the 2–3 GeV mass region by describing them as belonging to the $[\mathbf{56},4^+]$ multiplet. This is the first study of such excited states based on the $1/N_c$ expansion of QCD. A better description should include multiplet mixing, following the lines developed, for example, in Ref. [27].

We support previous assertions that better experimental values for highly excited non-strange baryons as well as more data for the Σ^* and Ξ^* baryons are needed in order to understand the role of the operator $\bar{B}_2$ within a multiplet and for the octet-decuplet mixing. With better data the analytic work performed here will help to make reliable predictions in the large N_c limit formalism.

5. PERSPECTIVES

As clearly stated in Ref. [18] the study of excited baryons is not free of difficulties. The use of the spin-flavor symmetry at zero-th order can be justified only from practical point of view, due to the smallness of the spin-orbit effects and of configuration mixings. More fundamentally, the excited baryons are unstable states for which the consistency condition is more difficult to ensure in large N_c QCD [28].

Moreover the analysis of the multiplet $[\mathbf{56}, \ell^+]$ is simplified by the fact that a distinction between excited and core quarks is not necessary. This is not the case for mixed representations.

Our next objective is to analyse the $[\mathbf{70}, 0^+]$ and $[\mathbf{70}, 2^+]$ multiplets. The simplicity of the $[\mathbf{56}, \ell^+]$ does not hold anymore. In addition, one can not assume that the excitation is associated to a single quark which can be decoupled from a core free of excitations, as in the case of $[\mathbf{70}, 1^-]$.

This can be illustrated by writing, for example, the two independent wave functions associated to $[\mathbf{70}, 2^+]$ in the form where the center of mass motion has been removed. For $N_c = 3$ we have

$$|\mathbf{70}, 2^+\rangle_{\rho,\lambda} = \sqrt{\frac{1}{3}} |[21]_{\rho,\lambda}(0s)^2(0d)\rangle + \sqrt{\frac{2}{3}} |[21]_{\rho,\lambda}(0s)(0p)^2\rangle. \quad (12)$$

The coefficients of the linear combination above are independent of N_c (we could prove this assertion for $N_c = 4, 5$ and 6). This implies that the two terms contribute to the same order and we have to consider both of them in the calculations. The first is common to $[\mathbf{56}, 2^+]$ and will be treated accordingly. For the second we have to include excitations both in the core and the decoupled quark. We shall analyse the role of an excited core in a future publication.

ACKNOWLEDGMENTS

The work of one of us (N. M.) was supported by the Institut Interuniversitaire des Sciences Nucléaires (Belgium).

REFERENCES

1. G. 't Hooft, Nucl. Phys. B **72** (1974) 461.
2. E. Witten, Nucl. Phys. B **160** (1979) 57.

3. R. Dashen and A. V. Manohar, Phys. Lett. B **315** (1993) 425; Phys. Lett. B **315** (1993) 438.
4. R. Dashen, E. Jenkins, and A. V. Manohar, Phys. Rev. D **49** (1994) 4713.
5. R. Dashen, E. Jenkins, and A. V. Manohar, Phys. Rev. D **51** (1995) 3697.
6. C. D. Carone, H. Georgi and S. Osofsky, Phys. Lett. B **322** (1994) 227.
   M. A. Luty and J. March-Russell, Nucl. Phys. B **426** (1994) 71.
   M. A. Luty, J. March-Russell and M. White, Phys. Rev. D **51** (1995) 2332.
7. E. Jenkins, Phys. Lett. B **315** (1993) 441.
8. E. Jenkins and R. F. Lebed, Phys. Rev. D **52** (1995) 282.
9. J. Dai, R. Dashen, E. Jenkins, and A. V. Manohar, Phys. Rev. D **53** (1996) 273.
10. J. L. Goity, Phys. Lett. B **414**, (1997) 140.
11. D. Pirjol and T. M. Yan, Phys. Rev. D **57** (1998) 1449.
    D. Pirjol and T. M. Yan, Phys. Rev. D **57** (1998) 5434.
12. C. E. Carlson, C. D. Carone, J. L. Goity and R. F. Lebed, Phys. Lett. B **438** (1998) 327; Phys. Rev. D **59** (1999) 114008.
13. C. D. Carone, H. Georgi, L. Kaplan and D. Morin, Phys. Rev. D **50** (1994) 5793.
14. C. E. Carlson and C. D. Carone, Phys. Lett. B **441** (1998) 363; Phys. Rev. D **58** (1998) 053005.
15. C. L. Schat, J. L. Goity and N. N. Scoccola, Phys. Rev. Lett. **88** (2002) 102002; J. L. Goity, C. L. Schat and N. N. Scoccola, Phys. Rev. D **66** (2002) 114014.
16. D. Pirjol and C. Schat, Phys. Rev. D **67** (2003) 096009.
17. C. E. Carlson and C. D. Carone, Phys. Lett. B **484** (2000) 260.
18. J. L. Goity, C. L. Schat and N. N. Scoccola, Phys. Lett. B **564** (2003) 83.
19. T. D. Cohen *et al.*, Phys Rev D **69** (2004) 056001.
20. Fl. Stancu, *Group Theory in Subnuclear Physics*, Clarendon Press, Oxford, 1996, Ch. 10.
21. N. Isgur and G. Karl, Phys. Rev. D **19** (1979) 2653.
22. R. Sartor and Fl. Stancu, Phys. Rev. D **34** (1986) 3405.
23. S. Eidelman et al., Phys. Lett. B **592** (2004) 1.
24. P. Stassart and Fl. Stancu, Z. Phys. A **351** (1995) 77.
25. N. Matagne and Fl. Stancu, Phys. Rev. D **71** (2005) 014010.
26. L. Y. Glozman, Phys. Lett. B **541** (2002) 115.
27. J. L. Goity, arXiv:hep-ph/0405304.
28. T. D. Cohen and R. F. Lebed, Phys. Rev. D **67** (2003) 096008; Phys. Rev. Lett **91** (2003) 012001; Phys. Rev. D **68** (2003) 056003.

Dynamics of pentaquarks in constituent quark models: recent developments

Fl. Stancu

*Physique théorique fondamentale, Département de Physique, Université de Liège,
allée du 6 Août 17, bât. B5, B-4000 Liège 1, Belgium
E-mail: fstancu@ulg.ac.be*

Abstract. Some recent developments in the study of light and heavy pentaquarks are reviewed, mainly within constituent quark models. Emphasis is made on results obtained in the flavor-spin model where a nearly ideal octet-antidecuplet mixing is obtained. The charmed antisextet is reviewed in the context of an SU(4) classification.

Keywords: Pentaquarks, parity, spin and representation mixing in constituent quark models
PACS: 12.38.-t,12.39.-x,14.20.-c,14.65.-q

1. INTRODUCTION

The existence of pentaquarks has been discussed for more than 30 years. Light and heavy pentaquarks have alternatively been predicted and searched for. The new wave of interest in light pentaquarks was triggered by the prediction of a narrow width antidecuplet, made by Diakonov, Petrov and Polyakov [1] in the framework of the chiral soliton model, although predictions for the mass of Θ^+ have been around for nearly 20 years (see e. g. [2] and [3] and references therein).

The observation of the exotic pentaquarks Θ^+, Ξ^{--} and Θ_c^0 still remains controversial, the number of positive results being, in each case, about the same as that of null evidence. However new efforts are currently being made to confirm the previous positive results of LEPS and CLAS Collaborations for Θ^+, of NA49 Collaboration for Ξ^{--} and of H1 Collaboration for Θ_c^0 [4].

Theoretically there is a large variety of approaches to describe pentaquarks: Skyrme and chiral soliton models, large N_c studies, constituent quark models, QCD sum rules, lattice calculations, etc. (for recent reviews see for example [5, 6, 7, 8]).

Regarding the light antidecuplet the main issues are: the mass of Θ^+, the spin and parity of the antidecuplet members, the splitting between isomultiplets, the influence of the representation mixing on the masses and on the strong decay widths, etc.

After discussing the light antidecuplet main issues, the charmed antisextet is shortly reviewed in the context of an SU(4) classification.

2. CONSTITUENT QUARK MODELS

Constituent quark models describe a large variety of observables in baryon spectroscopy. It seems thus natural to inquire about their applicability to exotic hadrons. Any constituent quark model Hamiltonian has a spin-independent part (free mass term + kinetic energy + confinement) and a short-range hyperfine interaction. The most common constituent quark models used in pentaquark physics have either a color-spin (CS) or a flavor-spin (FS) interaction. There are also studies in the so-called hybrid models, which contain a superposition of CS and FS interactions [9]. Attempts to describe Θ^+ by using an instanton induced interaction have also been made [10].

In the following we shall present results in the FS model and compare them with corresponding results from other models.

3. THE LIGHT ANTIDECUPLET

In $SU(3)_F$ $q^4\bar{q}$ multiplets can be obtained from the direct product of four quarks and an antiquark irreducible representation as

$$3_F \times 3_F \times 3_F \times 3_F \times \bar{3}_F = 3(1_F) + 8(8_F) + 3(27_F) + 4(10_F)$$
$$+ 2(\overline{10}_F) + 35_F.$$

which shows that the antidecuplet $\overline{10}_F$ is one of the possible multiplets. The $SU(3)_F$ breaking induces representation mixing. One expects an important mixing between octet members and antidecuplet members with the same quantum numbers. This will be discussed below.

3.1. Parity and Spin

The parity and spin can be found by looking first at a q^4 subsystem to which an antiquark is subsequently coupled.

In the FS model the lowest negative parity state of a $q^4\bar{q}$ system with total spin $S = 1/2$ results from a q^4 subsystem which has the structure $|[4]_O[211]_C[211]_{OC};[211]_F[22]_S[31]_{FS}\rangle$, where O, C, F and S stand for orbital, color, flavor and spin degrees of freedom. The symmetry $[4]_O$ implies that there is no orbital excitation and the parity of the pentaquark is negative, i. e. the same as the intrinsic parity of the antiquark. But if one quark is excited to the p-shell the parity becomes positive and the lowest symmetry allowed for the orbital part of the wave function is $[31]_O$. Then the Pauli principle requires the q^4 subsystem to have the structure $|[31]_O[211]_C[1111]_{OC};[22]_F[22]_S[4]_{FS}\rangle$ in its lowest state, which has $I = 0$ and $S = 0$. Although this state contains one unit of orbital excitation the attraction brought by the FS interaction is so strong that it overcomes the excess of kinetic energy and generates a positive parity state below the negative parity one [11]. After coupling q^4 to $\bar{q}$ the total angular momentum is 1/2 or

3/2. Calculations based on the realistic FS Hamiltonian of Ref.[12], have been performed for Θ^+ in Ref. [13] and for the whole antidecuplet in Ref. [14]. Similar variational calculations were made earlier for heavy pentaquarks. The positive parity pentaquarks turned out to be lighter by several hundreds MeV [11] than the negative parity ones with the same quark content [15].

Recently, more involved calculations for Θ^+, performed in the framework of a semi-relativistic version of the FS model proved once more that in the FS model the lowest state has positive parity [16].

Based on semi-schematic estimates, in Ref. [17] it was claimed that in the CS model the lowest state for Θ^+ has also positive parity. As above, this implies an excess of kinetic energy due to an extra unit of orbital angular momentum. Then, according to the Pauli principle, the lowest symmetry of the wave function in the relevant degrees of freedom is $[31]_{CS}$. This symmetry brings less attraction than $[4]_{FS}$ in the FS model, which is insufficient to overcome the excess of kinetic energy. The realistic study of Ref. [16] proves that this is the case, so that in the CS model the lowest resonant state has $J^P = 3/2^-$. The $J^P = 1/2^-$ state is even lower, but in the continuum. The same study shows that the hybrid models favor negative parity, in agreement with Ref. [9].

3.2. The Antidecuplet Mass Spectrum in the FS Model

There are several reasons to study pentaquarks in the FS model. This model reproduces the correct sequence of positive and negative parity levels in the low-energy spectra of nonstrange and strange baryons. In addition it is supported by lattice calculations [18] and the flavor-spin symmetry is consistent with the large N_c limit of QCD [19].

The results for the mass spectrum of $\overline{10}_F$ pentaquarks based on the Graz parametrization of Ref. [12] are shown in Fig. 1a. Here, as in any other model including the chiral soliton, one cannot determine the absolute mass of Θ^+. This mass has been fitted to the presently accepted experimental value of 1540 MeV. Reasons to accommodate such a value are given in Ref. [13]. The pure $\overline{10}_F$ spectrum, Fig. 1a, can approximately be described by the linear mass formula $M \simeq 1829 - 145\,Y$ where Y is the hypercharge. The FS model result is quite close to the presently estimated level spacing in the chiral soliton model [20] where the parameters were allowed to vary considerably for well justified reasons. In the CS model the level spacing is much smaller. To a good approximation one has $M \simeq M_0 - 58\,Y$ [21], where M_0 can be fixed by the mass of Θ^+.

To construct all the antidecuplet members the masses of the following systems have been calculated in the Graz parametrization [12]: $M(uudd\bar{d}) = 1452$ MeV, $M(uudd\bar{s}) \equiv M(\Theta^+) = 1540$ MeV, $M(uuds\bar{d}) = 1723$ MeV, $M(uuds\bar{s}) = 1800$ MeV, $M(ddss\bar{u}) \equiv M(\Xi^{--}) = 1962$ MeV and $M(uuss\bar{s}) = 2042$ MeV. The antidecuplet members with $Y = 1$ and $Y = 0$ were obtained according to their wave functions (see Ref. [14]) as

$$M(N_{\overline{10}}) = \frac{1}{3}M(uudd\bar{d}) + \frac{2}{3}M(uuds\bar{s}) = 1684 \text{ MeV},$$
$$M(\Sigma_{\overline{10}}) = \frac{2}{3}M(uuds\bar{d}) + \frac{1}{3}M(uuss\bar{s}) = 1829 \text{ MeV}. \quad (1)$$

The octet members with $Y = 1$ and $Y = 0$ were obtained in a similar way

$$M(N_8) = \frac{2}{3}M(uudd\bar{d}) + \frac{1}{3}M(uuds\bar{s}) = 1568 \text{ MeV},$$
$$M(\Sigma_8) = \frac{1}{3}M(uuds\bar{d}) + \frac{2}{3}M(uuss\bar{s}) = 1936 \text{ MeV}. \quad (2)$$

3.3. Representation Mixing in the FS Model

As a consequence of the $SU(3)_F$ breaking the representations $\overline{10}_F$ and 8_F mix. The existing data require mixing.

There are some phenomenological studies where, by fitting the mass and the width of known resonances, one can obtain the mixing angle between the antidecuplet and one (or more) octets. In such studies the number of quarks and antiquarks is arbitrary in every baryon.

In phenomenological models where one assumes mixing between states having $J^P = 1/2^+$, it turns out that the selected masses require a large mixing angle [28, 29] and the widths a small mixing angle [29]. A compromise was found when the antidecuplet and octet states which mix had $J^P = 3/2^-$, in which case a large mixing angle consistent with both masses and widths was obtained [30]. As mentioned above, the $J^P = 3/2^-$ state is the lowest resonant Θ^+ state in the CS model [16]. Its negative parity corresponds to an $\ell = 2$ relative partial wave which can produce a rather large centrifugal barrier, thus a small width.

The mixing of the antidecuplet with three octets with $J^P = 1/2^+$ has also recently been investigated phenomenologically in the chiral soliton model where it was assumed that the mixing is small [31]. It was found that this can reduce the size of the widths of the antidecuplet members without much affecting the masses. However, another chiral soliton study [3], based on an "exact" treatment (not only the first order) of $SU(3)_F$ breaking advocates large $8 + \overline{10}$ mixing from the mass analysis. Thus the representation mixing seems to remain a controversial problem in the chiral soliton model.

The mixing takes place between octet and antidecuplet members with the same quantum numbers, i.e. for $Y = 1$, $I = 1/2$ and $Y = 0$, $I = 1$ states. Here we suppose that there is mixing with the lowest pentaquark octet only. Then there is only one mixing angle, introduced by the physical states, defined as

$$|N^*\rangle = |N_8\rangle \cos\theta_N - |N_{\overline{10}}\rangle \sin\theta_N,$$
$$|N_5\rangle = |N_8\rangle \sin\theta_N + |N_{\overline{10}}\rangle \cos\theta_N, \quad (3)$$

for N and similarly for Σ.

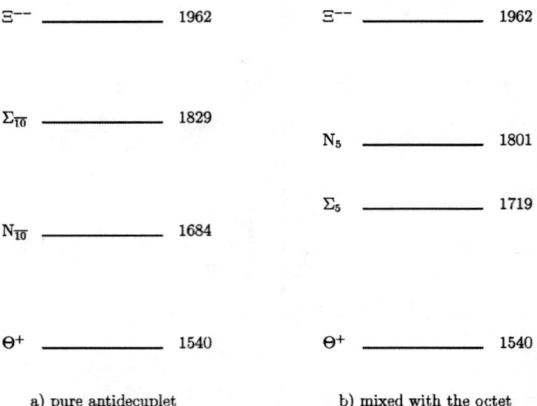

FIGURE 1. The pentaquark antidecuplet masses (MeV) in the FS model: (a) pure antidecuplet and (b) after mixing with the pentaquark octet.

In the FS model the mixing angles θ_N and θ_Σ were calculated dynamically in Ref. [14]. The states which mix are all $q^4\bar{q}$ states, i. e. the number of quarks and antiquarks is fixed, contrary to the above phenomenological studies or to the spirit of the chiral soliton model. The mixing is determined by the the coupling matrix element V of $\overline{10}_F$ and 8_F states. This has contributions from every part of the Hamiltonian which breaks $SU(3)_F$ symmetry: the free mass term, the kinetic energy and the hyperfine interaction. It reads

$$V = \begin{cases} \frac{2\sqrt{2}}{3}(m_s - m_u) + \frac{\sqrt{2}}{3}\left[S(uuds\bar{s}) - S(uudd\bar{d})\right] = 166 \text{ MeV} & \text{for N} \\ \frac{2\sqrt{2}}{3}(m_s - m_u) + \frac{\sqrt{2}}{3}\left[S(uuss\bar{s}) - S(uuds\bar{d})\right] = 155 \text{ MeV} & \text{for } \Sigma \end{cases} \quad (4)$$

where the first term is the free mass term which alone generates an ideal mixing and S represents the combined contribution of the kinetic energy and hyperfine interaction

$$S = \langle T \rangle + \langle V_\chi \rangle . \quad (5)$$

The expressions (4) result from the wave functions of N and Σ respectively and reflect their quark content. One can see that the numerical values of V are similar for N and Σ.

The mixing angle derived from the definitions (3) is

$$\tan 2\theta_N = \frac{2V}{M(N_{\overline{10}}) - M(N_8)} \quad (6)$$

and similarly for Σ. The resulting numerical values are $\theta_N = 35.34^0$ and $\theta_\Sigma = -35.48^0$. Each value is very close to the ideal mixing angle $\theta_N^{ideal} = 35.26^0$ and

$\theta_\Sigma^{ideal} = -35.26^0$ respectively. This implies that the "mainly antidecuplet" state N_5 is 67 % antidecuplet and 33 % octet, which represents a large mixture. The content of the "mainly octet" state N^* is the other way round, i. e. 67 % octet and 33 % antidecuplet. Then, for example, for positive charge pentaquarks with $Y=1, I=1/2$ one has

$$|N^*\rangle \simeq \frac{1}{2} |\,(ud-du)(ud-du)\bar{d}\,\rangle,$$
$$|N_5\rangle \simeq \frac{1}{2\sqrt{2}} |\,[(ud-du)(us-su)+(us-su)(ud-du)]\bar{s}\,\rangle, \tag{7}$$

i. e. the "mainly octet" state has no strangeness and the "mainly antidecuplet" state contains the whole available amount of (hidden) strangeness. The physical masses, obtained from the diagonalization of a 2 × 2 matrix, are

$$M(N_5) = M(N_{\overline{10}}) + V\tan\theta_N = 1801 \text{ MeV},$$
$$M(N^*) = M(N_8) - V\tan\theta_N = 1451 \text{ MeV}. \tag{8}$$

In the Σ sector the situation is opposite. The mixing angle, $\theta_\Sigma = -35.48^0$ ($\sin\theta_\Sigma \simeq -1/\sqrt{3}$, $\cos\theta_\Sigma \simeq \sqrt{2/3}$) minimizes the number of strange quarks (antiquarks) in Σ_5 and maximizes it in Σ^*. This can be readily seen from the analogues of Eqs. 3 which give

$$|\Sigma_5\rangle \simeq \frac{1}{2\sqrt{2}} |\,[(ud-du)(us-su)+(us-su)(ud-du)]\bar{d}\,\rangle,$$
$$|\Sigma^*\rangle \simeq \frac{1}{2} |\,(us-su)(us-su)\bar{s}\,\rangle, \tag{9}$$

so that

$$M(\Sigma_5) = M(\Sigma_{\overline{10}}) + V\tan\theta_\Sigma = 1719 \text{ MeV},$$
$$M(\Sigma^*) = M(\Sigma_8) - V\tan\theta_\Sigma = 2046 \text{ MeV}. \tag{10}$$

i. e. $M(\Sigma_5) < M(\Sigma^*)$. As a consequence, the order of N and Σ is interchanged by the mixing, as illustrated in Fig. 1. The N_5 state is 70 Mev higher than the option for a 1730 MeV resonance in the new analysis of Ref. [22]. The Σ_5 is about 30 MeV far off the upper end of the experimental mass range 1630 - 1690 MeV of the three star $\Sigma(1660)$ resonance and 10 MeV below the lowest experimental edge of the one star $\Sigma(1770)$ resonance (see the PDG [23] full listings).

The N^* state is located in the Roper resonance mass region 1430 - 1470 MeV. However this is a $q^4\bar{q}$ state, i. e. different from the q^3 radially excited state obtained in Ref. [12] at 1493 MeV. A mixing of the q^3 and the $q^4\bar{q}$ states could possibly be a better description of the reality.

3.4. The Decay Width

If the pentaquark Θ^+ exists, its width is expected to be small [1]. The positive experiments have reported upper limits. In particular the LEPS Collaboration at

SPring-8, which reported the first observation of Θ^+ [24], gives $\Gamma < 25$ MeV. Some recent analysis of the K^+N scattering gives an even smaller limit $\Gamma < 1$ MeV [25].

In quark models an option to reduce the width is to introduce rearrangements, as for example, diquark correlations in the orbital and/or color-flavor-spin spaces.

In Ref. [26] Θ^+ was described as a bound state $J^P = 1/2^+$ of two extended ud diquarks and an antiquark $\bar{s}$. The size parameters of the wave function were varied and a decay width of about 1 MeV was obtained for an asymmetric "peanut" structure with $\bar{s}$ in the center and the diquarks rotating around it. However, there is no dynamics justifying the range of values of the size parameters.

In Ref. [27] both parities were considered. But the width of the lowest positive parity state was expected to be smaller than that of the lowest negative parity state by a factor of 50 due to the centrifugal barrier. It was also suggested that the width can be lowered down by adequately reducing the coupling constant $g_{KN\Theta}$ as compared to $g_{\pi NN}$, due to large N_c arguments. However the estimates of the widths have been made in a limit where the emitted meson is a point-like particle, like for ordinary baryon decay. It is hoped that more dynamical studies will better settle the width issue in the future.

4. HEAVY PENTAQURKS

In most models which accommodate Θ^+ and its antidecuplet partners, heavy pentaquarks $q^4\bar{Q}$ ($\bar{Q} = \bar{c}$ or $\bar{b}$) can be accommodated as well. From general arguments they are expected to be more stable against strong decays than the light pentaquarks [32]. In an SU(4) classification, including the charm C, in Ref. [33] it has been shown that $\overline{10}+8$ discussed above and having $C = 0$ and a charm antisextet $\bar{6}$ with $C = -1$, belong to the same SU(4) irreducible representation of dimension **60**. This implies that Θ^0_c, the $I = 0$ member of this antisextet, is obtained from Θ^+ by replacing $\bar{s}$ by $\bar{c}$. The other SU(3) representations belonging to **60** of SU(4) are $\overline{15}+6$ with $C = 1$ and **15** with $C = 2$. Although SU(4) is badly broken, such a classification may be as useful as that of ordinary baryons [34].

TABLE 1. Masses (MeV) of the positive parity antisextet charmed and beauty pentaquarks.

Pentaquark	I	Content	FS model Ref. [11]	CS model Ref. [35]
Θ^0_c	0	u u d d $\bar{c}$	2902	2835±30
N^0_c	1/2	u u d s $\bar{c}$	3161	
Ξ^0_c	1	u u s s $\bar{c}$	3403	
Θ^+_b	0	u u d d $\bar{b}$	6176	6180±30
N^+_b	1/2	u u d s $\bar{b}$	6442	
Ξ^+_b	1	u u s s $\bar{b}$	6683	

The masses of the charmed antisextet calculated in the FS model [11] are presented in Table 1. [1]

For completeness, in Table 1, the masses of a beauty antisextet, calculated in the FS model [11], are presented as well. The results are consistent with the heavy quark limit. They are compared to the available estimates of Ref. [35] in the CS model. Close similarity is observed.

The experimental observation of charmed pentaquarks is contradictory so far. While the H1 collaboration [36] confirmed evidence for a narrow resonance at about 3100 MeV [4], there is still null evidence from the CDF collaboration [4].

For an orientation, it is interesting to calculate excited charmed pentaquarks Θ_c^{0*}. In the FS model the first excited state having $I = 1$ and $S = 1/2$ is located 200 Mev above Θ_c^0 which has $I = 0$. This value is close to the mass observed by the H1 collaboration [36]. In addition it supports the large spacing result obtained approximately in Ref. [35] in the FS model.

5. CONCLUSIONS

The recent research activity on pentaquarks could bring a substantial progress in understanding the baryon structure. It shows that the N and Σ partners of Θ^+ lie in the midst of low-lying positive parity baryonic states. Thus it suggests that the simple description of a baryon resonance as a q^3 configuration is insufficient. The addition of higher Fock components to the nucleon wave function may perhaps help to improve the description of strong decays of baryons.

Furthermore, in light hadrons it is necessary to clarify the role of the spontaneous breaking of chiral symmetry, the basic feature of the chiral soliton model [1] which motivated this new wave of interest in pentaquarks. Recent lattice calculations [18] suggest that the order reversal of the Roper and the negative parity $S_{11}(1535)$ resonance, compared to heavy quark systems, is caused by the flavor-spin interaction and conclude that the spontaneous breaking of chiral symmetry dictates the dynamics of light quarks.

REFERENCES

1. D. Diakonov, V. Petrov and M. Polyakov, Z. Phys. **A359**, 305 (1997) [arXiv:hep-ph/9703373].
2. M. Praszalowicz, Acta. Phys. Polon. **B35**, 1624 (2004) [arXiv:hep-ph/0402038].
3. H. Weigel, Eur. Phys. J. **A21**, 133 (2004) [arXiv:hep-ph/0404173].
4. For an overview of the recent status see presentations by K. Hicks (LEPS and CLAS Collaborations), T. Anticic (NA49 Collaboration), K. Lipka (H1 Collaboration) and M. Neubauer

---

[1] Actually in Ref. [11], instead of masses, binding energies were presented as $\Delta E = M(pentaquark) - E_T$ where $E_T = M_{baryon} + \overline{M}_{meson}$ is the threshold energy involving the average mass $\overline{M}_{meson} = (M + 3M^*)/4$, with M the pseudoscalar and M^* the vector meson mass respectively. In Table 1 the absolute value $M(pentaquark)$ is indicated. The lowest physical threshold is $N + D = 2808$ MeV for Θ_c^0 and $N + B = 6219$ MeV for Θ_b^+.

(CDF Collaboration) at the International Workshop *Exotic Hadrons* held at ECT* Trento, February 21-24, 2005, http://www.tp2.rub.de/talks/trento05.index.html.
5. Fl. Stancu, *Int. J. Mod. Phys.* **A20**, 209 (2005) [arXiv:hep-ph/0408042].
6. R. L. Jaffe, *Phys. Rept.* **409**, 1 (2005) [arXiv:hep-ph/0409065].
7. K. Goeke, H.-C. Kim, M. Praszalowicz and G.-S. Yang., hep-ph/0411195.
8. D. Diakonov, hep-ph/0412272.
9. F. Huang, Z. Y. Zhang, Y. W. Yu and B. S. Zou, *Phys. Lett.* **B586**, 69 (2004).
10. N. I. Kochelev, H.-J. Lee and V. Vento, *Phys. Lett.* **B594**, 87 (2004); C. Semay, F. Brau and B. Silvestre-Brac, *Phys. Rev. Lett.* **94**, 062001 (2005).
11. Fl. Stancu, *Phys. Rev.* **D58**, 111501 (1998).
12. L. Ya. Glozman, Z. Papp and W. Plessas, *Phys. Lett.* **B381**, 311 (1996).
13. Fl. Stancu and D. O. Riska, *Phys. Lett.* **B575**, 242 (2004).
14. Fl. Stancu, *Phys. Lett.* **B595** 269 (2004); ibid. Erratum, **B598**, 295 (2004).
15. M. Genovese, J. -M. Richard, Fl. Stancu and S. Pepin, *Phys. Lett.* **B425** 171 (1998).
16. S. Takeuchi and K. Shimizu, hep-ph/0411016.
17. B. Jennings and K. Maltman, *Phys. Rev.* **D69**, 094020 (2004).
18. N. Mathur, Y. Chen, S. J. Dong, T. Draper, F. X. Lee, K. F. Liu and J. B. Zhang, *Phys. Lett.* **B605**, 137 (2005).
19. E. Jenkins and A. Manohar, hep-ph/0402024 and references therein.
20. J. Ellis, M. Karliner and M. Praszalowicz, *JHEP* **0405** 002 (2004) [arXiv:hep-ph/0401127].
21. V. Dmitrasinovic and Fl. Stancu, in preparation.
22. R. A. Arndt, Ya. I. Azimov, M. V. Polyakov, I. I. Strakovsky, and R. L. Workman, *Phys. Rev.* **C69**, 035208 (2004).
23. S. Eidelman et al., *Phys. Lett.* **B592**, 1 (2004).
24. T. Nakano et al., LEPS Collaboration, *Phys. Rev. Lett.* **91**, 012002 (2003).
25. R. N. Cahn and G. H. Trilling, *Phys. Rev.* **D69**, 011501 (2004).
26. D. Melikhov, S. Simula and B. Stech, *Phys. Lett.* **B594**, 265 (2004).
27. A. Hosaka, M. Oka and T. Shinozaki, hep-ph/0409102.
28. M. Diakonov and V. Petrov, *Phys. Rev.* **D69**,(2004).
29. S. Pakvasa and M. Suzuki, *Phys. Rev.* **D70**, 036002 (2004).
30. T. Hyodo and A. Hosaka, hep-ph/0502093, to appear in *Phys. Rev.* **D**.
31. V. Guzey and M. Polyakov, hep-ph/0501010 and these proceedings
32. Fl. Stancu, *Few Body Syst. Suppl.* **10**, 399 (1999).
33. Bin Wu and Bo-Qiang Ma, *Phys. Rev.* **D70**, 034025 (2004).
34. Fl. Stancu, *Group Theory in Subnuclear Physics*, Clarendon Press, Oxford, 1996, Ch. 8.
35. K. Maltman, *Phys. Lett.* **B604**, 175 (2004).
36. A. Aktas, H1 Collaboration, *Phys. Lett.* **B588**, 17 (2004).

Review of Experimental Aspects of Pentaquark Physics

I.I. Strakovsky*, R.A. Arndt†, Ya.I. Azimov**, M.V. Polyakov‡ and R.L. Workman†

*Center for Nuclear Studies, Department of Physics, The George Washington University, Washington, D.C. 20052, USA
E-mail: igor@gwu.edu

†Center for Nuclear Studies, Department of Physics, The George Washington University, Washington, D.C. 20052, USA

**Petersburg Nuclear Physics Institute, Gatchina, St. Petersburg 188300, Russia

‡Physique théorique fondamentale, Département de Physique, Université de Liège, allée du 6 Août 17, bât. B5, B-4000 Liège 1, Belgium

Abstract. Given the existing empirical evidence for an exotic Θ^+ baryon, we analyze possible properties of its $SU(3)_F$-partners, paying special attention to the nonstrange member of the antidecuplet N^*. A modified πN partial-wave analysis results in two candidate masses, 1680 MeV and 1730 MeV. In both cases, the N^* should be rather narrow and highly inelastic. Our results suggest several directions for experimental studies that may clarify properties of the antidecuplet baryons, and structure of their mixing with other baryons. Recent experimental evidence from the GRAAL and STAR Collaborations could be interpreted as observations of a candidate for the Θ^+ nonstrange partner. We also briefly discuss recent negative results regarding the Θ-baryon.

Keywords: Exotic Hadrons, Pentaquarks, Partial-Wave Analysis
PACS: 14.20.Gk, 11.80.Et, 13.30.Eg

The problem of observing multiquark (exotic and/or "cryptoexotic") states is as old as quarks themselves. The first experimental results on searches for exotics [1] were published soon after invention of quarks [2]. The initial straightforward motivation of "Why not?" was later supported by duality considerations [3] (duality was understood in those times as a correspondence between the sum over resonances and the sum over reggeons). However, several years of experimental uncertainty generated the question: "Why are there no strongly bound exotic states, such as those of two quarks and two antiquarks or four quarks and one antiquark?" [4].

Results from a wide range of recent experiments are consistent with the existence of an exotic S=+1 resonance, the $\Theta^+(1540)$ with a narrow width and a mass near 1540 MeV [5]. Direct width determinations have been hindered by the limitations of experimental resolution, resulting in upper bounds of order 10 MeV. The quantum numbers of this state remain unknown, though the prediction of $J^P = 1/2^+$ was obtained in the work [6] that provided motivation for the original search.

Additional information related to the assignment of unitary partners is due to a more recent experimental result [7] giving evidence for one further explicitly exotic particle $\Xi_{3/2}^{--}$, with the mass 1862 ± 2 MeV and width < 18 MeV (*i.e.*, less than resolution). Such a particle had been expected to exist as a member

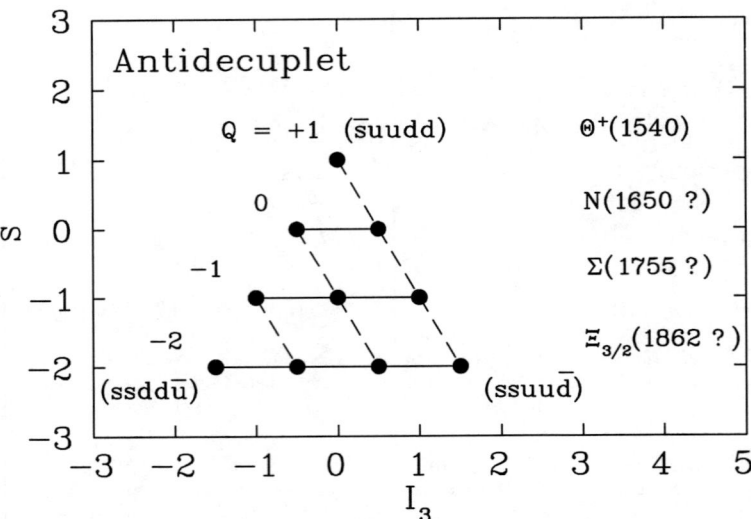

FIGURE 1. Tentative unitary anti-decuplet with Θ^+. Isotopic multiplet (constant values of the charge) shown by solid (dashed) lines.

of an antidecuplet, together with the Θ^+. However, the soliton calculation of the mass difference requires some assumptions. In particular, it depends on the value of the σ-term, which is the subject of controversy. Its value, taken according to the latest πN data analysis [8], leads to an antidecuplet mass difference of about 110 MeV, instead of the originally predicted 180 MeV [9]. So, if the states $\Xi_{3/2}$ [7] and Θ are indeed members of the same antidecuplet, then, according to the Gell-Mann-Okubo rule, the mass difference of any two neighboring isospin multiplets in the antidecuplet should be constant and equal $(M_{\Xi_{3/2}} - M_\Theta)/3 \approx 107$ MeV, which fits the GW SAID σ-term result very well [8]. This shift also affects the masses of other unitary partners of the Θ^+: nucleon-like and Σ-like. The supposed antidecuplet, with Σ- and N-masses determined by the Gell-Mann-Okubo rule, looks today as shown on Fig. 1.

Due to $SU(3)_F$-violating mixing with lower-lying nucleon-like octet states, M_{N^*} may shift upward, and reach about 1680 MeV [9]. Mixing with higher-lying nucleon-like members of exotic 27- and 35-plets may also play a role.

The state $N(1710)$, though listed in the PDG Baryon Summary Table [10] as a 3-star resonance and used as input in the Θ^+ prediction [6], is not seen in the latest analysis of pion-nucleon elastic scattering data (see Table 1). Studies which have claimed to see this state have given widely varying estimates of its mass and width (from ~ 1680 MeV to ~ 1740 MeV for the mass and from ~ 90 MeV to ~ 500 MeV for the total width). Branching ratios have also been given with large uncertainties (10–20% for $N\pi$, 40–90% for $N\pi\pi$, and so on), apart from one which has been presented with much greater precision ($6 \pm 1\%$ for $N\eta$). In any case, the PDG width of $N(1710)$ seems to be too large for the partner of the narrow Θ^+. It would be more natural for members of the same unitary multiplet to have comparable widths.

TABLE 1. Comparison of N(1710) properties.

Collaboration	Mass (MeV)	Width (MeV)	Ref
DPP	1710 (input)	<40	[6]
KH	1723±9	120±15	[11]
CMU	1700±50	90±30	[12]
KSU	1717±28	480±230	[13]
GWU	~1700	not seen	[8]

As has been suggested recently (see Refs. [14, 15]), any standard PWA by itself tends to miss narrow resonances due to a very small πN branching ratio or small total width $\Gamma < 30$ MeV. For this reason, we have considered [14] a modified PWA, assuming the existence of a narrow resonance, and comparing the quality of fit with and without such structures (a more detailed description is given in Ref. [15]). Such an approach was used initially to look for light nucleon resonances [14].

This method, applied to studies of the $\Theta^+(1540)$ [16], though unable to determine the Θ^+ quantum numbers, places a tight limit on the width, in full agreement with the results of other approaches [17]. We have used this method [15] to search πN scattering data for a narrow nucleon-like state assumed to be a member of the antidecuplet, accompanying the $\Theta^+(1540)$. The two candidate states, with masses 1680 MeV and 1730 MeV, would necessarily be quite inelastic with $\Gamma_{el} < 0.5$ MeV and 0.3 MeV, respectively. Some support for a narrow structure in this mass region has recently been obtained in a preliminary report based on direct measurements by the STAR [18] and GRAAL [19] Collaborations. Thus, the modified PWA may be a useful instrument in the search for narrow resonances.

However, there are also collaborations that have not (yet) found the Θ^+ in their data. Some of these negative results have been formally published (see, *e.g.*, Refs. [20, 21, 22, 23, 24, 25]). Others exist either as rumors, or as conference slides. Nevertheless, all such claims cast doubt on the evidence for the Θ^+. Note that the negative results mainly correspond to higher energies than positive ones. New dedicated experiments of several groups are soon expected to provide decisive conclusions on the existence of this and other exotic hadrons.

Meanwhile, more detailed analysis shows that though the present non-observation data require the exotics production to be small, as compared to conventional hadrons, they do not exclude existence of the Θ^+ and/or its companions/analogs. For example, analysis of the BES data [20] in Ref. [26] tends to demonstrate some suppression of the Θ production. This restriction, however, seems to be not very strong; an essentially stronger suppression of the exotics could still have a quite natural explanation. Similar conclusions may be obtained for some other data as well (see, *e.g.*, [27, 28]).

There is also a statement that the observed peak of Θ^+ could be due to a kinematical reflection of some of known resonances. A particular consideration has been suggested by Dzierba *et al.* [29] addressed to the CLAS analysis [30]. The specific model used by Dzierba *et al.* has, however, been criticized [31, 32], and may not be a serious concern for the CLAS results.

We should emphasize here that if the present evidence for the Θ turns out to be incorrect, we would have to answer a different, but also difficult, question: why do we *not* see exotic hadrons? Here we take the position that the Θ does exist, but its production may be governed by various mechanisms. Though we essentially agree with suggestions of Karliner and Lipkin [33] for ways to clarify the problem, we think that, first of all, it is important to reliably confirm the existence of the Θ in the processes where it has been reported. New data are being collected for this purpose, by several collaborations, and one would hope for a definitive answer within a year.

That is why, at the moment, we will assume that the Θ^+ (as well as other multi-quark hadrons) exist, and will discuss some consequences of this assumption (for details, see Ref. [15]).

To summarize, given our current knowledge of the Θ^+, the state commonly known as the $N(1710)$ is not the appropriate candidate to be a member of the antidecuplet together with the Θ^+. Instead, we suggest candidates with nearby masses, $N(1680)$ (more promising) and/or $N(1730)$ (less promising, but not excluded). Our analysis suggests that the appropriate state should be rather narrow and very inelastic. Similar considerations have been applied to the $\Xi_{3/2}(1862)$, assumed to be also a member of the same antidecuplet. It should also be quite narrow.

How reliable are our theoretical predictions? They have, indeed, essential theoretical uncertainties. We have yet to establish the existence of the (narrow) state originally associated with the $N(1710)$. Moreover, we have assumed the presence of only one state with $J^P = 1/2^+$, either $N(1680)$ or $N(1730)$. If both exist with the same spin and parity, our conclusions should be reconsidered.

Futhermore, we use the mixing angle ϕ, taken from Ref. [6], which was actually determined through formulas containing the σ-term (just as the mass difference in the antidecuplet). If we use parameters corresponding to more recent information, for both the σ-term and the mass difference, we obtain larger mixing, up to $\sin\phi \approx 0.15$. With our formulas, this would most strongly influence the partial width $N^* \to \pi\Delta$, increasing it to about 15 MeV. Other partial widths of N^* change not so dramatically, and the total width appears to remain not higher than ~ 30 MeV. Such a width could well be measured, but not in elastic scattering, because of an expected very small elastic branching ratio. Note, however, that the above large value for $\sin\phi$ may appear problematic, since the formulas of Ref. [6] assume linearization with respect to $SU(3)_F$-violation, and need to be reconsidered if the violation appears to be large.

Nevertheless, even having in mind all theoretical uncertainties, we can suggest several directions for experimental studies. First of all, one should search for possible new narrow nucleon state(s) in the mass region near 1700 MeV. Searches may use various initial states, (*e.g.*, πN collision or photoproduction). We expect the largest effect in the $\pi\pi N$ final state (mainly through $\pi\Delta$, though it is forbidden by $SU(3)_F$). The final states ηN and $K\Lambda$ may also be interesting and useful, especially the ratio of ηN and πN partial widths, as the latter is very sensitive to the structure of the octet–antidecuplet mixing. Another interesting possibility to separate antidecuplet and octet components of N^* is provided by comparison of photoexcitation amplitudes for neutral and charged states of this resonance, the point being that the antidecuplet contribution to the photoexitation of the charged N^* is strongly suppressed (see details in Ref. [34]).

On the other hand, such a relatively simple picture of mixing cannot reproduce our small value(s) of Γ_{el}. We assumed in our analysis that this could result from more complicated mixing with several other multiplets [15]. Such a possibility was recently confirmed [35].

For $\Xi_{3/2}$, attempts to measure the total width are necessary, though it could possibly be even smaller than Γ_{Θ^+}. Branching ratios for $\overline{K}\Sigma$ and $\pi\Xi(1530)$, in relation to $\pi\Xi$, are very interesting. These may give important information on the mixing of antidecuplet baryons with octets and higher $SU(3)_F$-multiplets.

ACKNOWLEDGMENTS

The work was partly supported by the U. S. Department of Energy Grant DE–FG02–99ER41110, by the Jefferson Laboratory, by the Southeastern Universities Research Association under DOE Contract DE–AC05–84ER40150, by the Russian State Grant SS–1124.2003.2.

REFERENCES

1. R. L. Cool et al., Phys. Rev. Lett. **17**, 102 (1966).
   R. J. Abrams et al., Phys. Rev. Lett. **19**, 259 (1967).
   J. Tyson et al., Phys. Rev. Lett. **19**, 255 (1967).
2. M. Gell-Mann, Phys. Lett. **8**, 214 (1964).
   G. Zweig, CERN preprints TH-401, TH-412 (1964).
3. J. Rosner, Phys. Rev. Lett. **21**, 950 (1968).
4. H. J. Lipkin, Phys. Lett. **45B**, 267 (1973).
5. See for example, T. Nakano, in *Proceedings of the Workshop on the Physics of Excited Nucleons (NSTAR2004), Grenoble, France, March, 2004*, edited by J.-P. Bocquet, V. Kuznetsov, and D. Rebreyend (World Scientific, 2004), p. 3.
6. D. Diakonov, V. Petrov, and M. Polyakov, Z. Phys. A **359**, 305 (1997).
7. C. Alt et al., Phys. Rev. Lett. **92**, 042003 (2004).
8. R.A. Arndt, W.J. Briscoe, I.I. Strakovsky, R.L. Workman, and M.M. Pavan, Phys. Rev. C **69**, 035213 (2004).
9. D. Diakonov and V. Petrov, Phys. Rev. D **69**, 094011 (2004).
10. S. Eidelman et al., Phys. Lett. B **592**, 1 (2004).
11. R. Koch, Z. Phys. C **29**, 597 (1985); G. Höhler, *Pion–Nucleon Scattering*, Landolt–Börnstein Vol. I/9b2, edited by H. Schopper (Springer Verlag, 1983).
12. R.E. Cutkosky et al., in *Proceedings of 4th International Conference on Baryon Resonances, Toronto, Canada, July 1980*, published in Baryon 1980:19 (QCD161:C45:1980); R.E. Cutkosky and S. Wang, Phys. Rev. D **42**, 235 (1990).
13. D.M. Manley and E.M. Saleski, Phys. Rev. D **45**, 4002 (1992).
14. Ya.I. Azimov, R.A. Arndt, I.I. Strakovsky, and R.L. Workman, Phys. Rev. C **68**, 045204 (2003).
15. R.A. Arndt, Ya.I. Azimov, M.V. Polyakov, I.I. Strakovsky, and R.L. Workman, Phys. Rev. C **69**, 035208 (2004).
16. R.A. Arndt, I.I. Strakovsky, and R.L. Workman, Phys. Rev. C **68**, 042201 (2003).

17. R.N. Cahn and G.H. Trilling, *Phys. Rev. D* **69**, 011501 (2004).
18. S. Kabana, hep-ex/0406032.
19. V. Kuznetsov, in *Proceedings of the Workshop on the Physics of Excited Nucleons (NSTAR2004), Grenoble, France, March, 2004*, edited by J.-P. Bocquet, V. Kuznetsov, and D. Rebreyend (World Scientific, 2004), p. 197; hep-ex/0409032.
20. J.Z. Bai et al., *Phys. Rev. D* **70**, 012004 (2004).
21. I. Abt et al. [HERA-B Collaboration], *Phys. Rev. Lett.* **93**, 212003 (2004).
22. C. Pinkenburg (for the PHENIX Collaboration), *J. Phys. G* **30**, S1201 (2004).
23. Yu. M. Antipov et al. [SPHINX Collaboration], *Eur. Phys. J. A* **26**, 455 (2004).
24. M.J. Longo et al. [HyperCP Collaboration], *Phys. Rev. D* **70**, 111101 (2004).
25. S. Schael et al. [ALEPH Collaboration], *Phys. Lett. B* **599**, 1 (2004).
26. Ya.I. Azimov and I.I. Strakovsky, *Phys. Rev. C* **70**, 035210 (2004).
27. K. Hicks, Plenary talk at 1st Meeting of the APS Topical Group on Hadronic Physics, Batavia, IL, USA, October, 2004; hep-ex/0412048.
28. B. Aubert et al. [BABAR Collaboration], submitted to Phys. Rev. Lett.; hep-ex/0502004.
29. A. Dzierba et al., *Phys. Rev. D* **69**, 051901 (2004).
30. S. Stepanyan et al., *Phys. Rev. Lett.* **91**, 252001 (2003).
31. K. Hicks, V. Burkert, A. Kudryavtsev, I. Strakovsky, and S. Stepanyan, Submitted to *Phys. Rev. D*, hep-ph/0411265.
32. Y. Oh, K. Nakayama, and T.-S.H. Lee, hep-ph/0412363.
33. M. Karliner and H. Lipkin, *Phys. Lett. B* **597**, 309 (2004).
34. M.V. Polyakov and A. Rathke, *Eur. Phys. J. A* **18**, 691 (2003).
35. V. Guzey and M.V. Polyakov, hep-ph/0501010

DIAGONAL AND OFF-DIAGONAL STRUCTURE

Probing the partonic structure of exotic particles in hard electroproduction

I.V. Anikin[*,†], B. Pire[†], L. Szymanowski[**,‡], O.V. Teryaev[§] and S. Wallon[*]

LPT, Université Paris-Sud[1], F-91405 Orsay, France
†CPhT, École Polytechnique,[2] F-91128 Palaiseau, France
**Physique théorique fondamentale, Département de Physique, Université de Liège,
allée du 6 Août 17, bât. B5, B-4000 Liège 1, Belgium
E-mail: lechszym@fuw.edu.pl
‡SINS, Warsaw, Poland
§BLTP, Joint Institute for Nuclear Research, Dubna, Russia

Abstract. We argue that the electroproduction of exotic particles is a useful tool for study of their partonic structure. In the case of hybrid mesons, the magnitude of their cross sections shows that they are accessible for measurements in existing electroproduction experiments.

Keywords: Hybrid meson electroproduction
PACS: 12.38.Bx 12.39.Mk

1. INTRODUCTION

Searching for exotic particles whose structure cannot be explained in the framework of the constituent quark model is at present a very lively subject of studies. One of their main streams concentrates on finding a firm evidence for the existence (or absence) of pentaquarks, with the lowest Fock state containing $uudd\bar{s}$ quarks. This is particularly important in view of some contradictory results of experimental searches, reviewed at this conference by I. Strakovsky [1]. Because of the apparently very small decay width of pentaquarks ($\approx 1\,\text{MeV}$) their theoretical description represents a big challenge, as discussed by M. Polyakov [2]. Another family of exotic particles which are at present a subject of intense studies are glueballs and hybrid mesons, for their review see, e.g. [3]. The theoretical description of these particles requires to include in the lowest Fock state the gluonic degrees of freedom. All these facts stress the importance of studies of the structure of exotic particles within Quantum Chromodynamics (QCD) for understanding the mechanism of quark and gluon confinement.

The hard exclusive electroproduction of particles permits to probe the structure of particles at the parton level. The scattering amplitude factorizes in this case into perturbatively calculable coefficient function, the non-perturbative distribution

---

[1] Unité mixte 8627 du CNRS
[2] Unité mixte 7644 du CNRS

amplitude (DA) of a produced particle and the non-perturbative generalized parton distribution (GPD) describing the transition probability amplitude of the target. The DA and GPD encode full information about parton distribution inside particles participating in a collision, for a review see, e.g. [4].

Recently it was emphasized [5] that hard electroproduction can also be used to reveal the partonic structure of exotic particles. In particular, for the processes involving pentaquarks

$$\gamma^* p \to \bar{K}^0 \Theta^+ , \qquad \gamma^* n \to \bar{K}^- \Theta^+ , \qquad (1)$$

as well as for their analogs with vector mesons K^*, the necessary formalism based on the GPD of $p(n) \to \Theta^+$ was developed which has permitted next to evaluate some contributions to the cross section in the Born approximation [5]. Unfortunately, realistic cross section estimates based on these expressions are not possible at the moment due to our ignorance of these $p(n) \to \Theta^+$ GPDs.

Another example of processes in which the partonic structure of exotic particles can be probed is supplied by the hard electroproduction of $J^{PC} = 1^{-+}$ hybrid meson $\pi_1(1400)$

$$\gamma^* p \to \pi_1 p . \qquad (2)$$

The process (2) is the subject of recent studies in Refs. [6] and [7]. Below we present in some details our results.

2. THE PUZZLE WITH DA OF $J^{PC} = 1^{-+}$ HYBRID MESON

In Ref. [8] Jaffe *et al.* analyse the particle spectrum (including its exotic sector) by construction of lowest-dimensional, gauge invariant, colorless *local* operators. As a result of this study a hybrid meson with exotic quantum numbers $J^{PC} = 1^{-+}$ is described by a *local* composite operator of dimension 5 built from quark and gluonic fields, e.g. $\bar{\psi}\gamma^\mu G_{\mu\nu}\psi$, where $G^{\mu\nu}$ is the field strength tensor of a gluon. Such local operator has a twist equal to 4. Naively one could think therefore that this result implies that the leading twist-2 DA of a hybrid meson vanishes. Consequently, the scattering amplitude for hard hybrid meson electroproduction would be suppressed (at large photon virtuality Q^2) in comparison with the amplitude for production in the same process of non-exotic ρ−meson ($J^{PC} = 1^{--}$). The analysis of Ref. [6] shows that this naive conclusion is not correct for longitudinally polarized hybrid meson for which the leading twist-2 DA is not zero. Thus, the electroproduction of such hybrid meson doesn't need to be strongly suppressed in comparison with a ρ−meson production.

The analysis of Ref. [6] consists in exploiting the fact that DAs of particles are defined by a *non local* composite operators. In the case of a longitudinally polarised hybrid meson $H(p,0)$ of momentum p the DA has the form

$$\langle H(p,0)|\bar{\psi}(-z/2)\gamma_\mu[-z/2;z/2]\psi(z/2)|0\rangle = if_H M_H e_{L\mu}^{(0)} \int_0^1 dy\, e^{i(\bar{y}-y)p\cdot z/2} \phi_L^H(y) , \quad (3)$$

where $[-z/2; z/2]$ on the l.h.s. denotes the path-ordered gluonic exponential along the straight line connecting the initial and final points $(-z/2, z/2)$ which provides the gauge invariance for bilocal operator and equals unity in a light-like (axial) gauge. In Eq. (3) f_H, M_H, $e_\mu^{(0)}$ denote the hybrid meson coupling constant, its mass and the longitudinal polarisation vector, respectively. Because of the positive charge parity of the hybrid meson, its DA, ϕ^H, is an antisymmetric function: $\phi^H(y) = -\phi^H(1-y)$. This last property implies in particular that

$$\int_0^1 dy\, \phi^H(y) = 0,$$

which means that the first term of the Taylor expansion in z of l.h.s of (3) vanishes and generally only terms with odd powers of z contribute to this expansion

$$\langle H(p,\lambda)|\bar{\psi}(-z/2)\gamma_\mu[-z/2; z/2]\psi(z/2)|0\rangle = \qquad (4)$$
$$\sum_{n\, odd} \frac{1}{n!} z_{\mu_1}..z_{\mu_n} \langle H(p,\lambda)|\bar{\psi}(0)\gamma_\mu \overleftrightarrow{D}_{\mu_1} .. \overleftrightarrow{D}_{\mu_n} \psi(0)|0\rangle,$$

in which D_μ is the usual covariant derivative and $\overleftrightarrow{D}_\mu = \frac{1}{2}(\overrightarrow{D}_\mu - \overleftarrow{D}_\mu)$. The first non-vanishing term of the expansion (4) corresponds to $n = 1$ and its twist-2 contribution involves operator

$$\mathcal{R}_{\mu\nu} = \mathcal{S}_{(\mu\nu)} \bar{\psi}(0) \gamma_\mu \overleftrightarrow{D}_\nu \psi(0), \qquad (5)$$

where $\mathcal{S}_{(\mu\nu)}$ denotes the symmetrization operator ($\mathcal{S}_{(\mu\nu)} T_{\mu\nu} = 1/2(T_{\mu\nu} + T_{\nu\mu})$). Let us note that $\mathcal{R}_{\mu\nu}$ is proportional to the quark energy-momentum tensor, $\mathcal{R}_{\mu\nu} = -i\Theta_{\mu\nu}$. Its matrix element of interest is

$$\langle H(p,\lambda)|\mathcal{R}_{\mu\nu}|0\rangle = \frac{1}{2} f_H M_H \mathcal{S}_{(\mu\nu)} e_\mu^{(\lambda)} p_\nu \int_0^1 dy (1-2y) \phi^H(y). \qquad (6)$$

Examining symmetry properties of the operator $\mathcal{R}_{\mu\nu}$ and the matrix element (6) reveals indeed that the C and P parities of $\mathcal{R}_{\mu\nu}$ are (-1), equal to those of the hybrid meson. This proves that f_H is non zero and allows to determine its value through non perturbative methods, such as, e.g. the QCD sum rules method [9]:
a) using the equations of motion one derives that $\partial^\mu \Theta_{\mu\nu} = g\bar{\psi}\gamma^\mu G_{\mu\nu}\psi$,
b) the value of the coupling constant f_H is determined by the correlator of two $\bar{\psi}\gamma^\mu G_{\mu\nu}\psi$ operators.
This results in the estimate $f_H \approx 50\,\text{MeV}$.

The description of the DA of hybrid meson is complete by fixing the form of $\phi^H(y)$. This DA satisfies usual non-singlet evolution equations and, forgetting the slowly varying logarithmic scaling violation factor, we assume in our estimates that it is given by the asymptotic expression [12]: $\phi^H(y)_{as} = 30y(1-y)(1-2y)$.

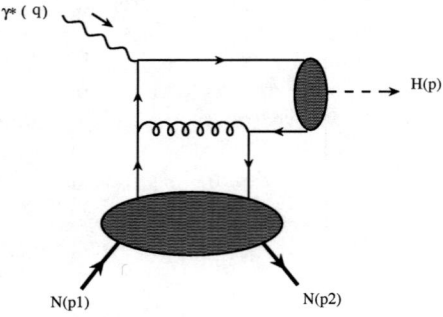

FIGURE 1. Typical diagram describing the electroproduction of a meson at lowest order. The grey blobs are the non-perturbative meson distribution amplitude and nucleon generalized parton distribution.

3. THE HYBRID MESON PRODUCTION

3.1. The scattering amplitude

Knowing the DA of the hybrid meson (3) one can determine the scattering amplitude at the leading order in strong coupling constant α_s. The calculations proceed in a full analogy as in the ρ-meson case. The electroproduction process (2) occurs in the scaling regime where the virtuality of the photon $Q^2 = -q^2$, see Fig. 1, is large and scales with the energy of the process, and the momentum transfer t is small, $-t \ll Q^2$. In such kinematics the conditions for the QCD factorization are fulfilled and the scattering amplitude at leading twist is at given factorisation scale μ expressed as a convolution

$$\mathcal{A} = \int_0^1 du \int_0^1 dx \, \phi(u,\mu^2) H(x,u,Q^2,\mu^2,\mu_R^2) F(x,\mu^2) \equiv \phi \otimes H \otimes F \qquad (7)$$

of the DA $\phi(u,\mu^2)$ of meson, the GPD of the target nucleon $F(x,\mu^2)$ and the perturbatively calculable coefficient function $H(x,u,Q^2,\mu^2,\mu_R^2)$. The hard part of the scattering amplitude depends also on the renormalisation scale μ_R. The expression for $H(x,u,Q^2,\mu^2,\mu_R^2)$ is obtained by adding the contributions of 4 diagrams, one of which is shown in Fig. 1. Taking into account the contributions of all diagrams one arrives to the scattering amplitude of the form

$$\mathcal{A}_{\gamma_L^* p \to H_L^0 p} = \frac{e\pi\alpha_s f_H C_F}{\sqrt{2}\, N_c Q} \left[e_u \mathcal{H}_{uu}^- - e_d \mathcal{H}_{dd}^- \right] \mathcal{V}^{H-}, \qquad (8)$$

where

$$\mathcal{H}_{ff}^{\pm} = \int\limits_{-1}^{1} dx \quad \left[\overline{U}(p_2)\hat{n}U(p_1)H_{ff}(x,\xi) + \overline{U}(p_2)\frac{i\sigma_{\mu\alpha}n^\mu\Delta^\alpha}{2M}U(p_1)E_{ff}(x,\xi)\right]$$
$$\times \left[\frac{1}{x+\xi-i\epsilon} \pm \frac{1}{x-\xi+i\epsilon}\right], \qquad (9)$$

and

$$\mathcal{V}^{M,\pm} = \int\limits_{0}^{1} dy \phi^H(y)\left[\frac{1}{y} - \frac{1}{1-y}\right].$$

The functions H and E are standard leading twist GPD's having well known properties. Eq. (9) also show definitions of $\mathcal{H}_{ff}^+$ and $\mathcal{V}^{M,+}$ necessary for comparison with the production of ρ-meson.

3.2. Scale fixing and numerical predictions.

The QCD factorisation of the scattering amplitude as given by Eq. (7) introduces dependence of the coefficient function H, the soft DA ϕ and the target GPD F on the factorization scale μ and on the renormalisation scale μ_R, which in principle should be treated as two independent parameters. Since in the coefficient function H only first terms of its perturbative expansion are at best known, the dependence of the amplitude $\mathcal{A}$ on μ and μ_R can be large and leads to significant theoretical uncertainties of results. In order to minimize this uncertainty a scale fixing procedure has to be invoked.

Inspired by results obtained in calculations of the pion form-factor we adapted the convention that both scales are equal, $\mu = \mu_R$ [7]. Fixing of the scale μ is then done by applying a modified version of the Brodsky-Lepage-McKenzie (BLM) procedure: the scale μ is chosen in such a way which leads to a vanishing of large terms proportional to the β-function (which governs the μ-behaviour of the strong coupling constant $\alpha_s(\mu^2)$) in the square of the scattering amplitude known with at least the next-to-leading order accuracy, see [10] for details. In more physical terms the BLM procedure consists in absorbing numerically large terms originating from the renormalisation into redefinition of the argument of the strong coupling constant $\alpha_S(\mu^2)$.

Results of the numerical analysis of hybrid meson electroproduction are shown in Fig. 2. The solid line represents the cross-section of the non-exotic ρ-meson electroproduction (quark exchange contribution only) obtained with the BLM scale fixing. The dashed line describes the cross section for production of the hybrid meson $\pi_1(1400)$. The comparison of these two curves leads to the conclusion that the production process of hybrid meson has a sizeable cross section, so that it should be already now accessible to measurements. Fig. 2 also contains the comparison of our results with the predictions of Ref. [11] on the ρ-meson production which takes into account the intrinsic transverse momenta of partons without invoking the BLM procedure.

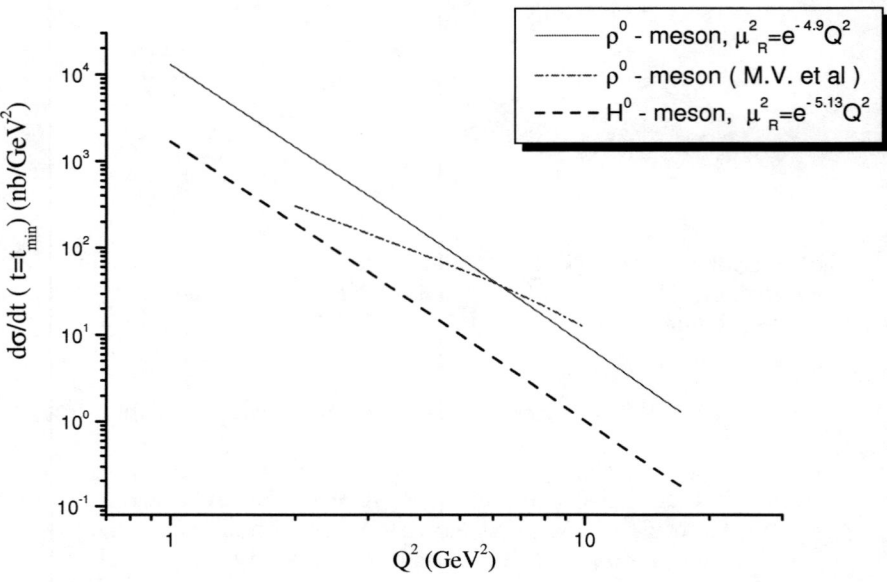

FIGURE 2. Differential cross-section for exotic hybrid meson electroproduction (dashed line) with $\mu_R^2 = e^{-5.13}Q^2$ compared with the quark contribution to ρ^0 electroproduction (solid line) with $\mu_R^2 = e^{-4.9}Q^2$, as a function of Q^2, for $x_B \approx 0.33$. The dash-dotted line is the result of Vanderhaegen et al [11] for ρ electroproduction.

4. HYBRID MESON PROBED THROUGH THE ELECTROPRODUCTION OF $\pi\eta$ PAIRS

In the case where there is no recoil detector which allows to identify the hybrid meson production events through a missing mass reconstruction, one has to base an identification process through the possible decay products of the hybrid meson H^0. Since the particle $\pi_1(1400)$ has a dominant $\pi\eta$ decay mode one can use the electroproduction process (see Fig. (3)),

$$e(k_1) + N(p_1) \to e(k_2) + \pi^0(p_\pi) + \eta(p_\eta) + N(p_2) \tag{10}$$

or $\gamma^*(q) + N(p_1) \to \pi^0(p_\pi) + \eta(p_\eta) + N(p_2)$, (see Fig. (4)), to probe the hybrid meson properties. The computation of the process in Fig. (4) requires a knowledge of the

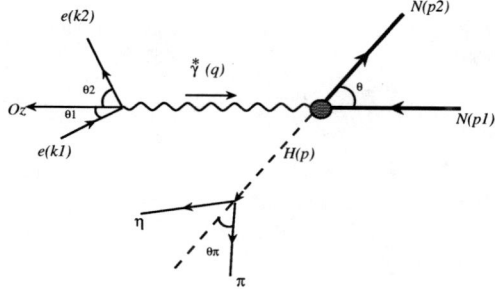

FIGURE 3. The kinematics of the electroproduction of $\pi\eta$ pair

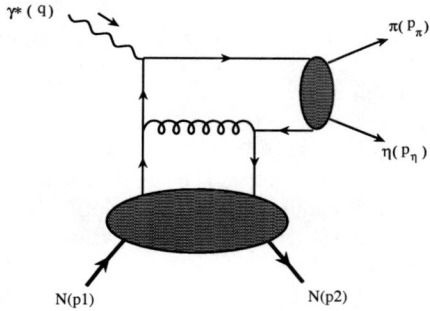

FIGURE 4. Typical diagram describing the electroproduction of $\pi\eta$ pair. The grey blobs are the non-perturbative $\pi\eta$-pair GDA and the nucleon GPD.

generalized distribution amplitude (GDA) for the $\pi\eta$ pair [13]:

$$\langle \pi^0(p_\pi)\eta(p_\eta)|\bar{\psi}_{f_2}(-z/2)\gamma^\mu[-z/2;z/2]\tau^3_{f_2 f_1}\psi_{f_1}(-z/2)|0\rangle =$$

$$p^\mu_{\pi\eta}\int_0^1 dy\, e^{i(\bar{y}-y)p_{\pi\eta}\cdot z/2}\Phi^{(\pi\eta)}(y,\zeta,m^2_{\pi\eta}), \qquad (11)$$

where the total momentum of $\pi\eta$ pair is $p_{\pi\eta} = p_\pi + p_\eta$ while $m^2_{\pi\eta} = p^2_{\pi\eta}$. The $\pi\eta$ distribution amplitude $\Phi^{(\pi\eta)}$ describes non resonant as well as resonant contributions.

It does not possess any symmetry properties concerning the $\tilde{\zeta}$-parameter

$$\tilde{\zeta} = \frac{p_\pi^+}{(p_\pi + p_\eta)^+} - \frac{m_\pi^2 - m_\eta^2}{2m_{\pi\eta}^2}, \quad 1 - \tilde{\zeta} = \frac{p_\eta^+}{(p_\pi + p_\eta)^+} + \frac{m_\pi^2 - m_\eta^2}{2m_{\pi\eta}^2}, \quad (12)$$

which describes roughly the fraction of total '+' momentum of the pair carried by the π-meson (in case of particles with equal masses $\tilde{\zeta} = p_\pi^+/p^+$) and which is related to the angle θ_{cm}, defined as the polar angle of the π meson in the center of mass frame of the meson pair:

$$2\tilde{\zeta} - 1 = \beta \cos\theta_{cm}, \qquad \beta = \frac{2|\mathbf{p}|}{m_{\pi\eta}}, \quad (13)$$

In the reaction under study, the $\pi\eta$ state may have total momentum, parity and charge-conjugation in the following sequence $J^{PC} = 0^{++}, 1^{-+}, 2^{++}, ...$, that corresponds to the following values of the $\pi\eta$ orbital angular momentum L: $L = 0, 1, 2, ...$, respectively. Thus a resonance with a $\pi\eta$ decay mode for odd orbital angular momentum L should be considered as an exotic meson. The mass region around 1400 MeV is dominated by the strong $a_2(1329)\,(2^{++})$ resonance, it is therefore natural to look for the interference of the amplitudes of hybrid and a_2 production, resulting in the angular asymmetry in $\pi\eta$ production.

Asymmetries are often a good way to get a measurable signal for a small amplitude by means of its interference with a larger one. In the asymmetry a small amplitudes enters linearly rather than quadratically as in the cross section which increases chances for sizeable effects. In our case, since the hybrid production amplitude may be rather small with respect to a continuous background, we use the supposedly large amplitude for a_2 electroproduction as a magnifying lens to unravel the presence of the exotic hybrid meson. Since these two amplitudes describe different orbital angular momentum of the π and η mesons, the asymmetry which is sensitive to their interference is an angular asymmetry defined by

$$A(Q^2, y_l, \hat{t}, m_{\pi\eta}) = \frac{\int \cos\theta_{cm}\, d\sigma^{\pi^0\eta}(Q^2, y_l, \hat{t}, m_{\pi\eta}, \cos\theta_{cm})}{\int d\sigma^{\pi^0\eta}(Q^2, y_l, \hat{t}, m_{\pi\eta}, \cos\theta_{cm})} \quad (14)$$

as a weighted integral over polar angle θ_{cm} of the relative momentum of π and η mesons. The variable y_l is the longitudinal fraction of the electron momentum k_1 carried by the virtual photon.

Our estimation of the asymmetry (14) is shown on Fig. 5 and it has a sizable magnitude. The structure of this asymmetry is very natural: the first positive extremum is located at $m_{\pi\eta}$ around the mass of a_2 meson while the second negative extremum corresponds to the hybrid meson mass.

5. SUMMARY

We advocate to use hard electroproduction to uncover the partonic structure of exotic particles. In the case of the exotic $J^{PC} = 1^{-+}$ hybrid meson $\pi(1400)$ we

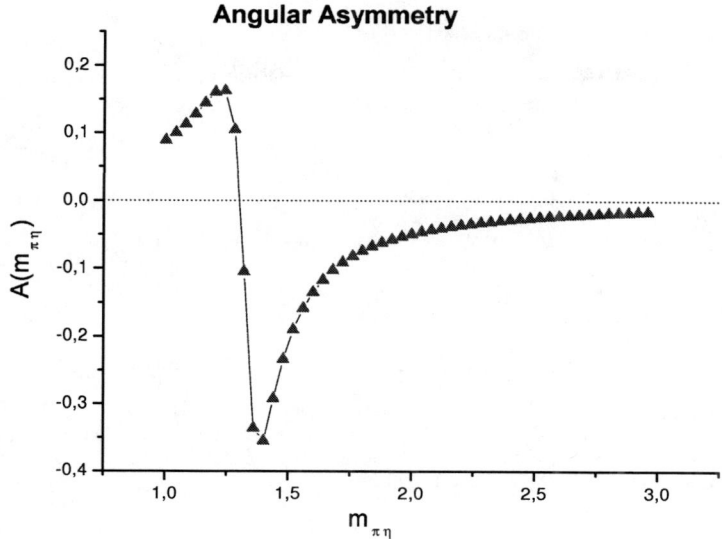

FIGURE 5. The angular asymmetry as a function of $m_{\pi\eta}$.

presented quantitative estimates of the leading twist contribution to the electroproduction amplitude, which lead to sizeable effects. The resulting order of magnitude of cross section is smaller than the ρ electroproduction but similar to the π electroproduction. Thus the exotic hybrid meson effects should be measurable at dedicated experiments at JLab, Hermes or Compass.

ACKNOWLEDGMENTS

This work is supported by the Polish Grant 1 P03B 028 28, the Joint Research Activity "Generalized Parton Distributions" of the european I3 program Hadronic Physics, contract RII3-CT-506078 and the French-Polish scientific agreement Polonium. The work of I.V.A. and O.V.T. is supported in part by INTAS (Project 00/587) and RFBR (Grant 03-02-16816). I. V. A. thanks NATO for a Grant. L.Sz. is a Visiting Fellow of the Fond National pour la Recherche Scientifique (Belgium).

REFERENCES

1. I. Strakovsky, talk at this conference.
2. M.V. Polyakov, talk at this conference.
3. S. Godfrey, "The phenomenology of glueball and hybrid mesons," arXiv:hep-ph/0211464, talk at *Workshop on Future Physics at COMPASS*, Geneva, Switzerland, 26-27 Sep 2002.

4. M. Diehl, *Phys. Rept.* **388** 41 (2003).
5. M. Diehl, B. Pire and L. Szymanowski, *Phys. Lett.* B **584** 58 (2004), "Exclusive production of pentaquarks in the scaling regime," arXiv:hep-ph/0501074, Proceedings of *10th International Conference on Structure of Baryons (Baryons 2004)*, Palaiseau, France, 25-29 Oct 2004.
6. I. V. Anikin, B. Pire, L. Szymanowski, O. V. Teryaev and S. Wallon, *Phys. Rev. D* **70** 011501 (2004).
7. I. V. Anikin, B. Pire, L. Szymanowski, O. V. Teryaev and S. Wallon, *Phys. Rev. D* **71** 034021 (2005).
8. R. L. Jaffe, K. Johnson and Z. Ryzak, *Annals Phys.* **168** 344 (1986).
9. I. I. Balitsky, D. Diakonov and A. V. Yung, *Z. Phys. C* **33** 265 (1986); I. I. Balitsky, D. Diakonov and A. V. Yung, *Sov. J. Nucl. Phys.* **35** 761 (1982).
10. I. V. Anikin, B. Pire, L. Szymanowski, O. V. Teryaev and S. Wallon, "On BLM scale fixing in exclusive processes," arXiv:hep-ph/0411408.
11. M. Vanderhaeghen, P. A. M. Guichon and M. Guidal, *Phys. Rev. D* **60**, 094017 (1999).
12. M. K. Chase, *Nucl. Phys. B* **174** 109 (1980).
13. M. Diehl, T. Gousset, B. Pire and O. Teryaev, *Phys. Rev. Lett.* **81**, 1782 (1998); M. Diehl, T. Gousset, B. Pire and O. V. Terayev, arXiv:hep-ph/9901233; M. V. Polyakov, *Nucl. Phys. B* **555** 231 (1999); B. Lehmann-Dronke *et al*, *Phys. Lett. B* **475** 147 (2000); B. Pire and L. Szymanowski, *Phys. Lett. B* **556**, 129 (2003).

A toy model for generalised parton distributions

J.R. Cudell *,†, F. Bissey**, J. Cugnon*, M. Jaminon*, J.P. Lansberg* and P. Stassart*

*Physique théorique fondamentale, Département de Physique, Université de Liège, allée du 6 Août 17, bât. B5, B-4000 Liège 1, Belgium
†JR.Cudell@ulg.ac.be
** Institute of Fundamental Sciences, Massey University, P.B. 11 222, Palmerston North, New Zealand

Abstract. We give the results of a simple model for the diagonal and off-diagonal valence quark distributions of a pion. We show that structure can be implemented in a gauge-invariant manner. This explicit model questions the validity of the momentum sum rule, and gives an explicit counter-example to the Wandzura-Wilczek ansatz for twist-3 GPD's.

Keywords: parton distributions, GPD, bound states, gauge invariance.
PACS: 13.60.Hb, 13.60.Fz, 14.40.Aq, 12.38.Aw.

1. INTRODUCTION

We review the results on structure functions and generalised parton distributions (GPD's) that we obtained some time ago in [1]. Our purpose is to explore the simplest model of a hadron in order to have an explicit representation and some intuition for GPD's. Hence we have considered the simplest hadron, *i.e.* the pion, and concentrated on its valence-quark content. In order to further simplify the model, we took a π^0, although our twist-2 and twist-3 results remain true for $\pi^\pm$, as the diagrams involving the direct $\gamma\pi^\pm$ coupling are suppressed by powers of Q^2.

2. DIAGONAL STRUCTURE

2.1. Structureless pions

We must first consider the diagonal structure functions. In order to calculate them explicitly, we must make a model for the pion. The minimal requirement is that it is a pseudo-scalar made of a quark and an antiquark. Hence we must select the proper spin states, which is easily done through the use of a γ_5 vertex. The simplest assumption is then that the pion can be treated as a point-like particle, coupling to quarks via an effective 3-point vertex (shown in Fig.1):

$$\Gamma_3 = ig\gamma_5, \qquad (1)$$

with g a coupling constant.

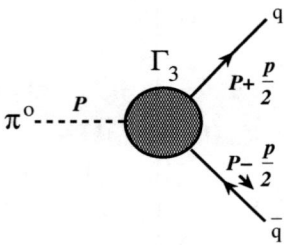

FIGURE 1. The lowest-order vertex $\pi^0 q\bar{q}$.

The lowest-order approximation for the structure functions F_1 and F_2 then comes from the discontinuity disc($\mathcal{A}_{\mu\nu}$) of the diagrams of Fig. 2. In order to make the calculation infrared finite (or at least to avoid the poles on the vertical lines of the diagrams of Fig. 2), we assume that the quarks are sufficiently massive:

$$2m_q > m_\pi \tag{2}$$

(We shall take $m_q \approx 300$ MeV in the following.) The discontinuity is then obtained by putting the intermediate states on-shell. The answer one gets is explicitly gauge-invariant, and one may obtain the structure functions via the usual formula:

$$\frac{1}{4\pi}\text{disc}(\mathcal{A}_{\mu\nu}) = \left(-g_{\mu\nu} + \frac{q_\mu q_\nu}{q^2}\right) F_1 + \left(P_\mu - q_\mu \frac{P.q}{q^2}\right)\left(P_\nu - q_\nu \frac{P.q}{q^2}\right)(P.q)F_2, \tag{3}$$

with P the momentum of the pion and q that of the photon.

In the Bjorken limit, $Q^2 \to \infty$, $x = Q^2/(2\,P.q)$ fixed, one obtains

$$F_1 = \frac{5g^2}{24\pi^2}\left[\log\left(\frac{(1-x)Q^2}{xM^2}\right) - \frac{m_\pi^2}{M^2}x(1-x)\right], \tag{4}$$

with $M^2 = m_q^2 - m_\pi^2 x(1-x)$.

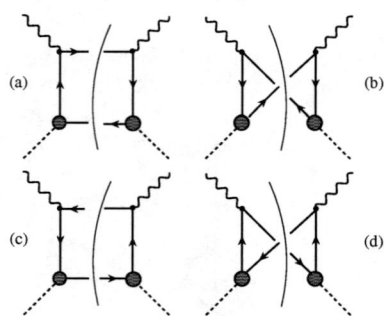

FIGURE 2. The four cut diagrams contributing to structure functions.

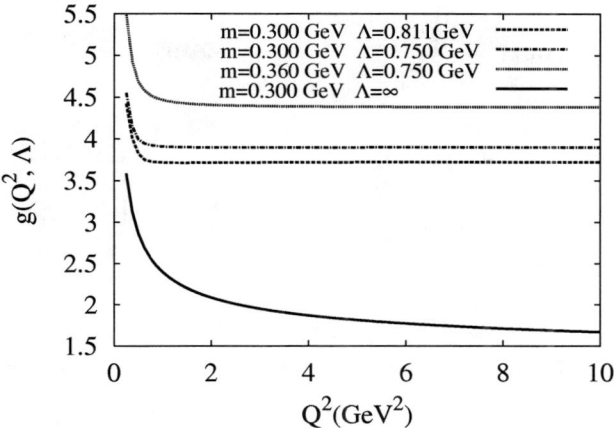

FIGURE 3. The coupling constant of the vertex.

One still needs however to determine the coupling g of the vertex (1). In principle, this can be done via the electromagnetic form factor at $t=0$. We find it easier to use the Adler sum rule, which should be equivalent for the leading twist:

$$\int_0^1 dx\ F_1(x,Q^2) = \frac{5}{18}. \tag{5}$$

It amounts to saying that our pion is made of $u\bar{u}$ and $d\bar{d}$ with equal probability. It leads to a coupling g given by the lower curve of Fig. 3. One obtains the Callan-Gross relation $F_2(x,Q^2) = 2xF_1(x,Q^2)$, and a prediction for F_1.

However, we see that, because F_1 has a logarithmic growth at fixed g, the normalisation condition actually makes g run down as $1/\log Q^2$. This is not consistent with the definition (1) of the vertex: g can depend only on the variables entering the vertex, i.e. P and p, and cannot depend directly on Q^2.

Furthermore, although the diagrams of Fig. 2.a and 2.c have a probabilistic interpretation, they are not enhanced by a power of Q^2 with respect the interference graphs of Fig. 2.b and 2.d: we cannot define parton distributions. The reason for this is obvious: our pion does not have a structure.

2.2. Structure and gauge invariance

Guided again by simplicity, we shall assume that it is a good approximation to represent confinement effects by a cut on the square of the relative 4-momentum of the partons. The vertex of Fig. 1 now becomes a step function of p^2,

$$\Gamma_3(P,p) = ig(Q^2,\Lambda^2)\gamma_5\theta(|p^2|<\Lambda^2) \tag{6}$$

and we shall take $\Lambda \approx 0.8$ GeV (or $r_\pi \approx 0.25$ fm). But the introduction of structure

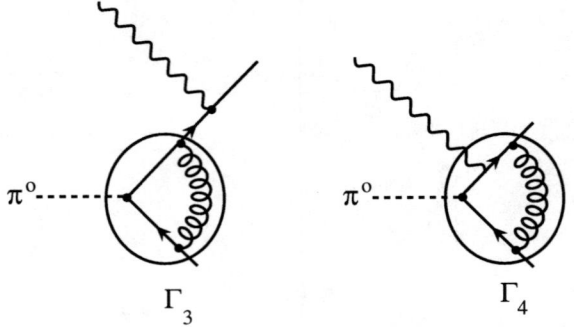

FIGURE 4. Illustration of the necessity of a 4-point vertex.

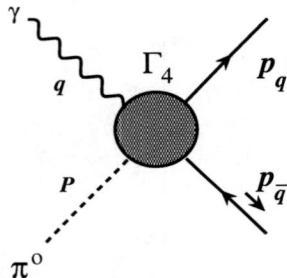

FIGURE 5. Representation of the 4-point vertex.

has a dire consequence: this simple modification is not sufficient as it breaks gauge invariance. This can easily be understood if we represent the vertex function by an exchange, as in Fig. 4. To get a complete set of diagrams leading to a gauge-invariant answer, on must include both diagrams in Fig. 4. The first diagram is analogous to a simple modification of the 3-point vertex, such as (6), whereas the second one can only be included in a 4-point vertex, as in Fig. 5.

The latter is unknown, and we shall use a simple trick to model it. To analyse the problem, it is enough to consider one half of the cut amplitude, as in Fig. 6. We can define the usual Mandelstam variables as $\hat{t} = (P - p_q)^2$ and $\hat{u} = (P - p_{\bar{q}})^2$. The cut on the relative momentum then amounts to a cut on t for diagram 6.a ($p^2 = -m_\pi^2 + 2m_q^2 + 2t$), whereas it is a cut on u for diagram 6.b ($p^2 = -m_\pi^2 + 2m_q^2 + 2u$). Hence both diagrams have different physical cuts, which gives rise to the gauge-invariance problem. The solution is then simple: one must invent a 4-point vertex such that both graphs are cut in the same way. Hence, we need to multiply the sum of the two graphs of Fig. 6 by the same function and, to obtain (6), we must choose

$$F(t,u) = [\theta(|-m_\pi^2 + 2m_q^2 + 2\hat{t}| < \Lambda^2) + \theta(|-m_\pi^2 + 2m_q^2 + 2\hat{u}| < \Lambda^2)]. \quad (7)$$

The first term is the 3-point vertex (6), whereas the second term can be interpreted as a contribution from a new 4-point vertex, shown in Fig. 5:

$$\Gamma_4(p_q, p_{\bar{q}}, P, q) = g(Q^2, \Lambda^2)\theta(|-m_\pi^2 + 2m_q^2 + 2\hat{u}| < \Lambda^2)\frac{\gamma_\mu(\gamma \cdot (P - p_{\bar{q}}) + m_q)\gamma_5}{(P - p_{\bar{q}})^2 - m_q^2} \quad (8)$$

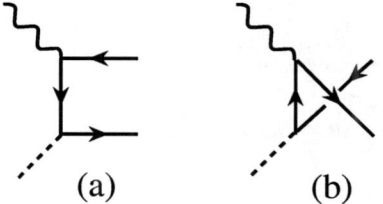

(a) (b)

FIGURE 6. Two amplitudes of the cut.

One can then re-calculate F_1 and F_2, and normalise them again. The coupling constants are shown in Fig. 3. We see that g can now be taken as a constant for values of Q^2 large enough for the Adler sum rule to hold. The curves for F_2 are given in Fig. 7, for various choices of the cut-off Λ and of the quark masses[1].

One can see that F_2 is stable w.r.t. Q^2, so that the structure function seems a possible candidate for the initial valence quark distribution in a pion $u_v(x) = \bar{u}_v(x) = d_v(x) = \bar{d}_v(x) = v(x)$, via the relation

$$F_1 = \frac{5}{9}v(x). \quad (9)$$

Note however that the condition $s > 4m_q^2$ leads to a cut on the values of x which are allowed: whereas for no cut-off, x can go to 1 in the limit $Q^2 \to \infty$, it is limited to $x < 1 - (m_q/\Lambda)^2$ for finite Λ. Although the elimination of the interval close to $x = 1$ is due to the fact that our cut-off is sharp, the suppression at large x is reminiscent of that obtained in the covariant parton model [2]. In this model, a vertex falling as $1/(p^2)^n$ at large p leads to a parton distribution that behaves like $(1-x)^{(n-1)}$.

This leads to one of the puzzles of this model. We can consider the average momentum carried by the quarks

$$2\langle x \rangle = 4\int_0^1 xv(x)dx = \frac{18}{5}\int_0^1 F_2(x)dx. \quad (10)$$

As we do not have gluons in the model, one would expect $2 < x >$ to be equal to 1 as $Q^2 \to \infty$. However, we find that it is in fact significantly lower, as shown in Fig. 8. In fact, the momentum sum rule is correct either for $\Lambda = \infty$ (*i.e.* for structureless pions), or in the case $m_q < m_\pi/2$ (as the infrared divergence can be re-absorbed in the normalisation).

---

[1] Please note that the curves are for $18/5F_2$. The factor $18/5$ was missed in the first paper of [1].

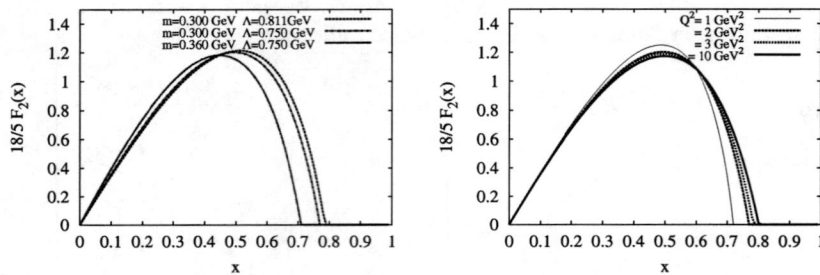

FIGURE 7. Results for the diagonal structure function. Left: for $Q^2 = 2$ GeV^2, right: $m_q = 0.3$ GeV, $\Lambda = 0.75$ GeV.

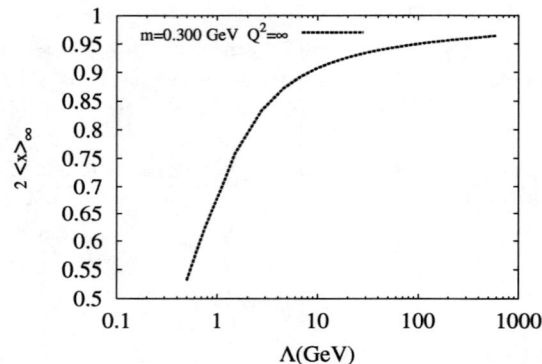

FIGURE 8. The average momentum carried by the quarks.

Because the momentum sum rule is obeyed in the structureless case, and because high momenta are cut off in our model (or in the covariant parton model), it is obvious that $2 < x >$ must be smaller than 1. There are two possible conclusions: the first is that the momentum sum rule holds only for free partons, so that it is not realised in our model, in which partons are always off-shell. But then physical quarks are always off-shell, so one may wonder if the sum rule is true. On the other hand, one may argue that the problem is that we did not consider cuts through the vertices, which would lead to a gluonic component. One must then understand why those cuts would have no effect on the sum rule if $m_q < m_\pi/2$, whereas they would increase it by a factor 2 for higher masses. Whatever the scenario, the conclusion is that our calculation is reasonable for valence quarks.

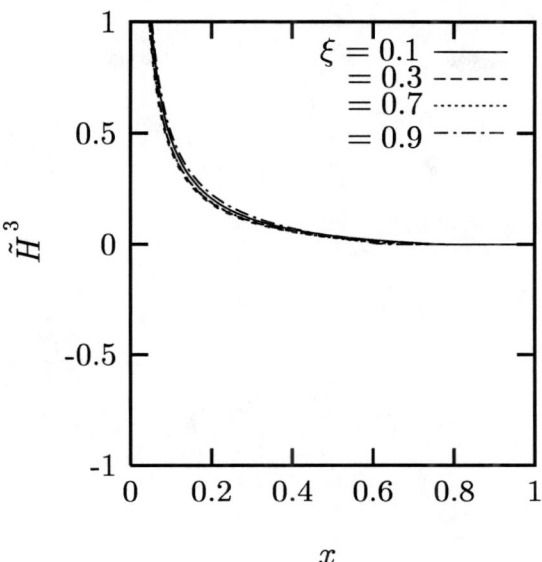

FIGURE 9. Our prediction for the valence component of $\tilde{H}^3$, for $Q^2 = 10$ GeV^2, $t = -0.1$ GeV^2 and $m_\pi = 0$.

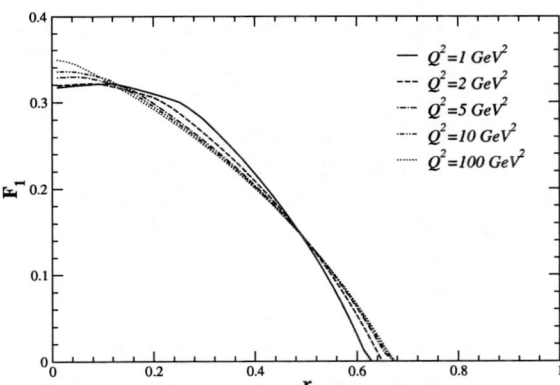

FIGURE 10. Our prediction for the valence component of H, in DVCS, as a function of Q^2, for $t = -0.1$ GeV^2, $\Lambda = 0.75$ GeV and $m_\pi = 0$.

2.3. Off-diagonal structure

It is easy to extend our previous calculation to the off-diagonal case. Here, we shall again calculate the discontinuity of the amplitude, but with an off-diagonal kinematics: the external photons have incoming momentum q_1 and outgoing mo-

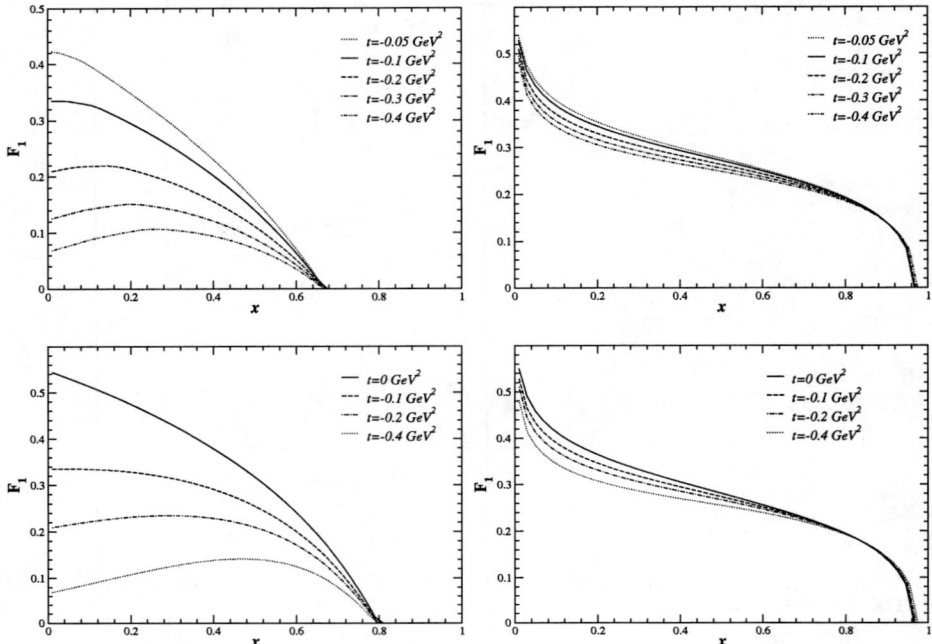

FIGURE 11. Our prediction for the valence component of H, in DVCS (top) and in elastic scattering (bottom), as a function of t, for $Q^2 = 10$ GeV^2 and $m_\pi = 0$. The left graphs are for $\Lambda = 0.75$ GeV and the right ones for structureless pions.

mentum q_2, whereas the pions have momenta P_1 and P_2. We define average momenta $P = (P_1 + P_2)/2$ and $q = (q_1 + q_2)/2$, and the momentum transfer $\Delta = P_2 - P_1$. The Lorentz invariants of the process are $t = \Delta^2$, $Q^2 = -q^2$, $x = Q^2/(2P.q)$, and $\xi = \Delta.q/(2P.q)$.

The calculation proceeds as in the diagonal case[2]: we calculate the discontinuity using the vertices Γ_3 and Γ_4 defined above. The answer is again gauge invariant, and can now be decomposed into 5 independent structures [3].

$$\begin{aligned}\frac{1}{4\pi}\text{disc}(\mathcal{A}_{\mu\nu}) &= -\mathcal{P}_{\mu\sigma}g^{\sigma\tau}\mathcal{P}_{\tau\nu}F_1 + \frac{\mathcal{P}_{\mu\sigma}p^\sigma p^\tau \mathcal{P}_{\tau\nu}}{p\cdot q}F_2 \\ &+ \frac{\mathcal{P}_{\mu\sigma}(p^\sigma(\Delta^\tau - 2\xi p^\tau) + (\Delta^\sigma - 2\xi p^\sigma)p^\tau)\mathcal{P}_{\tau\nu}}{2p\cdot q}F_3 \\ &+ \frac{\mathcal{P}_{\mu\sigma}(p^\sigma(\Delta^\tau - 2\xi p^\tau) - (\Delta^\sigma - 2\xi p^\sigma)p^\tau)\mathcal{P}_{\tau\nu}}{2p\cdot q}F_4 \\ &+ \mathcal{P}_{\mu\sigma}(\Delta^\sigma - 2\xi p^\sigma)(\Delta^\tau - 2\xi p^\tau)\mathcal{P}_{\tau\nu}F_5,\end{aligned} \quad (11)$$

---

[2] although we simplify the results by setting the pion mass to zero in the off-diagonal case.

where we have used the projector $\mathcal{P}_{\mu\nu} = g_{\mu\nu} - \frac{q_{2\mu}q_{1\nu}}{q_1 \cdot q_2}$. For neutral pions, the structure functions F_i that parameterize the discontinuity can be directly related to the GPD's H, H^3 and $\tilde{H}^3$ [3] to twist-3 accuracy:

$$\frac{1}{2\pi}F_1 = H, \tag{12}$$

$$\frac{1}{2\pi}F_2 = 2xH + \mathcal{O}(1/Q^2), \tag{13}$$

$$\frac{1}{2\pi}F_3 = \frac{2x}{x^2 - \xi^2}\left(H^3 x^2 + \tilde{H}^3 \xi x - H\xi\right) + \mathcal{O}(1/Q^2), \tag{14}$$

$$\frac{1}{2\pi}F_4 = \frac{2x}{x^2 - \xi^2}\left(H^3 \xi x + \tilde{H}^3 x^2 - Hx\right) + \mathcal{O}(1/Q^2), \tag{15}$$

$$\frac{1}{2\pi}F_5 = \mathcal{O}(1/Q^2). \tag{16}$$

Our calculation obeys Eqs. (13) and (16), and leads to definite predictions for H, H^3 and $\tilde{H}^3$. Before giving the explicit results, let us mention that we find explicit relations linking the twist-3 GPD's to H:

$$H^3 = \frac{(x-1)\xi}{x(\xi^2-1)}H + \mathcal{O}(1/Q^2) \tag{17}$$

$$\tilde{H}^3 = \frac{H^3}{\xi} + \mathcal{O}(1/Q^2) \tag{18}$$

Note that the polynomiality of the Mellin moments of H, H^3 and $\tilde{H}^3$, together with Eqs. (17) and (18), implies that H must be a polynomial $P_H(\xi)$ multiplying $\xi^2 - 1$. We show in Fig. 9 our results for $\tilde{H}^3$. The fact that it is almost independent of ξ shows that P_H is very close to a constant.

Let us point out that the relations (17, 18) are an explicit counter-example to the Wandzura-Wilczek ansatz [3, 4]. Not only are they numerically different, but our calculated H^3 and $\tilde{H}^3$ do not suffer from discontinuities at $x = \xi$, contrarily to what the Wandzura-Wilczek ansatz predicts.

We can finally turn to our predictions for H. Here we consider two regimes: deeply virtual Compton scattering ($\xi = -x$), and elastic scattering ($\xi = 0$). First of all, we show in Fig. 10 that our ansatz is stable w.r.t. Q^2. It can presumably be used as the initial parton distribution, as in the diagonal case.

We can also examine the role of hadronic structure by comparing our prediction at $\Lambda = 0.75$ GeV with that for a structureless pion ($\Lambda = \infty$). We do this both in the DVCS and in the elastic cases in Fig. 11. We see that the structure of the pion makes an enormous difference. The cut-off in x is in fact smaller than in the diagonal case, and the novelty is a rather large dependence on t. Hence we confirm that DVCS, and GPD's in general, will give us new information about hadronic structure.

3. CONCLUSIONS

We have built a very simple model for the pion, which goes beyond the spectator quark model. It implements all the symmetries of the problem, in particular gauge invariance. This toy model has enabled us to study explicitly initial valence quark (and antiquark) distributions, both in the diagonal and in the off-diagonal cases. We find that structure leads to corrections of order m_q/Λ to the momentum sum rule in the diagonal case. Such effects might be attributed to the contributions of cuts in the vertices. Using this model in the off-diagonal case, we have shown that the Wandzura-Wilczek ansatz, which can be used to relate twist-3 GPD's to twist-2, is likely to be wrong. We in fact obtain new relations between the GPD's. We have also shown that binding effects lead to a rich structure for GPD's, which is not present in the case of point couplings.

ACKNOWLEDGMENTS

We thank M.V. Polyakov and P.V. Landshoff for discussions. J.P.L. is a research fellow of the Institut Interuniversitaire des Sciences Nucléaires.

REFERENCES

1. F. Bissey, J. R. Cudell, J. Cugnon, J. P. Lansberg and P. Stassart, Phys. Lett. B **587** (2004) 189 [arXiv:hep-ph/0310184]; F. Bissey, J. R. Cudell, J. Cugnon, M. Jaminon, J. P. Lansberg and P. Stassart, Phys. Lett. B **547** (2002) 210 [arXiv:hep-ph/0207107].
2. P. V. Landshoff, J. C. Polkinghorne and R. D. Short, Nucl. Phys. B **28** (1971) 225; P. V. Landshoff and D. M. Scott, Nucl. Phys. B **131** (1977) 172.
3. A. V. Belitsky, D. Müller, A. Kirchner and A. Schäfer, Phys. Rev. D **64** (2001) 116002 [arXiv:hep-ph/0011314].
4. S. Wandzura and F. Wilczek, Phys. Lett. B **72** (1977) 195.

Deep inelastic lepton nucleus scattering and hadronization at HERMES energies

D. Grünewald

Universität Heidelberg, Institut für theoretische Physik, Philsophenweg 19,
D-69120 Heidelberg, Germany
E-mail: daniel@tphys.uni-heidelberg.de

Abstract. Semi-inclusive deep inelastic lepton nucleus scattering is studied. The possible hadron interactions inside the nucleus are taken into account by an absorption model which is based on flavor dependent hadron formation lengths, calculated in the framework of the LUND string fragmentation model. Additionally, the rescaling of parton distribution functions and fragmentation functions in the nuclear medium is considered, due to the hypothesis, that a quark in a bound nucleon has access to a larger region in space than in a free nucleon. The model predictions are compared with recent HERMES results for the multiplicity ratios normalized to deuterium on various hadron species and different nuclei. Beside the proton, a good agreement with the experimental data is found.

Keywords: Semi-Inclusive Deep Inelastic Scattering, Fragmentation, Particle Production in Nuclei
PACS: 12.38.-t, 13.60.Hb , 13.60.Le

1. INTRODUCTION

Recent HERMES results give precise data on hadron production in deep inelastic scattering (DIS) of 27.6 GeV positrons on D, He, N, Ne, and Kr nuclei [1, 2, 3]. The main observable is the multiplicity ratio, R_M, defined as the ratio between the hadron multiplicity on a nucleus and on deuterium. R_M has been studied as a function of the hadron fractional momentum z, of the virtual photon energy ν, of its virtuality Q^2, for different hadrons and for different nuclei. Without nuclear effects, this ratio would be equal to 1. The deviations of R_M from 1 give insights into the hadronization process itself, since the interactions with the surrounding medium which are responsible for these deviations during the evolution of the quark into the observed hadron probe the space-time evolution of hadronization. The nucleons play the role of very nearby detectors for the hadronization process. In contrast to heavy ion collisions, the starting point of hadronization in deep inelastic scattering events is well defined, since the virtual photon energy is given by the difference of the incoming and outgoing projectile energies and is therefore directly accessible in the experiment. Furthermore, the environment in which the struck quark propagates, i.e. cold nuclear matter, is well known. Therefore, deep inelastic lepton nucleus scattering is appropriate to study the processes appearing in heavy ion collision under "clean" conditions, which is crucial for the interpretation of ultra-relativistic proton-nucleus and nucleus-nucleus collisions.
In the following, we review the current status of our model [4, 5]. We start on the

basis of the deep inelastic lepton nucleon scattering cross-section. This cross-section is then extended to the case of a nuclear target. We relate the parton distribution functions and the fragmentation functions of a bound nucleon to the parton distribution and fragmentation functions of free nucleons in the deconfinement model, which is based on the hypothesis, that quarks in a bound nucleon have access to a larger region in space than in free nucleons. Furthermore, we take into account possible final state interactions of the produced hadrons by introducing an absorption factor in the cross-section. The color of the struck quark is neutralized after a short time given by the formation time of a prehadron, the predecessor of the final hadron. This formation time is calculated in the framework of the LUND string fragmentation model and turns out to be flavor dependent. The color neutral prehadron propagates through the nucleus and interacts with surrounding nucleons which absorbes it on its way out of the nucleus yielding a multiplicity ratio smaller than one.

2. BUILDING BLOCKS OF THE LEPTON NUCLEUS SCATTERING MODEL

The QCD factorization theorem predicts that the semi inclusive deep inelastic lepton nucleon scattering (SIDIS) cross-section can be written as a product of the parton distribution function $q_f(x, Q^2)$, the hard lepton quark scattering cross-section and the fragmentation function $D_f^h(z, Q^2)$:

$$\frac{d^2\sigma}{dxd\nu dz}\bigg|_{SIDIS} = \sum_f e_f^2 q_f(x, Q^2) \frac{d^2\sigma^{lq}}{dxd\nu} D_f^h(z, Q^2) \qquad (1)$$

Here f denotes the struck quark flavor, x is the momentum fraction of the nucleon momentum carried by the struck quark, Q^2 denotes the virtuality of the process and z is the energy fraction of the virtual photon carried by the produced hadron. If one considers deep inelastic lepton nucleus scattering instead of deep inelastic lepton nucleon scattering, one has to take into account several nuclear effects in the cross-section. First, we relate the parton distribution function of the nucleus to the one of the free nucleon by a rescaling of the physical scale $Q^2 \to \xi_A(Q^2)Q^2$ [6] due to the hypothesis, that a quark in a bound nucleon has access to a larger region in space than in a free nucleon, i.e. that the quarks in a bound nucleon are partially deconfined. The rescaling factor $\xi_A(Q^2)$ is given by

$$\xi_A(Q^2) = \left(\frac{\lambda_A}{\lambda_0}\right)^{\frac{\bar{\alpha}_s}{\alpha_s(Q^2)}} \qquad (2)$$

Here λ_A is the confinement scale in the nucleus and λ_0 in the free nucleon respectively, $\alpha_s(Q^2)$ is the QCD coupling constant computed in leading order with $\Lambda_{QCD} = 200 MeV$ and four quark flavors and $\bar{\alpha}_s = \alpha_s(\mu_A^2)$ with $\mu_A = \frac{\lambda_0}{\lambda_A}\mu_0$. $\mu_0^2 = 0.66 GeV^2$ denotes the starting scale of the DGLAP evolution in the free nucleon. Effectively, the DGLAP evolution of the nuclear structure function covers

a larger interval in momentum compared with the corresponding functions in the free nucleon at the same scale Q.
Similar considerations yield the same rescaling factor in the case of the fragmentation functions.
And second, we take into account possible final state interactions of the produced hadron with the nucleus by introducing an absorption factor $\mathcal{N}_{f;A}^h(z,\nu)$ which is explained in the following section. Therefore, the SIDIS cross-section for lepton nucleus scattering is given in our model by

$$\left.\frac{d^2\sigma}{dxd\nu dz}\right|_{SIDIS} = \sum_f e_f^2 q_f^A(x,\xi_A(Q^2)Q^2)\frac{d\sigma^{lq}}{dxd\nu}D_f^h(z,\xi_A(Q^2)Q^2)\mathcal{N}_{f;A}^h(z,\nu) \quad (3)$$

3. THE ABSORPTION MODEL

A schematic space time picture of hadronization in the Lab frame in the LUND [7] model is illustrated in fig. (1). The quark/anti-quark production points are

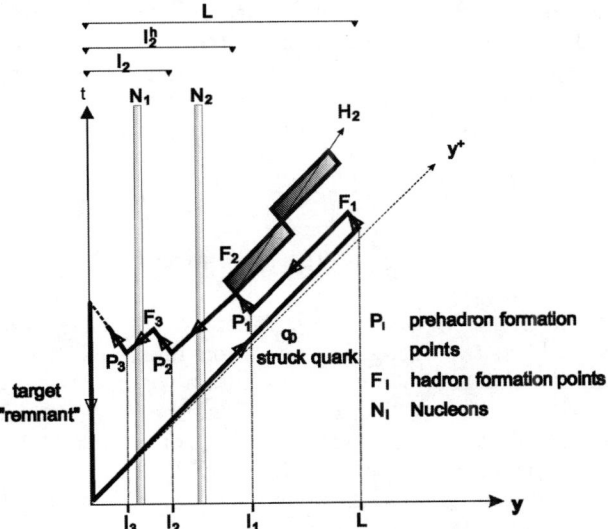

FIGURE 1. Schematic space time picture of hadronization in the LUND model for deep inelastic nucleus scattering in the laboratory frame, where the nucleus is at rest.

denoted by P_i and the points, where the quark and the anti-quark from two adjacent production points first intersect and therefore form a hadron, are called the hadron formation points F_i. There is an ordering of the hadrons, the rank. The distance between two quark-anti-quark pair production points is always space like [7], i.e. that the ordinary time ordering of the break up points is not Lorentz invariant. One therefore usually orders the hadrons along one of the two light cones, which is unique. Here, the positive light cone is chosen. The first rank hadron is per definition the one containing the struck quark. Inversely, only hadrons which contain the flavor of the struck quark are producible as rank 1 particles. Otherwise, they only chance to produce them, is as a rank 2 or any higher rank particle.

For leading hadrons the constituent produced at P_i is much earlier present than the other one. It forms a color neutral object with the struck quark until the second constituent is produced. Since this object already contains one of the hadron constituents and evolves to the observed hadron, we call it a pre-hadron (cf ref. [4]). Therefore, the production point P_n is interpreted as the formation point of the n-th rank pre-hadron, which propagates through the nuclear medium and may interact with the surrounding nucleons N_i.

The turning point of the struck quark, i.e. the point where the quark has lost its whole energy, is denoted by L. It is proportional to the energy transferred to the quark and is given by:

$$L = \frac{\nu}{\kappa} \qquad (4)$$

The rank independent average pre-hadron formation length can be computed (cf. Ref. [8]) as the average of the i-th rank pre-hadron formation point (P_i) distribution, which is the probability that the i-th rank pre-hadron is formed after the distance Δy after the virtual photon quark interaction, summed over all accessible hadron ranks for the desired hadron, i.e. starting from rank 1 or 2 respectively. The average pre-hadron formation length for particles, which are producible as rank 1 particles, is given by

$$\langle {}^1 l_r^* \rangle = \frac{1+C_r}{1+C+(C_r-C)z}(1-z)zL$$
$$\times [1 + \frac{1+C}{2+C_r}\frac{(1-z)}{z^{2+C_r}} {}_2F_1\left(2+C_r, 2+C_r; 3+C_r; \frac{z-1}{z}\right)], \qquad (5)$$

where ${}_2F_1$ is the Gauss hypergeometric function, the parameters C and C_r with $r=(q,qq)$ arise from the string fragmentation function $f(u) \propto (1-u)^{C_r}$, which is the probability that the string breaks into two pieces under the creation of a quark-anti-quark pair or a diquark-anti-diquark pair respectively. Furthermore, these parameters select whether a meson ($r=q$) or a baryon ($r=qq$) is produced. Their numerical values are given in Ref. [7] by $C=C_q=0.3$ and $C_{qq}=1.3$. The mean pre-hadron formation length for particles which can be produced only as a rank 2 or any higher rank particle is the following

$$\langle {}^2 l_r^* \rangle = [\frac{1+C_r}{2+C_r}\frac{(1-z)}{z^{2+C_r}} {}_2F_1\left(2+C_r, 2+C_r; 3+C_r; \frac{z-1}{z}\right)](1-z)zL, \qquad (6)$$

The average hadron formation length is given by

$$\langle {}^{1,2} l_r^h \rangle = \langle {}^{1,2} l_r^* \rangle + zL \qquad (7)$$

So far, the model is based on string fragmentation in $(1+1)$-dimensions. Realistic hadrons also have a transverse extension which depends on the number of quarks and their flavor. Empirically we know that baryons have a larger radius than π-mesons, which is larger than K-mesons radius. Therefore we renormalize the string

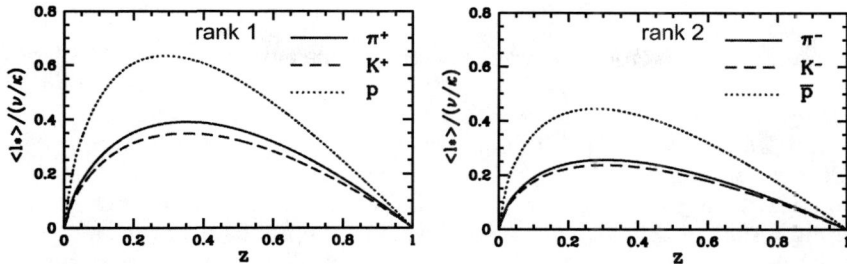

FIGURE 2. Computed pre-hadron formation lengths when an up quark is struck by the virtual photon. *Left:* When a π^+, K^+ or p is observed, the corresponding pre-hadron can be created at rank $n \geq 1$. *Right:* When a π^-, K^- or $\bar{p}$ is observed, the corresponding pre-hadron can be created only at rank $n \geq 2$.

tension, i.e. the quantity which sets the confinement scale in the Lund model [7, 8], to the confinement scales of realistic hadrons, whose values we take from Ref. [9]. Therefore, L is renormalized to

$$L_h = \frac{\nu}{\kappa} \frac{r_h^2}{r_\pi^2}. \tag{8}$$

This produces an increased formation time for baryons due to the larger size of the proton compared with pions.

The comparison between the computed formation lengths of pions, kaons, protons and anti-protons is shown in figure 2, where the hadron radii of [9] where used. In this figure, the effect of the flavor dependent formation length and the renormalization of the string tension is included already. The flavor-dependence of the formation time decreases the mean formation time of the negative mesons by about 30% compared to the positive ones, and by more than 40% for $\bar{p}$ compared to p. Taking into account the above considerations, the hadrons are not produced immediately after the virtual photon - quark interaction since they have a non vanishing formation length. The formation length can exceed nuclear radii at high energies. In lepton-nucleus reactions the reaction products, i.e. the pre-hadron and the hadron, therefore can interact with the surrounding nuclear medium. In order to take that effect into account, we introduce the nuclear absorption factor $\mathcal{N}_A$, which is defined as the probability that the final state does not interact with the nucleus during its evolution from the quark to the observed hadron. This factor approximates the evolution by treating absorption in the first scattering of the color neutral state with a target nucleon. In principal the observed hadron with energy fraction z might be the result of multiple elastic or inelastic re-scatterings whereby it degrades its energy fraction [10]. Since the fragmentation function is falling rapidly for $z \to 1$, multiple re-scattering processes are suppressed since they need a very energetic original particle at the origin at least for observed hadrons with sizable z. Therefore, the observed hadrons are dominated by the hadrons, which have not interacted during their evolution.

In order to calculate $\mathcal{N}_A$, we want to consider the evolution of the quark as a decay process as it is proposed in the simplistic Bialas-Chmaj [11] absorption formulae, but in contrast to the latter we fully couple the evolution of the quark into a pre-hadron and a hadron with the absorption processes. In the following we will not differentiate between time and space coordinates since $\Delta L = \Delta T$ for $c = 1$ and we neglect the quark flavor and rank indices of the formation lengths in order to simplify the notation. The quark decays into the pre-hadron with a mean lifetime equal to the average pre-hadron formation time $\langle l^* \rangle$, i.e the time it takes to form a pre-hadron after the $\gamma^* q$ interaction. The pre-hadron on its part, decays into the hadron. Its lifetime $\langle \Delta l \rangle$ is given by the difference of the average hadron and the average pre-hadron formation time:

$$\langle \Delta l \rangle = \langle l^h \rangle - \langle l^* \rangle \tag{9}$$

The possible interactions of the pre-hadron or the hadron respectively involve additional "lifetimes": The mean free path of the pre-hadron and the hadron, i.e. the average time between two interactions. The probabilities that the intermediate state at the longitudinal position y' is a quark $W_q(y, y')$, a pre-hadron $W_*(y, y')$ or a hadron $W_h(y, y')$ if the initial virtual photon quark interaction took place at longitudinal position y obey the following evolution equations

$$\begin{aligned}
\frac{\partial W_q(y, y')}{\partial y'} &= -\frac{W_q(y, y')}{\langle l^* \rangle} & , W_q(y, y' = y) = 1 \\
\frac{\partial W_*(y, y')}{\partial y'} &= \frac{W_q(y, y')}{\langle l^* \rangle} - \frac{W_*(y, y')}{\langle \Delta l \rangle} - \frac{W_*(y, y')}{\lambda_*(y')} & , W_*(y, y' = y) = 0 \\
\frac{\partial W_h(y, y')}{\partial y'} &= \frac{W_*(y, y')}{\langle \Delta l \rangle} - \frac{W_h(y, y')}{\lambda_h(y')} & , W_h(y, y' = y) = 0
\end{aligned} \tag{10}$$

Here $\lambda_{*,h}(y')$ is the local mean free path of the pre-hadron or the hadron respectively in a nucleus with mass number A. It is given by:

$$\lambda_{*,h}(y') = \frac{1}{\rho_A(y') \sigma_{*,h}} \tag{11}$$

where ρ_A is the nuclear density and $\sigma_{*,h}$ is the (pre-)hadronic cross-section. Note, that the dependence of ρ_A on the impact parameter b is suppressed to simplify the notation. The absorption factor $\mathcal{N}_A(b, y)$ for a fixed primary interaction point b, y is simply given as the probability of finding a hadron outside of the nucleus:

$$\mathcal{N}_A(b, y) = \lim_{y' \to \infty} W_h(y', y) \tag{12}$$

The final interaction point and impact parameter averaged expression is given by

$$\mathcal{N}_A = \lim_{y' \to \infty} \int d^2 b \int_{-\infty}^{\infty} dy \rho_A(b, y) W_h(y', y)$$

$$= \int d^2b \int_{-\infty}^{\infty} dy \rho_A(b,y) \int_y^{\infty} dx' \int_y^{x'} dx \frac{e^{-\frac{x-y}{\langle l^* \rangle}}}{\langle l^* \rangle} e^{-\sigma_* \int_x^{x'} ds A\rho_A(s)}$$

$$\times \frac{e^{-\frac{x'-x}{\langle \Delta l \rangle}}}{\langle \Delta l \rangle} e^{-\sigma_h \int_{x'}^{\infty} ds A\rho_A(s)} \qquad (13)$$

Note, that the absorption factor $\mathcal{N}_A$ implicitly depends on the energy fraction z of the produced hadron, the virtual photon energy ν, the flavor f of the struck quark and the hadron species h by the average formation length for pre-hadrons and hadrons and by the respective cross-sections. Therefore $\mathcal{N}_A = \mathcal{N}_{f;A}^h(z,\nu)$.

4. NUMERICAL RESULTS

In this section we discuss the results of our model for semi-inclusive charged hadron production in nuclei and compare them to the most recent HERMES experimental data, ref. [1, 2]. The main observable is the ratio R_M^h of the multiplicity for a nuclear target with mass number A and the deuterium multiplicity as a function of z for example:

$$R_M^h(z) = \frac{1}{N_A^{DIS}} \frac{dN_A^h(z)}{dz} \bigg/ \frac{1}{N_D^{DIS}} \frac{dN_D^h(z)}{dz} \qquad (14)$$

The upper ratio represents the number of produced hadrons of species h with energy fraction z, normalized to the total number of deep inelastic scattering events for a nuclear target with mass number A. The lower ratio contains the corresponding expression for a deuterium target.
Theoretically, the multiplicity for lepton-nucleus scattering as a function of z is the integrated SIDIS cross-section eqn. 3 divided by the total deep inelastic scattering (DIS) cross-section:

$$\frac{1}{N_A^{DIS}} \frac{dN_A^h(z)}{dz} = \frac{1}{\sigma^{lA}} \int_{\text{exp. cuts}} dx d\nu \sum_f e_f^2 q_f^A(x, \xi_A(Q^2)Q^2) \frac{d\sigma^{lq}}{dx d\nu}$$

$$\times D_f^h(z, \xi_A(Q^2)Q^2) \mathcal{N}_{f;A}^h(z,\nu) \qquad (15)$$

$$\sigma^{lA} = \int_{\text{exp. cuts}} dx d\nu \sum_f e_f^2 q_f^A(x, \xi_A(Q^2)Q^2) \frac{d\sigma^{lq}}{dx d\nu} \qquad (16)$$

Here, the total lepton nucleus DIS cross section is denoted by σ^{lA}. Note, that the experimental acceptance is explicitly accounted for in the integration limits in eqns. (15) and (16) and that the absorption factor intrinsically depends on the struck quark flavor f. In the numerical computations we use the leading order GRV98 parton distribution functions, which are given in ref. [12]. For the fragmentation functions, we use the leading-order Kretzer's parametrization of ref. [13] for $\pi^\pm$

and $K^{\pm}$. But, the set of Kretzer fragmentation functions does not provide the production of protons or anti-protons respectively. Therefore, we use for protons and anti-protons the Kramer, Poetter and Kniehl (KKP) parametrization [14] of the fragmentation function. In the computation of the nuclear absorption factor, we use as a density for the deuteron derived from the normalized sum of the Reid's soft core S- wave (denoted by $u(r_{np})$) and D-wave (denoted by $w(r_{np})$) functions squared, which are tabulated in ref. [15]. For heavier nuclei, we use a Fermi three parameter distribution which is tabulated in ref. [16] for several nuclei. For the hadron nucleon cross-sections, we take the isospin averaged total, i.e. the sum of elastic and inelastic cross-sections, hadron-proton σ_{hp} and hadron-neutron σ_{hn} cross-sections of ref. [17], where the energy dependence of σ_{hp} and σ_{hn} has been explicitly taken into account:

$$\sigma_h(E_h) = \frac{1}{A}[Z\sigma_{hp}(E_h) + (A-Z)\sigma_{hn}(E_h)] \qquad (17)$$

Predictions of our model for π, K, p and $\bar{p}$ multiplicity ratio on ^{4}He, ^{20}Ne, Kr and Xe are shown in Fig. 3 for the z dependence (Predictions for the ν dependence can be found in ref. [5]). In this plot comparisons with the HERMES multiplicity ratio for identified hadrons on Kr [2] and the preliminary data on ^{4}He and ^{20}Ne [3] are presented.

The only free parameter in our model is the prehadronic cross section. In ref. [5] we fitted a prehadron cross section equal to 2/3 of the hadron cross section which is realistic since the prehadron coming from gluon radiation of the struck quark is smaller than the final hadron. The z-dependence reported in Fig. 3 shows a nice agreement with the data for negative and positive pions for all the nuclei. The predictions for K^+ and K^- are in qualitatively agreement with the trend shown by the data except for the low-z region. This region contains contributions from both leading hadrons decelerated in nuclear rescattering and from secondary positive and negative kaons produced from initial pions and ρ-mesons. Both this contributions are not accounted in the purely absorption treatment of the final state interaction included in our model. While the z-dependence of the anti-protons data is described, within the statistical uncertainties, by our model, the protons multiplicity ratio is underestimated especially in the low z region.

If the disagreement between model and data in the low-z region can be ascribed to the fact that in the absorption model a hadron interacting with a nucleon is simply removed without accounting for a possible lowering of z in the interaction, a full transport theory [10] should reproduce the yield at low-z values. This is actually not the case. The discrepancy with the data may also point to a non negligible diquark contribution as also suggested in ref. [10].

5. SUMMARY AND CONCLUSIONS

Rescaling of the structure and fragmentation functions together with (pre)hadronic absorption is shown to be able to describe HERMES data. The hadron formation length turns out to be flavor dependent. Especially, negative kaons ($K^- = s\bar{u}$) as

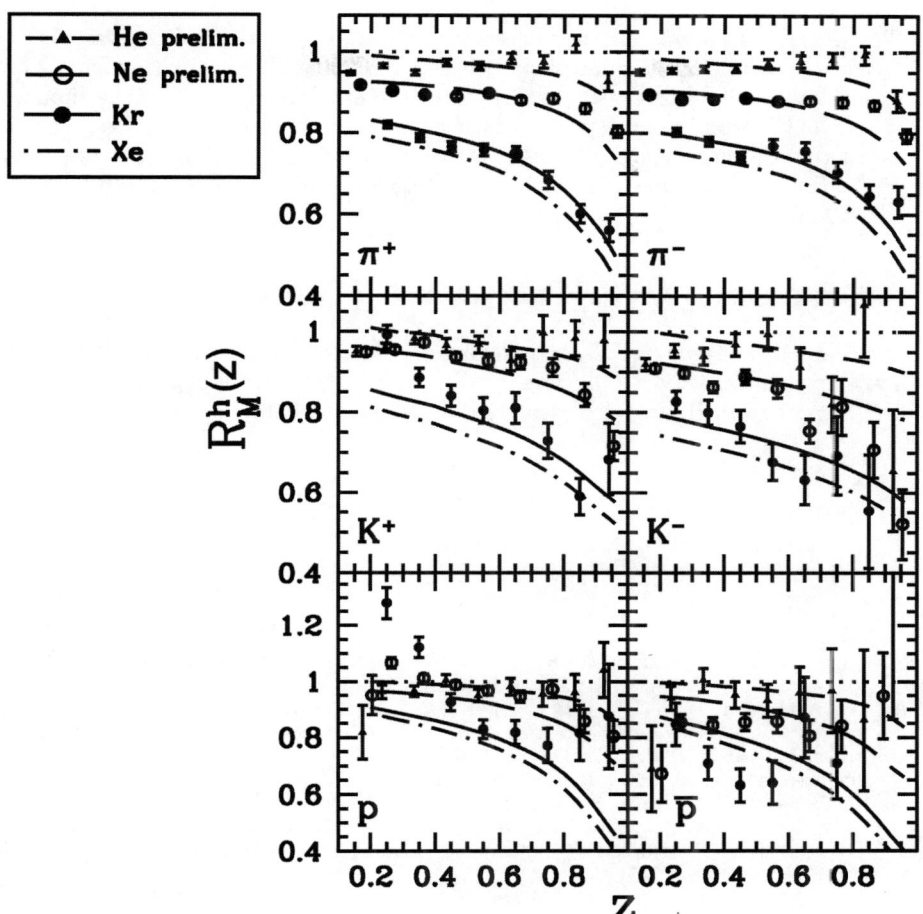

FIGURE 3. Computed multiplicity ratio for pions, kaons, protons and antiprotons as a function of z for ^{4}He (dashed line), ^{20}Ne (long-dashed line), Kr (solid line) and Xe (dot-dashed line). The HERMES data on Kr (solid circles) [2] and the preliminary data on ^{4}He (closed triangles) and on Ne (open circles) [3] are shown with their statistical uncertainties. To improve the readability of the figure, Ne and Kr experimental points have been shifted by 0.015 to the left and to the right, respectively.

well as antiprotons ($\bar{p} = \bar{u}\bar{u}\bar{d}$) cannot be formed by a struck valence quark which is the dominant contribution at HERMES energy. This yields a smaller formation length for K^- and antiprotons implying more hadron suppression compared to positive kaons or protons respectively. We extract the prehadronic cross-section $\sigma_* = 2/3\sigma_h$ from data on pion production on krypton. This shows that the prehadron does not have the full hadronic size yet.

The presented model correctly describes pion and kaon multiplicity ratios. A different magnitude of the multiplicity ratios for K^+ and K^- originating from a flavor dependent formation length is reflected in the experimental data. The model disagrees with experimental proton data at small z values. This discrepancy may be ascribed to a non negligible diquark fragmentation contribution in that region. Prehadron rescatterings and final state interactions are insufficient to fully account for the effect.

ACKNOWLEDGMENTS

We thank Jean-René Cudell, Joseph Cugnon, Jean-Marc Gérard and Jean-Philippe Lansberg for the very stimulating meeting in Spa. This work is partially funded by the EU program Hadron Physics (RII3-CT-2004-506078).

REFERENCES

1. A. Airapetian et al. [HERMES], Eur. Phys. J. C **20** (2001) 479.
2. A. Airapetian et al. [HERMES], Phys. Lett. B 577 (2003) 37.
3. G. Elbakian et al. [HERMES], Proceedings of the 11th International Workshop on Deep Inelastic Scattering (DIS 2003), St.Petersburg April 23 - 27, 2003; V.T. Kim and L.N Lipatov eds., p.597.
4. A. Accardi, V. Muccifora and H. J. Pirner, Nucl. Phys. A **720** (2003) 131.
5. A. Accardi, D. Grünewald, V. Muccifora and H. J. Pirner, arXiv:hep-ph/0502072.
6. O. Nachtmann and H. J. Pirner, Z. Phys. C **21** (1984) 277. R. L. Jaffe, F. E. Close, R. G. Roberts and G. G. Ross, Phys. Lett. B **134** (1984) 449; Phys. Rev. D **31** (1985) 1004. J. Dias de Deus, Phys. Lett. B **166** (1986) 98.
7. B. Andersson et al., Phys. Rep **97** (1983) 31.
8. A. Bialas and M. Gyulassy, Nucl. Phys. **291** (1987) 793.
9. S. R. Amendolia *et al.*, Phys. Lett. B **146** (1984) 116. S. R. Amendolia *et al.*, Phys. Lett. B **178** (1986) 435. F. Borkowski, P. Peuser, G. G. Simon, V. H. Walther and R. D. Wendling, Nucl. Phys. B **93** (1975) 461.
10. T. Falter, W. Cassing, K. Gallmeister and U. Mosel, Phys. Rev. C **70** (2004) 054609 and T. Falter Ph.D. Thesis Giessen University, 2004.
11. A. Bialas and T. Chmaj, Phys. Lett. **133** (1983) 241.
12. M. Glück, E. Reya and A. Vogt, Eur. Phys. J. C **5**, 461 (1998) [arXiv:hep-ph/9806404].
13. S. Kretzer, Phys. Rev. D **62** (2000) 054001.
14. B. A. Kniehl, G. Kramer and B. Potter, Nucl. Phys. B **582** (2000) 514.
15. R. V. Reid, Ann. Phys. **50**, 411 (1968).
16. H.R. Collard, L. R. B. Elton and R. Hofstader: Numerical data and functional relationships, Nuclear Radii volume 2, 1 edition (Springer , Heidelberg 1967)
17. K. Hagiwara *et al.* [Particle Data Group], Phys. Rev. D **66**, 010001 (2002).

Gluon transversity in the hard exclusive reactions

Nikolay Kivel

Institute for Theoretical Physics II, Ruhr University Bochum, Germany

Abstract. We discuss contributions of the gluon transversity in several hard exclusive reactions and suggest observables sensitive to such amplitudes.

Keywords: exclusive reactions, factorization, DVCS, higher twist
PACS: 13.88.+e, 13.65.+i

1. INTRODUCTION

The attractive property of the hard exclusive reactions is the possibility to measure quite rare configurations of the partons which can not be measured, for instance, in the inclusive channels. One of such examples is gluon transversity which is defined by the gluon twist-2 operator. The nice feature of this operator is decoupling from the quarks, i.e. the absence of the mixing with the quark antiquark state. Hence such operator probes pure gluonic structure of the scanned hadron. Note, for instance, that gluon admixture to meson wave functions is interesting for many reasons and has been subject to extensive and somewhat controversial discussion over many years, see e.g. [1]. On the other hand, separation of quark and gluon contributions in hard processes that are dominated by hadron wave functions at small transverse separations has a different meaning compared to the quark model; in a suitably chosen reaction quark contribution can be down compared to the gluon contribution by a power of the momentum transfer.

Below we consider several examples of the hard exclusive reactions which provides opportunity to probe such pure gluon state.

2. MATRIX ELEMENTS OF GLUON TRANSVERSITY

Let us first give few examples of the matrix elements which we are going use below.

Generalized Parton Distribution (GPD) from gluon transverse operator was introduced in Ji and Hoodbhoy [2] and farther studied by Diehl in [3]. Definition of the GPD gluon transversity can be written as:

$$F_{(\mu\nu)}(x,\xi,t) = \int_{-\infty}^{\infty} \frac{d\lambda}{2\pi} e^{-ix\lambda} \langle p'|SG^a_{n\mu}(\tfrac{1}{2}\lambda n) G^a_{n\nu}(-\tfrac{1}{2}\lambda n)|p\rangle, \qquad (1)$$

where we use standard notation $t=(p'-p)^2$, ξ is skewedness and n is light-cone vector $n^2=0$. Symbols **S** and $(\mu\nu)$ stand for symmetrisation of the two indices and removal of the trace: $SO_{\mu\nu} = \frac{1}{2}O_{\mu\nu} + \frac{1}{2}O_{\nu\mu} - \frac{1}{4}g_{\mu\nu}O^\alpha{}_\alpha$ and $G^a_{n\nu}$ is a shorthand notation for $n^\alpha G^a_{\alpha\nu}$. Using notation of [3] the gluonic matrix $F_{(\mu\nu)}(x,\xi,t)$ element can be parametrized in terms of four independent functions:

$$F^{(\mu\nu)}(x,\xi,t) = -S\frac{\Delta^\mu_\perp}{4m}\Big\{H^g_T(x,\xi,t)\langle\!\langle i\sigma^{+\nu}\rangle\!\rangle + \tilde{H}^g_T(x,\xi,t)\frac{\Delta^\nu_\perp}{m}\langle\!\langle 1/m\rangle\!\rangle$$

$$+ E^g_T(x,\xi,t)\left(\frac{\Delta^\nu_\perp}{2m}\langle\!\langle \gamma^+\rangle\!\rangle + \frac{1}{m}\langle\!\langle \gamma^\nu_\perp\rangle\!\rangle\right) + \tilde{E}^g_T(x,\xi,t)\frac{1}{m}\langle\!\langle \gamma^\nu_\perp\rangle\!\rangle\Big\} \quad (2)$$

where we used short notations $\langle\!\langle ...\rangle\!\rangle = \bar{N}(p')...N(p)$, $\Delta = p' - p$, $\gamma \cdot n = \gamma^+$ and m is nucleon mass. Note that in the forward limit matrix element tends to zero because it is proportional to momentum transfer Δ. Such nullification is expected because nucleon target can not compensate helicity flip on 2 units which is necessary in the case of the gluon transversity. This is the reason why such matrix elements can not be accessed in DIS with nucleon target.

As the second example, consider distribution amplitude (DA) of the tensor meson $f_2(1275)$ generated by transverse gluons:

$$f^T_g e_{\mu\nu}\phi^T_g(u) = \frac{1}{(p^+)^2}\int_{-\infty}^\infty \frac{d\lambda}{2\pi}e^{-iu\lambda}\langle f_2(P)|SG^a_{+\mu}(\lambda n)G^a_{+\nu}(-\lambda n)|0\rangle. \quad (3)$$

where polarization tensor $e_{\mu\nu}$ is symmetric and traceless, and satisfies the condition $e_{\mu\nu}P^\nu = 0$ The distribution amplitude $\phi^T_g(u)$ is symmetric to the interchange of $u \leftrightarrow -u$ and describes the momentum fraction distribution of the two gluons in the f_2-meson having the same helicity. The asymptotic distributions at large scales are equal to

$$\phi^{T,as}_g(u) = \frac{15}{16}(1-u^2)^2. \quad (4)$$

The normalization constant f^T_g is defined through the matrix element of the local two-gluon operator:

$$\langle f_2(P)|G^a_{\alpha\beta}(0)G^a_{\mu\nu}(0)|0\rangle = f^T_g\Big\{[(P_\alpha P_\mu - \frac{1}{2}M^2 g_{\alpha\mu})e^{(\lambda)}_{\beta\nu} - (\alpha\leftrightarrow\beta)] - (\mu\leftrightarrow\nu)\Big\}$$

$$+ \frac{1}{2}f^S_g M^2\Big\{[g_{\alpha\mu}e^{(\lambda)}_{\beta\nu} - (\alpha\leftrightarrow\beta)] - (\mu\leftrightarrow\nu)\Big\}. \quad (5)$$

where the second constant f^S_g describe normalization of the usual gluonic operator $G^a_{+\xi}G^a_{+\xi}$.

Let us remain again, that these are twist-2 matrix elements and their existence is direct consequence of QCD. Unfortunately, at present, we do not have any model or any estimate of the moments both for GPD's (2) and DA (3). These are completely unknown functions. Perhaps, we may expect only, that their contribution can not be so large as matrix elements of the usual gluons. Any arguments which claim that corresponding functions are very small don't exist either. Hence it can be interesting at least to prove using experimental data that the corresponding amplitudes exists and have expected scaling behaviour with respect to hard momentum Q^2. Then such observation can be considered as check on existence of the pure gluon configuration inside the target.

3. GLUON TRANSVERSITY IN REACTIONS $\gamma^*\gamma \to$ HADRONS

Gluon transversity appears in hard exclusive reactions $\gamma^*\gamma \to$ hadrons if the hadron final state carries orbital momentum $J = 2$ or large, for instance, two π-mesons or resonance with the spin 2 or higher [4, 5]. Here, we consider hard exclusive production of the tensor meson $f_2(1270)$ with the quantum numbers $J^{PC} = 2^{++}$, $I^G = 0^+$. This state is non-exotic, and, in quark model, can be constructed either from a quark and an antiquark, or from a pair of gluons. To get the spin projection $s = 2$ the quark and the antiquark have to be in a P-wave state, while the gluons can be in S-wave. On the light-cone, however, contribution of the orbital angular momentum is higher-twist and we will find that the form factor $\gamma + \gamma^* \to f_2(1270)$ corresponding to a pure helicity state $\lambda = \pm 2$ is dominated at large photon virtualities by the gluon contribution.

The amplitude of the process $\gamma^*(q) + \gamma(q') \to f_2(P)$ for one real $q'^2 = 0$ and one virtual $q^2 = -Q^2$ photon is related to the matrix element

$$T_{\mu\nu} = i \int d^4x\, e^{-ix(q-q')/2} \langle f_2(P,\lambda)|T\{j_\mu(x/2) j_\nu(-x/2)\}|0\rangle, \tag{6}$$

where $j_\mu = e_u \bar{u}\gamma_\mu u + e_d \bar{u}\gamma_\mu u + \ldots$ is the electromagnetic current. Neglecting contributions that vanish after the multiplication by photon polarizations, the amplitude $T_{\mu\nu}$ can be parametrized in terms of the three invariant form factors

$$\begin{aligned}
T_{\mu\nu} &= -g^\perp_{\mu\nu} e_{\alpha\beta} (q-q')^\alpha (q-q')^\beta \frac{m^2}{(2qq')^2} T_0(Q^2) + \\
&\quad -g^\perp_{\nu\alpha} e_{\alpha\beta} (q-q')^\beta \left(q - q'\frac{q^2}{(qq')}\right)_\mu \frac{m^2}{(2qq')^2} T_1(Q^2) \\
&\quad + \left(g^\perp_{\mu\alpha} g^\perp_{\nu\beta} + g^\perp_{\mu\nu} \frac{(q-q')^2}{(2qq')^2} (q-q')^\alpha (q-q')^\beta\right) e^{\alpha\beta} T_2(Q^2)\Big],
\end{aligned} \tag{7}$$

where we have introduced the metric tensor $g^\perp_{\mu\nu}$ that is transverse to the photon momenta

$$g^\perp_{\mu\nu} = g_{\mu\nu} - \frac{1}{(qq')}(q_\mu q'_\nu + q_\nu q'_\mu - \frac{q^2}{(qq')} q'_\mu q'_\nu). \tag{8}$$

The form factors T_0, T_1 and T_2 correspond to the three possible helicity amplitudes

$$\begin{aligned}
T_0 &: \quad \gamma^*(\pm 1) + \gamma(\pm 1) \to f_2(0), \\
T_1 &: \quad \gamma^*(0) + \gamma(\pm 1) \to f_2(\mp 1), \\
T_2 &: \quad \gamma^*(\pm 1) + \gamma(\mp 1) \to f_2(\pm 2).
\end{aligned} \tag{9}$$

We consider the region where the virtuality of one of the photons is large:

$$-q^2 \equiv Q^2 \gg M^2, \ q'^2 = 0. \tag{10}$$

In this kinematics, the form factors $\gamma^*(q) + \gamma(q') \to f_2(P)$ can be calculated by the light-cone expansion of the T-product of the electromagnetic currents in (6)

$$iT\{j_\mu(x)j_\nu(-x)\} = \frac{1}{(4\pi)^2 x^4} \sum e_q^2 \Big\{ \left[\bar{\psi}(-x)\gamma_\nu \not{x} \gamma_\mu \psi(x) - \bar{\psi}(x)\gamma_\mu \not{x} \gamma_\nu \psi(-x) \right]$$

$$-i\frac{\alpha_s}{4\pi} \int_{-1}^{1} du \int_{-1}^{1} dv\, (1+uv) S_{\mu\nu} G_{x\mu}(ux) G_{x\nu}(-vx) \Big\} + \ldots \tag{11}$$

where we have shown the contributions that will be relevant for what follows, and $G_{x\mu} = x_\xi G_{\xi\mu}$, etc. Note that we take into account the contribution of the two-gluon operator with helicity $\lambda = 2$ but do not consider the gluon contribution with $\lambda = 0$ that enters at the same level as the $\mathcal{O}(\alpha_s)$ corrections to the quark-antiquark operator.

Using (11) we calculate the most interesting case of the gluon-dominated amplitude T_2 also keeping the leading higher-twist correction. A simple calculation gives:

$$T_2 = 2\frac{\alpha_s}{\pi} \sum e_q^2 f_g^T \int_{-1}^{1} \frac{du}{1-u^2} \phi_g^T(u) +$$

$$+ \frac{5}{9} \frac{f_q M^2}{Q^2} \int_{-1}^{1} du \ln\left(\frac{1+u}{1-u}\right) g_v^{WW}(u), \tag{12}$$

where all distribution amplitudes and the coupling have to be taken at the scale $\mu^2 = Q^2$; $f_q(\mu = 1 \text{ GeV}) \simeq 56$ MeV is quark normalization constant [6] and $g_v^{WW}(u)$ is twist-4 DA in the, so-called, Wandzura-Wilczek approximation, see details in [4]. We expect that distribution amplitudes of the f_2-meson are not far from their asymptotic form given in the text. On this assumption, which is certainly acceptable for an estimate, we obtain for large Q^2

$$T_2 \simeq \frac{5}{3}\frac{\alpha_s}{\pi} f_g^T + \frac{50}{27} f_q \frac{M^2}{Q^2}, \tag{13}$$

where we have taken into account u, d and s-quark contribution in the quark loop in the box diagram. For a realistic value $Q^2 \sim 10$ GeV^2 the gluon contribution is comparable to the subleading quark contribution if the constants f_g^T and f_q are of the same order. It is interesting to note that $T_2(0)$ is much larger than both $T_0(0)$ and f_q. This may indicate that f_g^T is quite large.

Since f_2 decays in two pions with the branching ratio about 95%, one natural possibility to measure the form factors is via the hard exclusive two-pion production $ee \to ee\pi\pi$. Let k_1 and k_2 be the momenta of the two pions in the final state. The tensor form factor of interest $T_2(Q^2)$ is related to the two-pion helicity amplitude A_{+-} (we use notations of Ref. [7], see Eq. (75) there) in the region $(k_1 + k_2)^2 \sim m^2$

as

$$A_{+-} = \varepsilon_+^\mu \varepsilon_-'^\nu (k_1 - k_2)_\mu^\perp (k_1 - k_2)_\nu^\perp \frac{g_{f_2\pi\pi}}{m} \frac{T_2(Q^2)}{m^2 - (k_1 + k_2)^2}, \tag{14}$$

where $g_{f_2\pi\pi}$ is the corresponding decay constant:

$$\langle \pi(k_1)\pi(k_2)|f_2(P,\lambda)\rangle = \frac{g_{f_2\pi\pi}}{m} e_{\alpha\beta}^{(\lambda)} (k_1 - k_2)^\alpha (k_1 - k_2)^\beta \tag{15}$$

and $\varepsilon_+^\mu, \varepsilon_-'^\nu$ denote the transverse polarisation vectors of the virtual and the real photon, respectively. As was discussed in Ref. [5, 7] the amplitude A_{+-} can be separated using its nontrivial dependence on the azimuthal angle φ. Moreover, A_{+-} is symmetric to the interchange of the pion momenta and can be measured by the interference with the (much larger) contribution of the bremsstrahlung process that is antisymmetric to the interchange of the pion momenta: In the difference of cross sections $\sigma(k_1, k_2) - \sigma(k_2, k_1)$ only the interference term survives. The relevant expressions for the physical cross sections have been worked out in [7].

4. DEEPLY VIRTUAL COMPTON SCATTERING

Calculation of the tensor gluon contribution in DVCS is very similar. This is also two photon process but only in the cross channel. The amplitude sensitive to transverse gluon GPD's is double helicity flip $\gamma^*(+1)N(p) \to \gamma(-1)N(p')$ amplitude. To the next-to-leading twist accuracy this amplitude can be written:

$$A_{+-}^{\mu\nu} = \frac{1}{2}\left[g_\perp^{\mu i}\left(g_\perp^{\nu j} + \frac{P^\nu \Delta_\perp^j}{(Pq)}\right) + g_\perp^{\mu j}\left(g_\perp^{\nu i} + \frac{P^\nu \Delta_\perp^i}{(Pq)}\right) - \left(g_\perp^{\mu\nu} + \frac{P^\nu \Delta_\perp^\mu}{(Pq)}\right)g_\perp^{ij}\right]\left[A_{(ij)}^{tw2} + \frac{m^2}{Q^2}A_{(ij)}^{tw4}\right]. \tag{16}$$

Here $A_{(ij)}^{tw2}$ is the leading twist contribution:

$$A_{\mu\nu}^{tw2} = (\sum e_q^2)\frac{\alpha_s(Q^2)}{4\pi\xi}\int_{-1}^1 dx\, F_{(\mu\nu)}(x,\xi,t)\left(\frac{1}{x-\xi+i\varepsilon} - \frac{1}{x+\xi-i\varepsilon}\right), \tag{17}$$

depending on transverse gluon GPD's (2). The twist-4 part, arising from the handbag diagram through the WW mechanism can be parametrized as $\frac{m^2}{Q^2}A_{(ij)}^{tw4}$, with m being the nucleon mass. A convenient physical interpretation of this term follows from an observation that beyond the leading twist approximation one can imagine partons as carrying non-zero orbital angular momentum along the collision axis. That allows quarks to participate in the LO, i.e. through the handbag diagram, in the photon helicity-flip amplitude. As two units of angular momentum have to flow through the hard vertex, such an amplitude is suppressed by $1/Q^2$ and represents a twist-4 contribution. In the WW approximation this term have been calculated in [8].

The standard way in the analysis of DVCS observables is to use interference with the Bethe-Heitler process. This allows to measure asymmetries which are

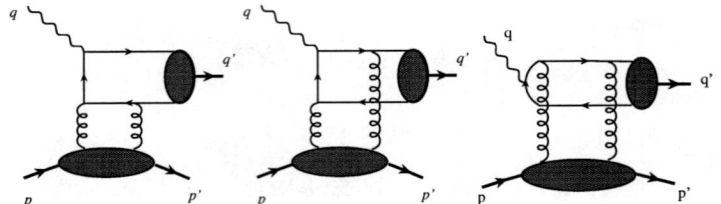

FIGURE 1. Diagrams for the electroproduction of ρ-meson with double-helicity-flip.

linear in DVCS amplitudes. Double helicity flip amplitude give rise to a $cos(3\varphi)$ angular distribution. For the case of unpolarised beam and target interference term proportional to the GPD's (2) appears in following combination [3]:

$$\frac{(t_{min}-t)^{3/2}}{8m3}\left(H_T^g F_2 - E_T^g F_1 - 2\tilde{H}_T^g(F_1 + \frac{t}{4m^2}F_2)\right)\cos(3\varphi) \qquad (18)$$

where F_1 and F_2 are the Dirac and Pauli form factors of the proton, respectively. Such combination can contribute for instance to the charge asymmetry measured already by HERMES collaboration [9]. Unfortunately, at present, the accuracy of the data do not allow to perform angular analysis in order to extract the value of the $cos(3\varphi)$ term. Different observables for the cases of polarized beam or proton have been presented in [10].

5. HARD EXCLUSIVE ELECTROPRODUCTION OF MESONS

Double helicity flip amplitude in the electroproduction of the ρ-meson, similar to DVCS, is also sensitive to the transverse gluons but such amplitude is of subleading twist. Remind that factorization was proved only for the process with longitudinal photon $\gamma_L^* N \to \rho_L N$. Now we are interested in helicity double flip amplitude $\gamma_T^*(+) N \to \rho_T(-) N$. For the transverse photon factorization can be broken and this question has to be studied. In [11] such amplitude have been investigated and it was fond that for the two-particle twist-3 meson wave function amplitude is free from end-point singularities which can spoil factorization. The relevant for the process diagrams are shown in the Fig.1.

The twist-3 meson wave function is approximated by two-particle state only. For simplicity, we skip all 3-particle contributions assuming that they are small. Then the answer for the amplitude reads:

$$A_{+-}^{\mu} = \frac{2}{N_c}\frac{f_\rho m_\rho}{Q^2}(4\pi\alpha_S)\int_{-1}^{1}du\frac{\phi_\rho(u)e_\nu^*}{1-u^2}\int_{-1}^{1}dx\frac{F^{(\mu\nu)}(x,\xi,t)\xi^2}{(x-\xi+i0)^2(x+\xi-i0)^2} \qquad (19)$$

where the $\phi_\rho(u)$ is twist-2 ρ-meson DA and f_ρ is corresponding normalization constant. The asymptotic DA is given by $\phi_\rho^{as}(u) = \frac{3}{4}(1-u^2)$. The formula (19) is valid if the convolution integrals are convergent. The integral of meson DA is well defined if the end-point behavior of $\phi_\rho(u)$ is not worse then the end-point behavior $\phi_\rho^{as}(u)$. Usually it is assumed like that. The GPD's integrals will be well defined if derivatives of the $F^{(\mu\nu)}(x,\xi,t)$ with respect to momentum fraction x are still continuous functions. That we expect to be valid too. Hence result (19) provides valid expression for the leading term of the double flip amplitude at large Q^2.

From phenomenological point of view, we expect that helicity non-conserving amplitudes are small relative to helicity conserving. Hence the best way to access A_{+-}^μ is due its to interference with large helicity conserving amplitudes. Such analysis is known and was carried out long time ago [12]. Different spin-density matrix elements for the hard ρ-meson electroproduction have been measured by HERMES and ZEUS in the different kinematical regions [14, 13]. The interesting for us matrix elements, according to HERMES results, are different from non-zero values only at the low Q^2 and three or four standard deviations from zero. Better statics is necessary to get more qualitative results. The ZEUS measurements are compatible with zero as expected for the small-x region. For better understanding of situation we need information about spin-density matrix elements from region of relatively large-x and large Q^2. Such measurements are also very important to get more clear information of the other amplitudes of the hard meson electroproduction and to extract information about GPD's of the nucleon.

REFERENCES

1. F. E. Close, Rept. Prog. Phys. **51** (1988) 833;
   F. E. Close, G. R. Farrar and Z. Li, Phys. Rev. **D55** (1997) 5749.
2. P. Hoodbhoy and X. Ji, Phys. Rev. **D58** (1998) 054006.
3. M. Diehl, Eur. Phys. J. C **19** (2001) 485 [hep-ph/0101335].
4. V. M. Braun and N. Kivel, Phys. Lett. B **501** (2001) 48 [arXiv:hep-ph/0012220].
5. N. Kivel, L. Mankiewicz and . V. Polyakov, Phys. Lett. B **467** (1999) 263 [arXiv:hep-ph/9908334].
6. T. M. Aliev and M. A. Shifman, Sov. J. Nucl. Phys. **36** (1982) 891.
7. M. Diehl, T. Gousset and B. Pire, Phys. Rev. **D62** (2000) 073014.
8. N. Kivel and L. Mankiewicz, Eur. Phys. J. C **21** (2001) 621 [arXiv:hep-ph/0106329].
9. F. Ellinghaus [HERMES Collaboration], Nucl. Phys. A **711**, 171 (2002) [arXiv:hep-ex/0207029].
10. A. V. Belitsky, D. Muller and A. Kirchner, Nucl. Phys. B **629**, 323 (2002) [arXiv:hep-ph/0112108].
11. N. Kivel, Phys. Rev. D **65** (2002) 054010 [arXiv:hep-ph/0107275].
12. K. Schilling and G. Wolf, Nucl. Phys. B **61** (1973) 381.
13. J. Breitweg et al. [ZEUS Collaborations], Eur. Phys. J. C **12** (2000) 393 [hep-ex/9908026].
14. K. Ackerstaff et al. [HERMES Collaboration], Eur. Phys. J. C **18** (2000) 303 [hep-ex/0002016].

New insight on global QCD fits using Regge theory

G. Soyez[1]

*CEA Saclay, Service de Physique Théorique, Orme des Merisiers Bât 774, F-91191
Gif-Sur-Yvette, France
E-mail: gsoyez@spht.saclay.cea.fr*

Abstract. In global QCD fits, one has to choose an initial parton distribution at $Q^2 = Q_0^2$. I shall argue that the initial condition choses in usual standard sets is inconsistent with analytic S-matrix theory. I shall show how one can combine these two approaches, leading to a Regge-compatible next-to-leading order global QCD fit. This allows one to extend the parametrisation in the low-Q^2 region. Finally, I shall discuss how it it possible to use the Dokshitzer-Gribov-Lipatov-Altarelli-Parisi (DGLAP) equation to obtain information on Regge models at high Q^2.

Keywords: DGLAP equation, Global QCD fit, Regge theory, triple-pole pomeron
PACS: 11.55.-m, 13.60.-r

1. INTRODUCTION

One of the most important results of QCD is the Dokshitzer-Gribov-Lipatov-Altarelli-Parisi (DGLAP) evolution equation [1] which gives the evolution of the parton densities with the virtuality Q^2 of the photon. The kernel of these equation, *i.e.* the splitting functions, has been computed recently [2, 3] at the three-loop level in perturbation theory (NNLO). From the phenomenological point of view, the large number of experimental data coming from the highly-accurate measurement of the structure functions, *e.g.* at HERA [4, 5], allows for the determination of the parton densities.

The technique is to choose a parametric distribution for the parton at an initial scale Q_0^2, to obtain the Parton Distribution Functions (PDF) at all Q^2 using the DGLAP equation and to construct the physical quantities by convolution with the coefficient functions. One can then adjust the initial condition in order to reproduce the data as well as possible. This way of getting a standard PDF set, usually called *a global QCD fit*, has been applied many times with many updates by various teams [6, 7, 8, 9, 10].

Beside this approach, it is well-known that one can reproduce the hadronic data using the analyticity properties of the S matrix. In this framework, the amplitudes are considered not as a function of the energy but in complex angular-momentum space. At high energy, the behaviour of the amplitude is then given

---

[1] FNRS Research Fellow (Chargé de Recherches). On leave from the fundamental theoretical physics group of the University of Liège.

by the leading singularities in the complex-j plane, corresponding to pomeron and reggeons exchanges. Therefore, in order to reproduce the data, one chooses a structure of singularity in the complex-j plane and adjusts their residues. Usually, we consider a pomeron contribution reproducing the growth of the cross-section at high energies, together with reggeon terms taking into account f-, a- and ω-meson trajectories exchanges. Within this framework, different models of pomeron are consistent with the present data [11]. As an example, one can consider the Donnachie-Landshoff two-pomeron model [12] with one simple pole at $j = 1.4$ ($F_2 \propto x^{-0.4}$) and a simple pole at $j = 1.08$ ($F_2 \propto x^{-0.08}$). Throughout these proceedings, we shall concentrate on another choice known as the triple-pole pomeron model [13], where the pomeron singularity is a triple pole at $j = 1$, corresponding to a cross-section growing like $\log^2(1/x)$.

2. REGGE-COMPATIBLE GLOBAL FIT

2.1. The initial condition problem

The DGLAP equation, being a renormalisation group equation, gives you the the Q^2-evolution of the PDF. This means that you still have to provide an initial condition at an initial scale Q_0^2. Let us consider, *e.g.*, the CTEQ 6M [6] initial distributions. At small x, they obtain

$$xq(x) \propto x^{-0.30} \quad \text{and} \quad xg(x) \propto x^{0.51}.$$

This result suffers from two problems: first, the corresponding j-plane singularities are not seen in soft amplitudes. Secondly, quarks and gluons, while being coupled, do not have the same singularity structure. These two problems are in contradiction with the fact that Regge theory requires [14] all hadronic amplitudes to have the same j-plane singularities.

2.2. Proposed solution

Let us show how to solve this problem [15]. We shall perform a global QCD fit starting at $Q_0^2 = 5$ GeV^2 with initial parton distributions compatible with Regge theory. In other words, one need to parametrise the initial parton distribution in agreement with S-matrix theory. In order to obtain a complete PDF set, one need to parametrise the valence quarks u_v and d_V, the sea quarks u_s, d_s, s_s and c_s, and the gluon density g. These expressions have to reproduce the pomeron and reggeon exchanges. Inserting a power of $1 - x$ in order to ensure that the parton

TABLE 1. Result of the global QCD fit together with the Regge small-Q^2 extension.

Data			LO global fit			NLO global fit		
Exp.	nop	$Q^2 \geq 5$	$Q^2 < 5$	total	$Q^2 \geq 5$	$Q^2 < 5$	total
F_2^p	1828	0.969	0.867	0.948	0.943	0.933	0.937
F_2^d	391	1.075	1.124	1.089	1.073	1.041	1.064
F_2^n/F_2^p	211	1.233	0.886	1.051	1.270	0.764	1.004
F_2^ν	84	2.385	5.063	2.767	2.096	4.353	2.418
xF_3^ν	111	0.440	1.420	0.652	0.670	0.613	0.658
Total	2615	1.026	1.017	**1.023**	1.008	0.975	**1.000**

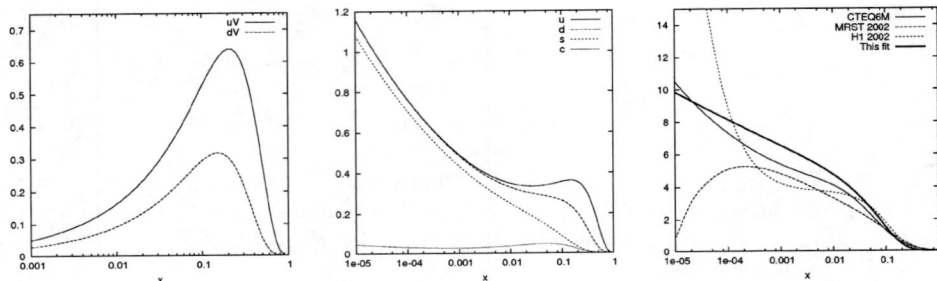

FIGURE 1. Valence quark, sea quark and gluon distributions for the NLO global QCD fit.

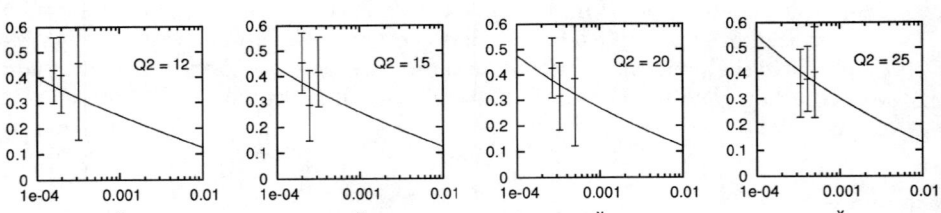

FIGURE 2. Predictions of our fit for the longitudinal structure function F_L.

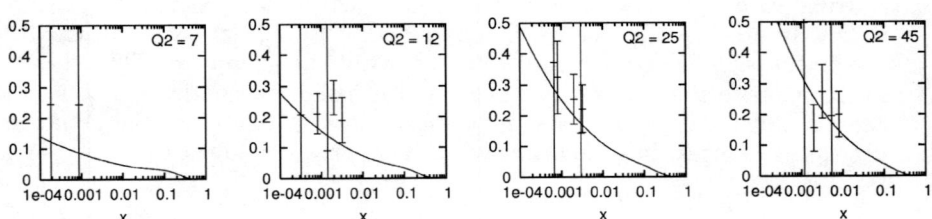

FIGURE 3. Predictions of our fit for the charm structure function F_c^2.

distributions go to 0 as x goes to 1, this gives

$$\begin{aligned}
xu_V &= \frac{2}{N_u^*} x^\eta (1+\gamma_u x)(1-x)^{b_u}, \\
xd_V &= \frac{1}{N_d^*} x^\eta (1+\gamma_d x)(1-x)^{b_d}, \\
xu_s &= \left[A\log^2(1/x) + B\log(1/x) + C + D_u x^\eta\right](1-x)^b, \\
xd_s &= \left[A\log^2(1/x) + B\log(1/x) + C + D_d x^\eta\right](1-x)^b, \\
xs_s &= N_s \left[A\log^2(1/x) + B\log(1/x) + C + D_s x^\eta\right](1-x)^b, \\
xc_s &= N_c \left[A\log^2(1/x) + B\log(1/x) + C + D_s x^\eta\right](1-x)^b, \\
xg &= \left[A_g\log^2(1/x) + B_g\log(1/x) + C_g^*\right](1-x)^{b+1},
\end{aligned} \qquad (1)$$

In these distributions, the $\log^2(1/x)$, the $\log(1/x)$ and the constant terms correspond to a triple-pole pomeron exchange while the fourth one, with $\eta = 0.4$ is a f-reggeon exchange. To build the initial conditions (1), we have applied the following physical arguments

- The pomeron is a flavour-singlet object. It therefore decouples from valence quarks and couples in the same way to all sea quarks.
- The reggeon is expected to represent quark exchanges. We assume that it is not coupled to gluons.
- The power of $1-x$ defining the large-x behaviour of the parton distributions is taken to be the same for all sea quarks. The exponent in the gluon distribution appears from a study of the DGLAP equation at large x [16].
- The mass effects has been introduced through the normalisation factors N_s and N_c. Due to the small strange mass, we expect $N_s \approx 1$. N_c is non-zero for initial scales Q_0^2 larger than $m_c^2 = 2$ GeV^2. Similarly, during the DGLAP evolution, the b quarks will be switched on at $Q^2 = m_b^2 = 20.25$ GeV^2.
- Quark number conservation fixes N_u and N_d. The momentum sum rule is used to constrain C_g.

We have performed a global QCD fit with this initial condition at leading and next-to-leading order (in the $\overline{MS}$ scheme). We have included the data for the proton [4, 5, 17, 18, 19] and deuteron [18, 20, 21] structure functions, the ratio F_2^n/F_2^p [21] and the neutrino structure functions [22, 23]. The initial scale Q_0^2 has been set to 5 GeV^2 and we have imposed $W^2 \geq 12.5$ GeV^2 in order to cut the region where we expect higher-twist effects. We obtain, as presented in Table 1, a chi-square per data point slightly larger than 1. This is at least as good as the results reached by standard global fits.

In Fig. 1 we show the valence and sea quark distributions. We also present the gluons obtained from our NLO fit compared with some distributions taken from various standard PDF sets at $Q^2 = 5$ GeV^2. Our distributions are compatible with the standard PDF sets.

Finally, one can consider predictions for the charm structure function F_2^c or for the longitudinal structure function F_L [24]. These are sensitive to the gluon

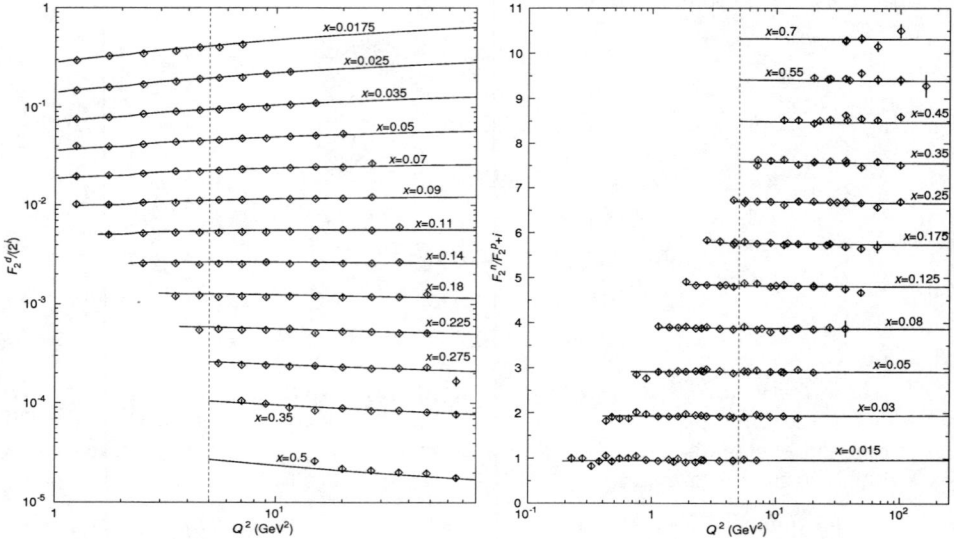

FIGURE 4. Result of our fit for the NMC data for F_2^d (left) and F_2^n/F_2^p (right). In each plot, the high-Q^2 part is described by DGLAP evolution and the low-Q^2 data by Regge theory.

distribution and, in addition, F_2^c [25] is a good test for the charm distribution. Our predictions are compatible with the data, as presented in Figs. 2 and 3. We also obtain a good description of the Tevatron Drell-Yan data with a K-factor between 1.3 and 1.4.

3. SMALL-Q^2 DESCRIPTION

One of the main advantages of restoring relevant analytic expressions for the initial distribution is that one can use [26] Regge-theory techniques to describe the data for $0 \leq Q^2 \leq Q_0^2$. We need to use the parametrisation (1) and let the coefficients of the fit being Q^2-dependent. For the Q^2 dependence, we have adopted for these dependences, the usual parametrisation

$$\phi(Q^2) = \phi(Q_0^2)\frac{Q^2}{Q_0^2}\left(\frac{Q_0^2 + Q_\phi^2}{Q^2 + Q_\phi^2}\right)^{\varepsilon_\phi},$$

where we have imposed matching with the parameters obtained from the global fit at $Q^2 = Q_0^2$. With a few simplifications, e.g., $Q_{D_u}^2 = Q_{D_d}^2 = Q_{D_s}^2 = Q_{D_c}^2$, it gives a total of 13 parameters.

At leading order, it is sufficient to consider extensions of the quark distributions. Due to the convolution with coefficient functions at NLO, one also needs small-Q^2 expressions for the gluon distribution and for the strong coupling constant α_s. Since

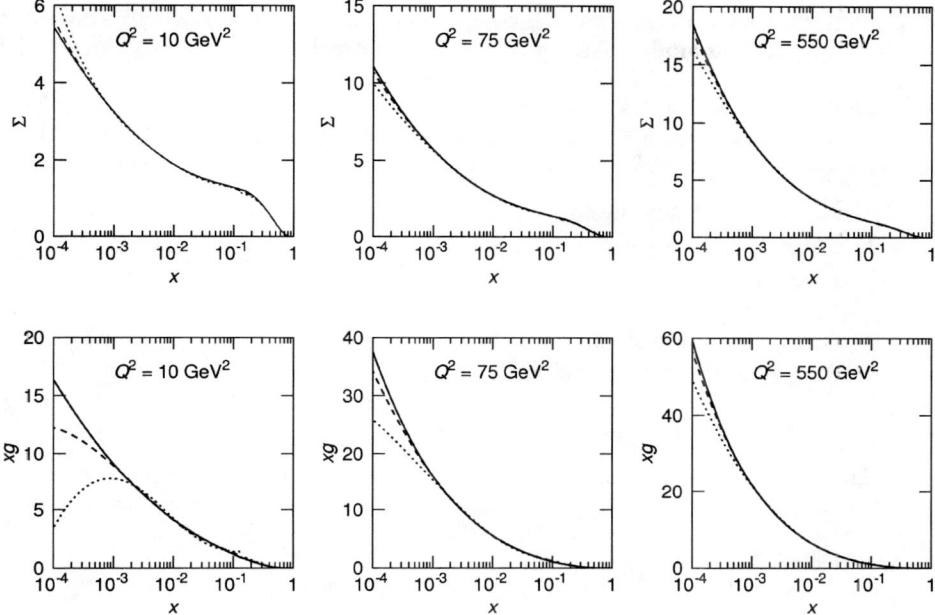

FIGURE 5. Sea quark (Σ) and gluon (xg) distributions at high Q^2. The solid (resp. dashed, dotted) line corresponds to the fit starting at $Q_0^2 = 10$ GeV^2 (resp. 75, 550 GeV^2). For example, the solid curve has a triple-pole pomeron behaviour at $Q^2 = 10$ GeV^2 and a essential singularity at other Q^2 values.

we expect our results to be relatively independent of this parametrisation, we have simply considered that the gluon density scales with Q^2 ($g(x,Q^2) = Q^2/Q_0^2 g(x,Q_0^2)$) and that the running coupling stays constant for $Q^2 \leq m_c^2$.

The χ^2 resulting from this fit is presented in Table 1 together with the global QCD fit. *We see that at NLO, one finally obtains a description of the DIS data, at all values of Q^2, for $W^2 \geq 12.5$ GeV^2 with a chi-square of 1 per data point.*

To illustrate this, we have shown in Fig. 4 the final result of our fit compared with the F_2^d and F_2^n/F_2^p measures from NMC.

4. REGGE THEORY AT HIGH-Q^2

We have shown that it is possible to have a description of the DIS data based on Regge theory at small Q^2 and on DGLAP evolution at large Q^2. However, it is well-known that DGLAP evolution generates an essential singularity at $j = 1$, which seems in disagreement with Regge theory.

It is proven that, in the small-x region, additional resummation needs to be done if we want to recover a Regge-like high-energy behaviour. Our statement is that, even if their analytic structure are not physical, the DGLAP-obtained parton densities have to be treated as numerical approximations. Our aim is to show that, if we assume that DGLAP evolution approximates a triple-pole pomeron, we can extract the residues of the pomeron at large Q^2 from the DGLAP equation. In order to find the residues we may therefore adopt the following strategy [27]:

1. choose an initial scale Q_0^2 at which one searches for the Regge residues,
2. choose a value for the parameters in the initial distribution,
3. compute the parton distributions for $Q_0^2 \leq Q^2 \leq Q_{\max}^2$ using forward DGLAP evolution and for $Q_{\min}^2 \leq Q^2 \leq Q_0^2$ using backward DGLAP evolution,
4. repeat 2 and 3 until the value of the parameters reproducing the F_2 data for $Q^2 > Q_{\min}^2$ and $x \leq 0.15$ is found.
5. This gives the residues at the scale Q_0^2 and steps 1 to 4 are repeated in order to obtain the residues at all Q^2 values.

This technique allows us to obtain the residues of a Regge fit directly from QCD evolution, without having to postulate an analytic expression for the Q^2 dependence. It is also motivated by the fact that physics is expected to be independent on the choice of Q_0^2.

Applying this method between 10 and 1000 GeV^2 and for $x \leq 0.15$, we have obtained the Q^2-dependent residues of the pomeron giving a fit to F_2^p with $\chi^2/nop = 1.02$. One may then check that the DGLAP result is a correct numerical approximation to a triple-pole pomeron model. In Fig. 5, we show the results of the DGLAP fit for 3 different choices of the initial scale. As seen from the upper part of the plot, in the case of the sea quark distribution, it is impossible to distinguish between the triple-pole pomeron behaviour $(Q^2 = Q_0^2)$ and the curves with a DGLAP-generated essential singularity $(Q^2 \neq Q_0^2)$. This validates our initial argument.

The situation is not as clear for the gluon distributions (lower part of Fig. 5). Since each of these fits reproduces equally well the F_2^p data, the differences between them have to be considered as errors on the gluon distribution. We thus predict large errors on the gluon density at small Q^2 and small x, which may be of prime importance for LHC physics.

5. CONCLUSIONS

We have shown that it was possible to use Regge theory to constrain the parametrisation of the initial parton distributions in a global QCD fit. We obtain a standard PDF set at LO and NLO which reproduces the usual features of global QCD fits with the advantage of being consistent with analytic S-matrix theory.

This allows us to extend the fit in the small-Q^2 region giving a complete description of the data, for $W^2 \geq 12.5$ GeV^2, over the whole Q^2 range.

Finally, we have discussed the compatibility of Regge theory and DGLAP evolution at large Q^2. We show that one can use the DGLAP equation to extract the residues of the triple-pole pomeron and that this predicts large errors on the gluon distribution at small x and small Q^2.

In the future, it might be interesting to repeat this Regge-constrained global QCD fit with other Regge models and to combine the low- and high-Q^2 fits in order to allow for a determination of the initial scale for DGLAP evolution.

The high-Q^2 study also invites one to look for QCD corrections to the DGLAP equation which stabilises the Regge behaviour.

ACKNOWLEDGMENTS

The author is funded by the National Funds for Scientific Research (FNRS, Belgium)

REFERENCES

1. V.N. Gribov and L.N. Lipatov, *Sov. J. Nucl. Phys.* **15** (1972) 438; G. Altarelli and G. Parisi, *Nucl. Phys.* **B126** (1977) 298; Yu.L. Dokshitzer, *Sov. Phys. JETP* **46** (1977) 641.
2. S. Moch, J. A. M. Vermaseren and A. Vogt, Nucl. Phys. B **688** (2004) 101 [arXiv:hep-ph/0403192].
3. A. Vogt, S. Moch and J. A. M. Vermaseren, Nucl. Phys. B **691** (2004) 129 [arXiv:hep-ph/0404111].
4. I. Abt et al. [H1 Collaboration], Nucl. Phys. B **407** (1993) 515; T. Ahmed et al., Nucl. Phys. B **439** (1995) 471 [arXiv:hep-ex/9503001]; S. Aid et al., Nucl. Phys. B **470** (1996) 3 [arXiv:hep-ex/9603004]; C. Adloff et al., Nucl. Phys. B **497** (1997) 3 [arXiv:hep-ex/9703012]; Eur. Phys. J. C **13** (2000) 609 [arXiv:hep-ex/9908059]; Eur. Phys. J. C **19** (2001) 269 [arXiv:hep-ex/0012052]; Eur. Phys. J. C **21** (2001) 33 [arXiv:hep-ex/0012053].
5. M. Derrick et al. [ZEUS Collaboration], Phys. Lett. B **316** (1993) 412; Z. Phys. C **65** (1995) 379; Z. Phys. C **69** (1996) 607 [arXiv:hep-ex/9510009]; Z. Phys. C **72** (1996) 399 [arXiv:hep-ex/9607002]; J. Breitweg et al., Phys. Lett. B **407** (1997) 432 [arXiv:hep-ex/9707025]; Eur. Phys. J. C **7** (1999) 609 [arXiv:hep-ex/9809005]; Eur. Phys. J. C **12** (2000) 35 [arXiv:hep-ex/9908012]; S. Chekanov et al., Eur. Phys. J. C **21** (2001) 443 [arXiv:hep-ex/0105090].
6. J. Pumplin, D. R. Stump, J. Huston, H. L. Lai, P. Nadolsky and W. K. Tung, JHEP **0207**, 012 (2002) [arXiv:hep-ph/0201195].
7. A. D. Martin, R. G. Roberts, W. J. Stirling and R. S. Thorne, Eur. Phys. J. C **23**, 73 (2002) [arXiv:hep-ph/0110215].
8. M. Gluck, E. Reya and A. Vogt, Eur. Phys. J. C **5**, 461 (1998) [arXiv:hep-ph/9806404].
9. A. M. Cooper-Sarkar, J. Phys. G **28**, 2669 (2002) [arXiv:hep-ph/0205153].
10. S. I. Alekhin, Phys. Rev. D **63**, 094022 (2001) [arXiv:hep-ph/0011002].
11. J. R. Cudell et al., Phys. Rev. D **65** (2002) 074024 [arXiv:hep-ph/0107219].
12. A. Donnachie and P. V. Landshoff, Phys. Lett. B **518** (2001) 63 [arXiv:hep-ph/0105088].
13. J. R. Cudell and G. Soyez, Phys. Lett. B **516** (2001) 77 [arXiv:hep-ph/0106307].
14. J. R. Cudell, E. Martynov and G. Soyez, Nucl. Phys. B **682** (2004) 391 [arXiv:hep-ph/0207196].
15. G. Soyez, arXiv:hep-ph/0407098.
16. C. Lopez and F. J. Yndurain, Nucl. Phys. B **171** (1980) 231.
17. A. C. Benvenuti et al. [BCDMS Collaboration], Phys. Lett. B **223** (1989) 485.
18. M. R. Adams et al. [E665 Collaboration], Phys. Rev. D **54** (1996) 3006.
19. M. Arneodo et al. [New Muon Collaboration], Nucl. Phys. B **483** (1997) 3 [arXiv:hep-ph/9610231].
20. A. C. Benvenuti et al. [BCDMS Collaboration], Phys. Lett. B **237** (1990) 595.
21. P. Amaudruz et al. [New Muon Collaboration], Nucl. Phys. B **371** (1992) 3.

22. W. G. Seligman *et al.*, Phys. Rev. Lett. **79** (1997) 1213.
23. B. T. Fleming *et al.* [CCFR Collaboration], Phys. Rev. Lett. **86** (2001) 5430 [arXiv:hep-ex/0011094].
24. S. Chekanov *et al.* [ZEUS Collaboration], Eur. Phys. J. C **21**, 443 (2001).
25. J. Breitweg *et al.* [ZEUS Collaboration], Phys. Lett. B **407**, 402 (1997) [arXiv:hep-ex/9706009]; Eur. Phys. J. C **12**, 35 (2000) [arXiv:hep-ex/9908012]; C. Adloff *et al.* [H1 Collaboration], Z. Phys. C **72**, 593 (1996) [arXiv:hep-ex/9607012].
26. G. Soyez, Phys. Lett. B **603** (2004) 189 [arXiv:hep-ph/0401177].
27. G. Soyez, Phys. Rev. D **69** (2004) 096005 [arXiv:hep-ph/0306113].

DIFFRACTION

Fluctuating pulled fronts & Pomerons

Edmond Iancu[1]

*Service de Physique Théorique, CEA/DSM/SPhT, Unité de recherche associée au CNRS,
CE Saclay, F-91191 Gif-sur-Yvette, France
E-mail: iancu@spht.saclay.cea.fr*

Abstract. I present a pedagogical discussion of the influence of particle number fluctuations on the high energy evolution in QCD. I emphasize the event–by–event description, and the correspondence with the problem of 'fluctuating pulled fronts' in statistical physics. I explain that the correlations generated by fluctuations reduce the phase–space for BFKL evolution up to saturation. Because of that, the evolution 'slows down', and the rate for the energy increase of the saturation momentum is considerably decreased. I also discuss the diagrammatic interpretation of the particle number fluctuations in terms of Pomeron loops.

1. INTRODUCTION

Much of the recent progress in our understanding of QCD evolution at high energy has been triggered by the observations that (i) the gluon number fluctuations play an important role in the evolution towards saturation and the unitarity limit [1, 2] and (ii) the QCD evolution in the presence of fluctuations and saturation is in the same universality class as a series of problems in statistical physics, the prototype of which being the 'reaction–diffusion' problem [3, 2, 4].

These observations have developed into a profound and extremely fruitful correspondence between high–energy QCD and modern problems in statistical physics, which relates topics of current research in both fields, and which has already allowed us to deduce some insightful results in QCD by properly translating the corresponding results from statistical physics.

At the same time, the recognition of the importance of fluctuations has revived the interest in the dilute regime of QCD at high energy, which has been somehow overlooked by the modern theory for gluon saturation, the Color Glass Condensate [5]. A more appropriate formalism in that respect is Mueller's 'color dipole' picture [6], which describes the gluon number fluctuations in the limit where the number of 'colors' N_c is large. In fact, it was in the context of this formalism that Salam has first noticed, through numerical simulations [7], the dramatic role played by fluctuations in the course of the evolution.

Thus, not surprisingly, the dipole picture occupies a central role in the recent developments aiming at the inclusion of the effects of particle number fluctuations in the non–linear evolution towards saturation [4, 8, 9]. Furthermore, the dipole picture will play a crucial role in the presentation we shall give here, and which is

---

[1] Membre du Centre National de la Recherche Scientifique (CNRS), France.

largely adapted from Refs. [1, 2, 4, 9].

2. THE BALITSKY–KOVCHEGOV EQUATION

The simplest physical context in which one can address the study of gluon saturation is the collision between a small *color dipole* (a quark–antiquark pair in a colorless state) and a high energy hadron (the "target"). At high energy, the target wavefunction is dominated by gluons, to which couple the quark and the antiquark in the dipole. Thus, by following the evolution of the dipole scattering amplitude towards the unitarity limit, one can obtain information about the evolution of the gluon distribution in the target towards saturation.

Since the projectile has such a simple structure, it is quite easy to deduce the equation describing the evolution of the corresponding S–matrix with increasing energy. We shall denote the elastic *scattering amplitude* as $\langle T(\boldsymbol{x},\boldsymbol{y})\rangle_\tau$, where $\boldsymbol{x}$ and $\boldsymbol{y}$ are the transverse coordinates of the quark and the antiquark, respectively, and $\tau \sim \ln s$ is the 'rapidity' variable, with s the total invariant energy squared. As we shall see, τ plays the role of an 'evolution time' for the quantum evolution with increasing energy. Now suppose we increase τ by a small amount $d\tau$. In order to compute the corresponding change in $\langle T\rangle_\tau$, it is more convenient to keep the rapidity of the target fixed and put the small change of rapidity into the elementary dipole. The latter then 'evolves', that is, it has a small probability of emitting a gluon due to this change of rapidity, which can be estimated as

$$dP = \frac{\alpha_s N_c}{2\pi^2}\mathcal{M}(\boldsymbol{x},\boldsymbol{y},\boldsymbol{z})\,d^2z\,d\tau, \qquad \mathcal{M}(\boldsymbol{x},\boldsymbol{y},\boldsymbol{z}) \equiv \frac{(\boldsymbol{x}-\boldsymbol{y})^2}{(\boldsymbol{x}-\boldsymbol{z})^2(\boldsymbol{y}-\boldsymbol{z})^2}, \qquad (2.1)$$

where N_c is the number of colors and $\boldsymbol{z}$ is the transverse coordinate of the emitted gluon. In the large–N_c limit, to which we shall restrict in what follows, the gluon can be effectively replaced by a zero–size $q\bar{q}$ pair, and the gluon emission appears as the splitting of the original dipole $(\boldsymbol{x},\boldsymbol{y})$ into two new dipoles $(\boldsymbol{x},\boldsymbol{z})$ and $(\boldsymbol{z},\boldsymbol{y})$.

If the emitted gluon is in the wavefunction of the dipole at the time it scatters on the target, then what scatters is a system of two dipoles. If the gluon is not in the wavefunction at the time of the scattering, it can be viewed as the "virtual" term which decreases the probability that the original quark–antiquark pair remain a simple dipole. The whole process can be summarized into the following *evolution equation*, which has been originally derived by Balitsky [10]:

$$\frac{\partial}{\partial\tau}\langle T(\boldsymbol{x},\boldsymbol{y})\rangle_\tau = \frac{\bar{\alpha}_s}{2\pi}\int_z \mathcal{M}(\boldsymbol{x},\boldsymbol{y},\boldsymbol{z}) \qquad (2.2)$$
$$\{-\langle T(\boldsymbol{x},\boldsymbol{y})\rangle_\tau + \langle T(\boldsymbol{x},\boldsymbol{z})\rangle_\tau + \langle T(\boldsymbol{z},\boldsymbol{y})\rangle_\tau - \langle T^{(2)}(\boldsymbol{x},\boldsymbol{z};\boldsymbol{z},\boldsymbol{y})\rangle_\tau\}.$$

This equation is illustrated with a few Feynman graphs in Fig. 1. (For simplicity, in this figure we represent the scattering between an elementary dipole and the target in the two–gluon exchange approximation.)

But although formally simple, Eq. (2.2) is not a closed equation — it relates a single–dipole scattering amplitude to a two–dipole one —, and the true difficulty

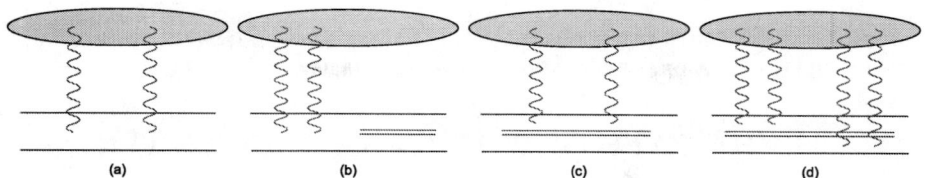

FIGURE 1. *Diagrams for the evolution of the dipole scattering amplitude, cf. Eq. (2.2): (a) the tree–level contribution; (b) the virtual correction* $-\langle T(\boldsymbol{x},\boldsymbol{y})\rangle$*; (c) the scattering of one child dipole,* $\langle T(\boldsymbol{x},\boldsymbol{z})\rangle$ *or* $\langle T(\boldsymbol{z},\boldsymbol{y})\rangle$*; (d) the simultaneous scattering of both child dipoles,* $\langle T^{(2)}(\boldsymbol{x},\boldsymbol{z};\boldsymbol{z},\boldsymbol{y})\rangle$.

refers to the evaluation of $\langle T^{(2)}\rangle_\tau$. To that aim, we need some information about the target. The simplest approximation is to assume factorization

$$\langle T^{(2)}(\boldsymbol{x},\boldsymbol{z};\boldsymbol{z},\boldsymbol{y})\rangle_\tau \approx \langle T(\boldsymbol{x},\boldsymbol{z})\rangle_\tau \langle T(\boldsymbol{z},\boldsymbol{y})\rangle_\tau, \qquad (2.3)$$

which is a *mean field approximation* (MFA) for the gluon fields in the target. This immediately yields a closed, non–linear, equation for $\langle T\rangle_\tau$:

$$\frac{\partial}{\partial \tau}\langle T(\boldsymbol{x},\boldsymbol{y})\rangle_\tau = \frac{\bar{\alpha}_s}{2\pi}\int_z \mathcal{M}(\boldsymbol{x},\boldsymbol{y},\boldsymbol{z}) \qquad (2.4)$$
$$\left\{-\langle T(\boldsymbol{x},\boldsymbol{y})\rangle_\tau + \langle T(\boldsymbol{x},\boldsymbol{z})\rangle_\tau + \langle T(\boldsymbol{z},\boldsymbol{y})\rangle_\tau - \langle T(\boldsymbol{x},\boldsymbol{z})\rangle_\tau\langle T(\boldsymbol{z},\boldsymbol{y})\rangle_\tau\right\}.$$

This is the equation originally derived by Kovchegov [11], and is commonly referred as the 'Balitsky–Kovchegov (BK) equation'. Remarkably, this equation predicts that the scattering amplitude should approach the unitarity bound $T=1$ in the high energy limit. By contrast, the linearized version of this equation (which is the BFKL equation [12]) predicts an exponential growth of the amplitude with τ, which would eventually violate unitarity. But, of course, the linear approximation breaks down when the average amplitude becomes of order one, since then the non–linear term becomes important and restores unitarity. As manifest on Fig. 1, the non–linear effects reflect *multiple scattering*.

In what follows we shall be primarily interested in the limitations of Eq. (2.5), coming from the factorization assumption Eq. (2.3). The latter may be a good approximation if the target is a large nucleus and for not very high energies, which is the situation for which Kovchegov has originally derived this equation. More generally, this should work reasonably well when the scattering is sufficiently strong, that is, when $\langle T\rangle_\tau$ is not much smaller than one, because in that case the external dipole scatters off a high–density gluonic system, and the *density fluctuations* are relatively unimportant. On the other, the MFA cannot be right if the scattering is *very weak*, because then the dipole is sensitive to the *dilute* part of the target wavefunction, where the fluctuations are, of course, essential. Still, given that our main interest when using Eq. (2.5) is in the *strong scattering* regime, one may expect the limitations of this equation in the dilute regime to be inessential for the problem at hand. However, this expectation turns out to be incorrect, and this is precisely what we would like to explain in what follows: The *particle number fluctuations* in the dilute regime have a strong influence, via their subsequent evolution, on the approach towards saturation and the unitarity limit.

3. THE FATE OF THE RARE FLUCTUATIONS

Since the fluctuations are a priori important in the weak scattering regime, we shall focus on the scattering of a *small dipole*, with transverse size $r \equiv |\boldsymbol{x}-\boldsymbol{y}| \ll 1/Q_s(\tau)$. We have introduced here the *saturation momentum* $Q_s(\tau)$, which is a characteristic scale of the gluon distribution in the target, and marks the scale at which a dipole scattering off the target makes the transition from weak ($r \ll 1/Q_s$) to strong ($r \gg 1/Q_s$) interactions. It is in fact common to *define* $Q_s(\tau)$ by the condition

$$\langle T(\boldsymbol{x},\boldsymbol{y})\rangle_\tau = 1/2 \qquad \text{for} \qquad r = 1/Q_s(\tau), \tag{3.1}$$

and to use this condition together with the solution to the BK equation (2.5) in order to compute the energy dependence of the saturation momentum. We shall discuss more about this in the next section.

Returning to our small external dipole, we would like to relate its scattering amplitude to the average gluon density in the target. This is indeed possible in the dilute regime, since then the dipole scatters only once. In fact, at large N_c we can achieve a more symmetric description by representing also the gluons in the target as color dipoles, with an *dipole number density* $n(\boldsymbol{u},\boldsymbol{v})$ (for dipoles with a quark at $\boldsymbol{u}$ and an antiquark at $\boldsymbol{v}$). The external dipole $(\boldsymbol{x},\boldsymbol{y})$ can scatter off any of the internal dipoles $(\boldsymbol{u},\boldsymbol{v})$ by exchanging two gluons. This gives:

$$\langle T(\boldsymbol{x},\boldsymbol{y})\rangle_\tau = \alpha_s^2 \int d^2\boldsymbol{u}\, d^2\boldsymbol{v}\, \mathcal{A}_0(\boldsymbol{x},\boldsymbol{y}|\boldsymbol{u},\boldsymbol{v})\, \langle n(\boldsymbol{u},\boldsymbol{v})\rangle_\tau, \tag{3.2}$$

where $\alpha_s^2 \mathcal{A}_0(\boldsymbol{x},\boldsymbol{y}|\boldsymbol{u},\boldsymbol{v})$ is the scattering amplitude for two elementary dipoles. Here, we shall not need its exact expression, but only the fact that it is *quasi–local* both with respect to the dipole *sizes* and with respect to their *impact parameters*. (The impact parameter of a dipole $(\boldsymbol{x},\boldsymbol{y})$ is its center–of–mass coordinate $\boldsymbol{b}=(\boldsymbol{x}+\boldsymbol{y})/2$.) This allows us to simplify Eq. (3.2) as:

$$\langle T(r,b)\rangle_\tau \simeq \alpha_s^2 \langle f(r,b)\rangle_\tau, \tag{3.3}$$

where the dimensionless quantity

$$f(r,b) \simeq r^2 \int_\Sigma d^2\boldsymbol{b}'\, n(\boldsymbol{r},\boldsymbol{b}') \tag{3.4}$$

is the *dipole occupation number* in the target, that is, the number of dipoles with size r (per unit of $\ln r^2$) within an area $\Sigma \sim r^2$ centered at b. Eq. (3.3) shows that a small dipole projectile is a very precise analyzer of the dipole distribution in the target: the external dipoles counts the numbers of internal dipoles having the same transverse size and impact parameter as itself.

Eq. (3.3) applies so long as $\langle T \rangle_\tau \ll 1$, but by extrapolation it shows that unitarity corrections in the dipole–target scattering become important when the dipole occupation factor in the target becomes of order $1/\alpha_s^2$. This is precisely the critical density at which *saturation effects* — i.e., non–linear effects in the target wavefunction leading to the saturation of the gluon distribution — are expected to occur [6]. This argument confirms that, by studying dipole scattering in the vicinity of the unitarity limit, one has access at the physics of gluon saturation.

Let us assume an initial condition like (3.3) at the initial rapidity τ_0 and follow the evolution of the scattering amplitude with increasing τ. At the beginning, the amplitude will rise very fast, according to the BFKL equation, but this rise will be eventually stopped by the non–linear term $\langle T^{(2)} \rangle_\tau \equiv \langle T^{(2)}(\bm{x}, \bm{z}; \bm{z}, \bm{y}) \rangle_\tau$ in Eq. (2.2), which in the linear regime rises even faster. We have, schematically,

$$\langle T \rangle_\tau \simeq T_0 \, e^{\omega_\mathbb{P}(\tau - \tau_0)}, \qquad \langle T^{(2)} \rangle_\tau \simeq T_0^{(2)} \, e^{2\omega_\mathbb{P}(\tau - \tau_0)}, \qquad (3.5)$$

where $\omega_\mathbb{P} = \text{const.} \times \bar{\alpha}_s$, $T_0 \equiv \langle T \rangle_{\tau_0} \simeq \alpha_s^2 f_0$ and $T_0^{(2)} \equiv \langle T^{(2)} \rangle_{\tau_0}$. ($f_0$ denotes the *average* occupation factor at $\tau = \tau_0$.) The unitarity limit is approached when $\langle T^{(2)} \rangle_\tau \sim \langle T \rangle_\tau$, which in turn implies $\tau \sim \tau_c$ with

$$e^{\omega_\mathbb{P}(\tau_c - \tau_0)} \sim T_0 / T_0^{(2)}. \qquad (3.6)$$

So, what is the ratio $T_0^{(2)}/T_0$? If one assumes the factorization property (2.3), then $T_0^{(2)} \approx (T_0)^2$, and therefore $T_0^{(2)}/T_0 \approx T_0 \simeq \alpha_s^2 f_0$. Then Eq. (3.6) implies:

$$\tau_c - \tau_0 \simeq \frac{1}{\omega_\mathbb{P}} \ln \frac{1}{\alpha_s^2 f_0} = \frac{1}{\omega_\mathbb{P}} \left(\ln \frac{1}{\alpha_s^2} + \ln \frac{1}{f_0} \right). \qquad (3.7)$$

But is the MFA (2.3) a reasonable approximation for a *dilute* initial condition? To answer this question, let us consider two physical situations: (i) $f_0 \gg 1$ (with $f_0 \ll 1/\alpha_s^2$ though) and (ii) $f_0 \ll 1$. Also, remember that $\langle T^{(2)}(\bm{x}, \bm{z}; \bm{z}, \bm{y}) \rangle_\tau$ is the scattering amplitude for two incoming dipoles $(\bm{x}, \bm{z})$ and $(\bm{z}, \bm{y})$ which have similar impact factors (since they have a common leg at $\bm{z}$) and also similar sizes (since the QCD evolution, Eq. (2.2), favors the splitting into dipoles with similar sizes).

(i) In the first case, the disk $\Sigma \sim r^2$ at b has a high occupancy, so the two external dipoles will predominantly scatter off *different* dipoles in that disk. Then, their scatterings are largely independent, and the MFA is reasonable. The result (3.7) can thus be trusted in this case.

(ii) The statement that the *average* occupation factor f_0 is much smaller than one requires an explanation. Clearly, in a *given configuration* of the target, the occupation number (3.4) is *discrete* : $f = 0, 1, 2, \ldots$; so, for its *average* value to be smaller than one, one needs to look at *rare configurations*. That is, if one considers the *statistical ensemble* of dipole configurations generated by the evolution up to rapidity τ_0, then for most of these configurations $f(r, b) = 0$, but for a small fraction among them, of order f_0, f is non–zero and of order one. Thus, f_0 is essentially the *probability* to find a dipole with the required characteristics (r, b) in the ensemble.

Consider now the scattering problem in such a *very* dilute regime: The fact that $T_0 \sim \alpha_s^2 f_0 \ll \alpha_s^2$ means that the incoming dipole (r, b) has a small probability

$f_0(r,b)$ to find a dipole with similar characteristics in the target, with which it then interacts with a strength α_s^2. Consider now *two* incoming dipoles, with similar sizes and impact parameters: there is a small probability $f_0(r,b)$ to find a corresponding dipole in the target, but whenever this happens, *both* external dipoles can scatter off it, with an overall strength α_s^4. This gives $T_0^{(2)} \sim \alpha_s^4 f_0 \sim \alpha_s^2 T_0$, which is much larger than the MFA prediction $T_0^{(2)} \sim (T_0)^2$. The scattering of the external dipoles is now *strongly correlated*. With this estimate for $T_0^{(2)}$, Eq. (3.6) implies

$$\tau_c - \tau_0 \simeq \frac{1}{\omega_\mathbb{P}} \ln \frac{1}{\alpha_s^2}. \tag{3.8}$$

For $f_0 \ll 1$, this is considerably smaller than the naive estimate (3.7) based on the MFA. Thus, *by enhancing the correlations in the dilute regime, the fluctuations in the particle number significantly reduce the rapidity window for BFKL evolution.*

Moreover, at the rapidity τ_c at which the unitarity corrections cut off the BFKL growth, Eqs. (3.5) and (3.8) imply $\langle T \rangle_{\tau_c} \sim \langle T^{(2)} \rangle_{\tau_c} \sim f_0 \ll 1$, in sharp contrast with the prediction of the MFA ! That is, the contribution that a *rare* fluctuation (r,b) at $\tau = \tau_0$ can give, through its subsequent evolution, to the average amplitude $\langle T(r,b) \rangle_\tau$ at $\tau > \tau_0$ saturates at a value *smaller* than one (of the order of the probability $f_0(r,b) \ll 1$ of the original fluctuation) [1]. Besides, this contribution violates the factorization assumption implicit in the BK equation [4].

We conclude that the correlations in the dilute regime significantly reduce the phase–space available for the BFKL evolution of the *average* amplitude towards saturation, by eliminating the rare fluctuations (r,b) for which $\langle f(r,b) \rangle_\tau < 1$, or, equivalently, for which $\langle T(r,b) \rangle_\tau < \alpha_s^2$ [1]. The limiting value α_s^2 is the elementary 'quantum' for the strength of T in the event–by–event description, that is, the minimal non–trivial value that a physical scattering amplitude can take in a particular event, where the dipole number is discrete [2].

In view of this, one expects the evolution to 'slow down' as compared to the MFA. This is confirmed by an original calculation by Mueller and Shoshi [1], which shows that the rate for the growth of saturation momentum with the energy is considerably reduced as compared to the MFA. In the next section, we shall recover the result of Ref. [1] from a broader perspective, which establishes a remarkable correspondence with modern results in statistical physics [2].

4. FLUCTUATING PULLED FRONTS

To perform a detailed study of the influence of fluctuations on the evolution towards high density, one needs a theory for correlations like $\langle T^{(2)} \rangle_\tau$ in the presence of fluctuations. Such a theory has been recently given (within the large–N_c approximation) [4, 9, 8], and we shall briefly comment on it in the last section. But before doing that, we would like to show that some very general results concerning the effects of fluctuations can be deduced without a detailed knowledge of the microscopic dynamics, by relying on universal results from statistical physics [2].

Specifically, the only assumptions that we shall need in order to derive these results are the following: (i) the *mean field description* of the dynamics of $\langle T \rangle_\tau$ is provided by the BK equation (2.5), and (ii) in the *event-by-event description*, the amplitude T is a discrete quantity, with step $\Delta T \sim \alpha_s^2$.

We start by summarizing those results about the BK equation that are needed for the present purposes. We shall neglect the impact parameter dependence of the amplitude, and write the corresponding solution as $\langle T(r) \rangle_\tau \equiv \overline{T}_\tau(\rho)$, where $\rho \equiv \ln(r_0^2/r^2)$ and r_0 is a scale introduced by the initial conditions at low energy. Note that small dipole sizes correspond to large values of ρ. Thus, the amplitude is small, $\overline{T}_\tau(\rho) \ll 1$, when ρ is sufficiently large: $\rho \gg \bar{\rho}_s(\tau)$, where $\bar{\rho}_s(\tau) \equiv \ln(r_0^2 \bar{Q}_s^2(\tau))$ and $\bar{Q}_s^2(\tau)$ denotes the saturation momentum extracted from the BK equation.

The solution $\overline{T}_\tau(\rho)$ can be visualized as a *front* which interpolates between $T = 1$ (the unitarity limit) at $\rho \to -\infty$ and $T = 0$ at $\rho \to \infty$ [3]. Note that $T = 1$ and $T = 0$ are stable and, respectively, unstable fixed points of the BK equation. The transition between the two regimes occurs at $\rho \sim \bar{\rho}_s(\tau)$; thus, the (logarithm of the) saturation momentum plays the role of the *position of the front*. With increasing τ, the saturation momentum rises very fast (exponentially in τ), so the front moves towards larger values of ρ. One finds [13, 14, 15, 3]:

$$Q_s^2(\tau) \simeq Q_0^2 \frac{e^{c\bar{\alpha}_s \tau}}{(\bar{\alpha}_s \tau)^{3/2\gamma_s}}, \tag{4.1}$$

where $Q_0^2 \propto 1/r_0^2$, and c and γ_s are numbers fixed by the BFKL dynamics: $c = 4.88...$ and $\gamma_s = 0.63...$ Eq. (4.1) implies the following expression for the *front velocity*:

$$\bar{\lambda}(\tau) \equiv \frac{d\bar{\rho}_s(\tau)}{d\tau} \simeq c\bar{\alpha}_s - \frac{3}{2\gamma_s}\frac{1}{\tau}. \tag{4.2}$$

Its asymptotic value at large τ represents the *saturation exponent* (the rate for the exponential growth of $Q_s^2(\tau)$), here estimated in the MFA: $\bar{\lambda}_{as} = c\bar{\alpha}_s$.

In the weak scattering (dilute) regime at $\rho \gg \bar{\rho}_s(\tau)$, the form of the amplitude can be obtained by solving the linearized version of Eq. (2.5), that is, the BFKL equation. One thus finds (up to an overall normalization factor) [14, 15, 3]:

$$\overline{T}_\tau(\rho) \simeq (\rho - \bar{\rho}_s) e^{-\gamma_s(\rho - \bar{\rho}_s)} \exp\left\{-\frac{(\rho - \bar{\rho}_s)^2}{2\beta\bar{\alpha}_s\tau}\right\}, \tag{4.3}$$

(with $\beta \simeq 48.2$). In particular, so long as the difference $\rho - \bar{\rho}_s$ remains much smaller than the *diffusion radius* $\sim \sqrt{2\beta\bar{\alpha}_s\tau}$, the Gaussian in Eq. (4.3) can be ignored, and the amplitude becomes purely a function of $\rho - \bar{\rho}_s(\tau)$:

$$\overline{T}_\tau(\rho) \simeq (\rho - \bar{\rho}_s(\tau)) e^{-\gamma_s(\rho - \bar{\rho}_s(\tau))} \qquad \text{for} \qquad \rho - \bar{\rho}_s \ll \sqrt{2\beta\bar{\alpha}_s\tau}. \tag{4.4}$$

This is the property referred to as 'geometric scaling' [16, 14]. It means that the front propagates without distortion, as a *traveling wave* [3].

Notice the mechanism leading to the front propagation: For a fixed $\rho \gg \bar{\rho}_s(\tau)$, the amplitude (4.3) rises rapidly with τ, due to the exponential factor $\exp(\gamma_s \bar{\rho}_s) \simeq e^{\omega_\mathbb{P} \tau}$

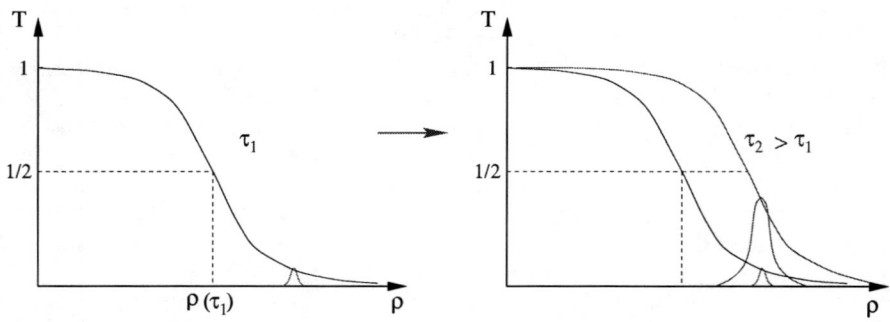

FIGURE 2. *Evolution of the continuum front of the BK equation with increasing rapidity τ.*

with $\omega_{\mathbb{P}} = \gamma_s \bar{\lambda}_{as}$; this is the BFKL instability (see Fig. 2). Thus the front is *pulled* by the unstable (BFKL) growth of its tail at large ρ. Besides, for a given (large) distance $\rho - \bar{\rho}_s$ ahead of the front, the amplitude increases through *diffusion* from smaller values of ρ, until it reaches the profile (4.4) of the traveling wave.

The fact that the front corresponding to the BK equation is a *pulled front* — it propagates via the growth and spreading of small perturbations around the unstable state $T = 0$ — is crucial for the problem at hand, as it shows that the front dynamics is driven by its *leading edge* (the front region where $T \ll 1$), and therefore it might be very sensitive to *fluctuations*. Although this property has been discussed here on the basis of the linear, BFKL, equation, it turns that this is an *exact* property of the non-linear BK equation [3]. Indeed, as shown by Munier and Peschanski, the BK equation is in the same universality class as the Fisher–Kolmogorov–Petrovsky–Piscounov (FKPP) equation [17], which appears as a mean field approximation to a variety of stochastic problems in chemistry, physics, and biology, and for which the pulled front property has been rigorously demonstrated (see [18, 19] for recent reviews and more references).

Let us now return to the actual microscopic dynamics, which is *stochastic* (it includes fluctuations in the number of dipoles in the target), and where the scattering amplitude (in a given event) is *discrete*. Then, as discussed in the previous section, one needs to consider a *statistical ensemble of configurations*, which correspond to different realizations of the same evolution. To any of these configurations one can associate a front $T_\tau(\rho)$, which characterizes the scattering between that particular configuration and external dipoles of arbitrary size ρ.

As in the mean field case, the evolution of a configuration is described as the propagation of the associated front towards larger values of ρ. What is however new is that, because of discreteness, a microscopic front looks like a *histogram*: both T_τ and ρ are now discrete quantities, with steps $\Delta T = \alpha_s^2$ and $\Delta \rho = 1$, respectively. Because of that, the front is necessarily *compact* — for any τ, there is only a finite number of bins in ρ ahead of ρ_s where T_τ is non-zero (see Fig. 3) —, and this property turns out to have dramatic consequences for the propagation of the front:

In the empty bins on the right of the front tip, the local, BFKL, growth is not possible anymore (this would require a seed !). Thus, the only way for the front to

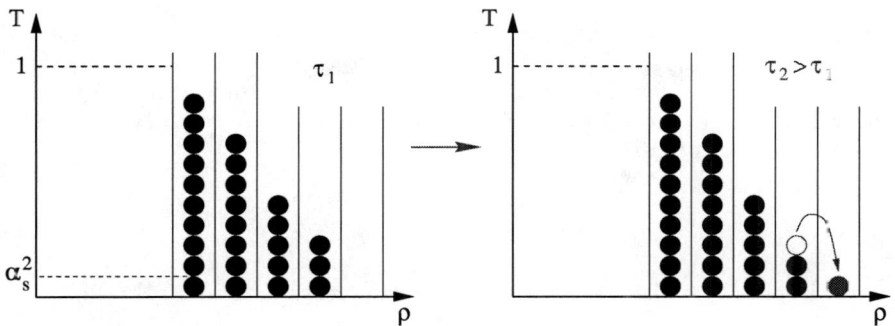

FIGURE 3. Evolution of the discrete front of a microscopic event with increasing rapidity τ. The small blobs are meant to represent the elementary quanta α_s^2 of T in a microscopic event.

progress there is via *diffusion*, i.e., via radiation from the occupied bins at $\rho < \rho_{\text{tip}}$ (see Fig. 3). But since diffusion is less effective than the local growth, we expect the velocity of the front — which is also the saturation exponent — to be reduced for the microscopic front as compared to the front of the MFA. The difference between the mechanisms for front propagation in the MFA and in a microscopic event can be also appreciated by comparing Figs. 2 and 3.

The extreme sensitivity of the pulled fronts to small fluctuations has been recognized in the context of statistical physics only in the recent years [20, 19]. The discrete particle version of a pulled front is generally referred to as a *"fluctuating pulled front"* [18, 19]. The most striking feature of such a system is that the convergence towards the mean field limit is extremely slow, *logarithmic* in the maximal occupation number N. (For QCD, $N \sim 1/\alpha_s^2$, as explained after Eq. (3.4).) Specifically, if λ_N denotes the velocity of the microscopic front for a finite value of N, and λ_∞ is the respective velocity in the MFA (which corresponds to the limit $N \to \infty$), then for $N \gg 1$ one finds: $v_N \simeq v_0 - \mathcal{C}/\ln^2 N$, where $\mathcal{C}$ is a constant.

An analytic argument which explains this slow convergence, and allows one to compute the coefficient $\mathcal{C}$, has been given by Brunet and Derrida [20]. Rather than reproducing the original derivation from Ref. [20], we prefer to present (directly for the case of QCD) a qualitative argument which explains the most salient feature of their result, namely its slow convergence to the mean field limit as $N \to \infty$.

This is related to the fact that, as mentioned before, the microscopic front has a compact width, and therefore its evolution is frozen in a state of 'pre-asymptotic velocity' [2]. The *width* of the front is the distance $\Delta \rho_f = \rho - \rho_s$ over which the amplitude $T_\tau(\rho)$ decreases from $T_\tau(\rho_s) \sim 1$ down the minimal allowed value $T \sim \alpha_s^2$. This can be estimated from Eq. (4.4) as $\Delta \rho_f \sim (1/\gamma_s)\ln(1/\alpha_s^2)$.

Now, from the discussion after Eq. (4.4), we know that the front sets in diffusively, and thus requires a formation 'time' : Eq. (4.3) shows that, for the front to spread over a given distance $\rho - \rho_s$, it takes a rapidity evolution

$$\bar{\alpha}_s \tau \sim \frac{(\rho - \rho_s)^2}{2\beta}. \qquad (4.5)$$

Through this evolution, the velocity of the front increases towards its asymptotic value according to Eq. (4.2). If the front is allowed to extend arbitrarily far away, as it was the case for the MFA, then the velocity will asymptotically approach the value $\bar{\lambda}$. However, when the front is compact, as for the discrete system, the formation time is finite as well, namely of the order

$$\bar{\alpha}_s \Delta \tau \sim \frac{(\Delta \rho_f)^2}{2\beta} \sim \frac{\ln^2(1/\alpha_s^2)}{2\beta \gamma_s^2}, \qquad (4.6)$$

which implies that the front velocity cannot increase beyond a value

$$\lambda_{as} \simeq \bar{\lambda}_{as} - \kappa \bar{\alpha}_s \frac{\gamma_s \beta}{\ln^2(1/\alpha_s^2)}. \qquad (4.7)$$

This estimate is valid when $\alpha_s^2 \ll 1$. The fudge factor κ cannot be determined by this qualitative argument, but this is computed in Ref. [20] as $\kappa = \pi^2/2$.

The first term in Eq. (4.7) is the mean field estimate $\bar{\lambda}_{as} \simeq 4.88 \bar{\alpha}_s$. But the second, corrective, term is particularly large, not only because it decreases very slowly with α_s^2, but also because its coefficient is numerically large: $\pi^2 \gamma_s \beta / 2 \approx 150$. Thus, although Eq. (4.7) becomes an *exact result* when α_s^2 is arbitrarily small, this result remains useless for practical applications.

5. POMERON LOOPS

So far, our discussion has been mostly qualitative, and the language used was essentially that of statistical physics. But it is also interesting to understand these results within the more traditional language of perturbative QCD, that is, in terms of Feynman graphs and evolution equations. This is especially important in view of the limitations of the correspondence with the statistical physics, which so far has only allowed us to obtain asymptotic results (valid when $\bar{\alpha}_s \tau \to \infty$ and $\alpha_s^2 \to 0$) like Eq. (4.7). To go beyond these results, we need the actual evolution equations in QCD in the presence of both fluctuations and saturation. These equations have been constructed in the large-N_c limit [4, 8, 9], by combining the Balitsky equations (or the CGC formalism) in the high density regime with the dipole picture in the dilute regime. To motivate the structure of these equations, we shall first discuss the diagrammatic interpretation of the particle number fluctuations.

We shall use, as before, the dipole picture for the target wavefunction in the dilute regime. Then, fluctuations in the dipole number appear because of the possibility that one dipole internal to the target splits into two dipoles in one step of the evolution. In the discussion of Eq. (2.2) we have already shown, in Fig. 1, the basic diagram for dipole splitting. In that discussion, the dipole appeared as the *projectile*, and the evolution was viewed as *projectile evolution* (that is, the small rapidity increment $d\tau$ was given to the projectile). Here, we would like to visualize the relevant fluctuations as splittings of the elementary dipoles inside the target, and to that aim we need to perform *target evolution*.

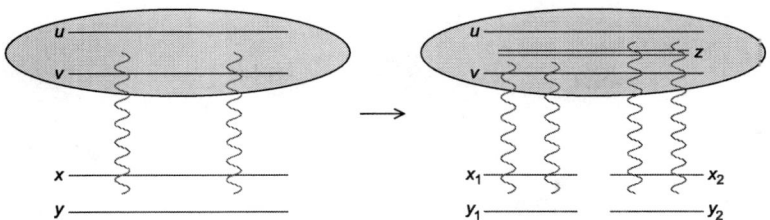

FIGURE 4. Target evolution in one step: the original dipole (u,v) splits into two new dipoles (u,z) and (z,v), which then scatter off two external dipoles.

In Fig. 4 we show one step in the evolution of the target, in which one of the dipoles there — the one with legs at u and v — has split into two new dipoles (with coordinates (u,z) and (z,v), respectively). As further illustrated there, the original dipole can be probed via scattering with *one* external dipole (x,y), in which case it provides a contribution to the scattering amplitude $\langle T(x,y)\rangle_\tau$ at the original rapidity τ. After evolution, the two child dipoles can be measured via the scattering with *two* external dipoles, thus giving a contribution to the respective amplitude $\langle T^{(2)}(x_1,y_1;x_2,y_2)\rangle_{\tau+d\tau}$ at rapidity $\tau+d\tau$. This is in agreement with the discussion in Sect. 3 where we have seen that one needs to scatter two external dipoles in order to be sensitive to fluctuations.

From the previous discussion, one can also understand what should be the role of the process in Fig. 4 in the evolution of the scattering amplitudes for external dipoles: This process generates a change in the two–dipole scattering amplitude $\langle T^{(2)}\rangle_\tau$ which is proportional to single–dipole amplitude $\langle T\rangle_\tau$.

Let us finally consider the evolution of the dipole scattering amplitude $\langle T\rangle$ after *two* steps. This involves several processes, but the most interesting among them is the one displayed in Fig. 5, which is sensitive to both fluctuations and saturation. Specifically, the first step of the evolution is the same as in Fig. 4: one dipole in the target wavefunction splits into two, which implies that $\langle T\rangle$ evolves into a $\langle T^{(2)}\rangle$. In the second step, the $\langle T^{(2)}\rangle$ evolves back into a $\langle T\rangle$, according to the non–linear term in Eq. (2.2). The latter process has been already represented from the perspective of projectile evolution in Fig. 1.d. In the lower half part of Fig. 5, this process is now represented as target evolution: From this perspective, it describes the merging of four gluons into two. Altogether, the two-step evolution depicted in Fig. 5 generates a *Pomeron loop*: This involves two 'vertices' — one for *dipole splitting*, the other one for *gluon merging* — and two 'propagators' — one for each dipole–dipole scattering amplitude $\alpha_s^2 \mathcal{A}_0$. An explicit expression for this loop can be found in Ref. [21].

In Sect. 4, we have mentioned that the BK equation — which, we recall, is a mean field approximation to the QCD evolution at high energy — is in the same universality class as the FKPP equation of statistical physics [3]. It is interesting to mention at this point that the complete evolution in QCD at large N_c is in the same universality class as the *stochastic FKPP equation* (sFKPP) [4] — a Langevin equation with a specific 'noise term' which simulates particle number fluctuations

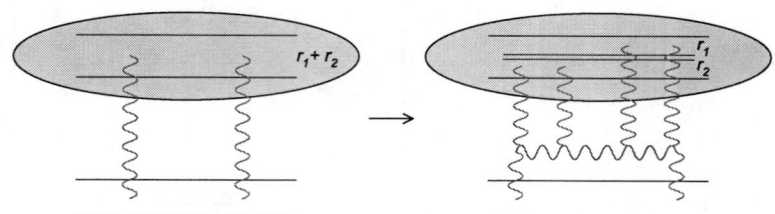

FIGURE 5. Two steps in the evolution of the average scattering amplitude of a single dipole: the original amplitude (left) and the Pomeron loop generated after two steps (right).

[19]. By further studying this equation (in particular, via numerical calculations), one should be able to go beyond the asymptotic results presented in Sect. 4 and find the behaviour of the scattering amplitudes in QCD for realistic values of the energy and of the coupling constant. This program is currently under way.

REFERENCES

1. A.H. Mueller and A.I. Shoshi, *Nucl. Phys.* **B692** (2004) 175.
2. E. Iancu, A.H. Mueller, and S. Munier, *Phys. Lett.* **B606** (2005) 342.
3. S. Munier and R. Peschanski, *Phys. Rev. Lett.* **91** (2003) 232001.
4. E. Iancu and D.N. Triantafyllopoulos, arXiv:hep-ph/0411405.
5. E. Iancu and R. Venugopalan, in *Quark-Gluon Plasma 3*, Eds. R.C. Hwa and X.-N. Wang, World Scientific, 2003 [arXiv:hep-ph/0303204].
6. A.H. Mueller, *Nucl. Phys.* **B415** (1994) 373; A.H. Mueller and B. Patel, *Nucl. Phys.* **B425** (1994) 471; A.H. Mueller, *Nucl. Phys.* **B437** (1995) 107.
7. G.P. Salam, *Nucl. Phys.* **B449** (1995) 589; *ibid.* **461** (1996) 512.
8. A.H. Mueller, A.I. Shoshi and S.M.H. Wong, arXiv:hep-ph/0501088.
9. E. Iancu and D.N. Triantafyllopoulos, arXiv:hep-ph/0501193 (to appear in Phys. Lett. **B**).
10. I. Balitsky, *Nucl. Phys.* **B463** (1996) 99.
11. Yu.V. Kovchegov, *Phys. Rev.* **D60** (1999), 034008; *ibid.* **D61** (2000) 074018.
12. L.N. Lipatov, *Sov. J. Nucl. Phys.* **23** (1976) 338;
    E.A. Kuraev, L.N. Lipatov and V.S. Fadin, *Zh. Eksp. Teor. Fiz* **72**, 3 (1977) (*Sov. Phys. JETP* **45** (1977) 199);
    Ya.Ya. Balitsky and L.N. Lipatov, *Sov. J. Nucl. Phys.* **28** (1978) 822.
13. L.V. Gribov, E.M. Levin, and M.G. Ryskin, *Phys. Rept.* **100** (1983) 1.
14. E. Iancu, K. Itakura, and L. McLerran, *Nucl. Phys.* **A708** (2002) 327.
15. A.H. Mueller and D.N. Triantafyllopoulos, *Nucl. Phys.* **B640** (2002) 331.
16. A.M. Stasto, K. Golec-Biernat and J. Kwiecinski, Phys. Rev. Lett. **86** (2001) 596.
17. R.A. Fisher, *Ann. Eugenics* **7** (1937) 355; A. Kolmogorov, I. Petrovsky, and N. Piscounov, *Moscou Univ. Bull. Math.* **A1** (1937) 1.
18. For a recent review, see W. Van Saarloos, *Phys. Rep.* **386** (2003) 29.
19. For a recent review, see D. Panja, Phys. Rep. **393** (2004) 87.
20. E. Brunet and B. Derrida, *Phys. Rev.* **E56** (1997) 2597; *Comp. Phys. Comm.* **121-122** (1999) 376; *J. Stat. Phys.* **103** (2001) 269.
21. J.-P. Blaizot, E. Iancu, K. Itakura, and D.N. Triantafyllopoulos, arXiv:hep-ph/0502221.

Odderon, Redone[1]

Kazunori Itakura

Service de Physique Théorique, CEA/Saclay, 91191 Gif-sur-Yvette, France
E-mail: itakura@dsm-mail.saclay.cea.fr

Abstract. We discuss the "odderon" exchange at high energy within the framework of the Color Glass Condensate (CGC). We explicitly construct gauge-invariant amplitudes for multiple odderon exchanges in the scattering between the CGC and two types of color-singlet projectiles: a 'color dipole' and three quarks.

Keywords: High energy scattering in QCD, Color Glass Condensate, Odderon
PACS: 12.38.Bx, 11.80.Fv, 11.80.La

1. INTRODUCTION

Recent years have seen a remarkable progress in understanding the high-energy scattering in QCD. The main trigger of this progress is the observation that, when the scattering energy is very large, the hadron or nucleus exhibits a *high density gluonic matter* which is now called the Color Glass Condensate (CGC) [1]. In short words, the CGC emerges as follows: When the scattering energy is increased, soft gluon emission ($g \to gg$) is induced, and its successive occurrence leads to multiple production of "small-x gluons" that have small fractions $x \ll 1$ of the total momentum ($x \propto 1/\sqrt{s}$ with $\sqrt{s}$ being the scattering energy). However, as the density of gluons becomes high, "*gluon recombination*" ($gg \to g$) starts to contribute and eventually leads to *saturation* where both processes are balanced. This state is the CGC. Technically, all these are described by a weak-coupling method ($\alpha_s \ll 1$ due to the presence of a hard scale $Q_s \gg \Lambda_{\rm QCD}$, see below) which is powered by resummation schemes with respect to $(\alpha_s \ln 1/x)^n$ and strong gauge fields.

The properties of CGC are specified by correlation functions of gluons, and the change of CGC with increasing energy is determined by "evolution" equations for these correlation functions. In particular, the *Balitsky-Kovchegov (BK) equation* is a nonlinear evolution equation for a *2-point correlation function* which physically corresponds to the *scattering amplitude* of a "color dipole" off the CGC, and reduces to the gluon density in the weak-field regime. The nonlinearity comes into play due to the recombination effects, whose contribution is naively proportional to square of the gluon density. As a result of extensive investigation of this equation both in analytic and numerical methods, it turned out that there exists saturation regime whose borderline in the kinematical plain is given by the *saturation momentum* $Q_s(x)$ in such a way that gluons having transverse momenta

---

[1] On the first slide of the talk, I showed an *odd*-looking lithograph "Smiling spider" drawn by a French painter, Odilon Redon.

lower than $Q_s(x)$ is saturated. Since the saturation momentum depends upon energy (or x) as $Q_s^2(x) \propto (1/x)^\lambda$ with $\lambda \simeq 0.3$, it grows with increasing energy (decreasing x) and the kinematical region for saturation expands. As most of the gluons have their transverse momenta around $Q_s(x)$, the weak-coupling treatment becomes better and better with increasing energy $\alpha_s(Q_s) \ll 1$, as mentioned above. Lastly, the solutions to the BK equation show new scaling phenomena called *geometric scaling* which naturally comes out due to the presence of a saturation momentum, and is also observed in experimental results in a beautiful way.

So far, studies of CGC have been focusing on the BK equation for a 2-point correlation function. However, in order to accurately understand the dynamics of CGC, we need to treat higher-point functions too. For example, the BK equation is obtained by a "mean-field" approximation of the more fundamental equation (the Balitsky equation), but this original equation involves 4-point function. Also, at high energy, we anticipate that multi-gluon exchanges between a projectile and a target, which are not expressed by multiple 2-gluon exchanges, will become important. All these should be addressed in the framework of the CGC.

The first nontrivial step beyond the BK equation in the direction of higher-point correlation functions should be the studies on "odderon" which requires an exchange of at least three reggeized gluons (in contrast to 2 reggeized gluons for a pomeron) and is odd under the charge conjugation operation (see Ref. [2] for a review of traditional description of odderon). This has been done in Ref. [3], and this talk gives a summary of the results presented there. Below, I first explain how to construct C-odd scattering amplitudes which correspond to "odderon" exchange in two different processes. Then, I show evolution equations for these amplitudes which are obtained by the application of the JIMWLK equation to them. Lastly, I compare the result with the BKP equation [4, 5] which is the evolution equation for the odderon exchanges in the traditional framework.

2. ODDERON OPERATORS IN THE CGC

Let us construct the relevant operators for multiple odderon exchanges in the scattering between the CGC and two types of simple projectiles: a color dipole and three quarks in a colorless state. The C-odd dipole-CGC scattering can be considered as a sub-process of the diffractive scattering of a virtual photon on the CGC into C-even mesons like η_c. The 3-quark system may be regarded as a crude model of a baryon.

2.1. The dipole-CGC scattering

Consider the high energy scattering of a $q\bar{q}$ dipole off the CGC which is treated as a random classical gauge field α. To obtain the S-matrix of this process, we first compute the S-matrix for a *fixed* configuration α, and then average over it. For the first step, we can use the eikonal approximation: $S(\mathbf{x}, \mathbf{y}; \alpha) = \langle \text{out} | \text{in} \rangle$, where the transverse positions of the quark (**x**) and the antiquark (**y**) are the same in the in-coming and the out-going states:

$$|\text{in}\rangle \sim \bar{\psi}_i^{in}(\mathbf{x})\psi_i^{in}(\mathbf{y})|0\rangle, \qquad |\text{out}\rangle \sim \bar{\psi}_i^{out}(\mathbf{x})\psi_i^{out}(\mathbf{y})|0\rangle.$$

The relation between the in-coming and the out-going fields is found by solving $(\partial_- - ig\alpha^a t^a)\psi = 0$ for a given gauge configuration α. Namely, $\psi_i^{out} = (V_\mathbf{x}^\dagger)_{ij}\psi_j^{in}$ with $V_\mathbf{x}^\dagger$ being the Wilson line in the fundamental representation along the trajectory of the quark:

$$V_\mathbf{x}^\dagger = \mathrm{P}\exp\left\{ig\int dx^- \alpha^a(x^-,\mathbf{x})t^a\right\}. \tag{2.1}$$

Then,

$$S(\mathbf{x},\mathbf{y};\alpha) = \langle \text{out}|\text{in}\rangle = \frac{1}{N_c}(V_\mathbf{x}^\dagger)^{ij}(V_\mathbf{y})^{ki}\delta^{kl}\delta^{jl} = \frac{1}{N_c}\mathrm{tr}(V_\mathbf{x}^\dagger V_\mathbf{y}).$$

The physical S–matrix is obtained after averaging over the random classical field:

$$S_\tau(\mathbf{x},\mathbf{y}) = \int \mathcal{D}\alpha\, W_\tau[\alpha] S(\mathbf{x},\mathbf{y};\alpha) = \frac{1}{N_c}\langle \mathrm{tr}(V_\mathbf{x}^\dagger V_\mathbf{y})\rangle_\tau. \tag{2.2}$$

Notice that this result depends upon the scattering energy or the rapidity τ defined by $\tau = \ln 1/x$ (x is the Bjorken variable) because the weight function $W_\tau[\alpha]$ which governs the randomness of the gauge field α changes with increasing energy. The change of $W_\tau[\alpha]$ under the change of rapidity is formulated as a renormalization group equation, which is called the JIMWLK equation.

So far, the scattering process is generic and the exchanged object can be either even, or odd, under the charge conjugation C. To single out C-even ('pomeron') or C-odd ('odderon') exchanges, one needs to project Eq. (2.2) onto in-coming and out-going states with appropriate C-parities. Since the charge conjugation for fermions is defined by $C\psi C^{-1} = -i(\bar{\psi}\gamma^0\gamma^2)^T$, and $C\bar{\psi}C^{-1} = (-i\gamma^0\gamma^2\psi)^T$, it is easy to check that the eigenstates of C in the dipole sector are given by $(\bar{\psi}(\mathbf{x})\psi(\mathbf{y}) \pm \bar{\psi}(\mathbf{y})\psi(\mathbf{x}))|0\rangle$, where $+(-)$ sign yields the C-even(odd) state. These are reasonable because the charge conjugation essentially works as the exchange of a quark and an antiquark.

Taking the C-odd dipole state as the in-coming state (a photon is C-odd), and the C-even dipole state as the out-going state, one obtains the C-odd contribution:

$$S_\tau^{odd}(\mathbf{x},\mathbf{y}) = \langle \text{out, even}|\text{in, odd}\rangle = \frac{1}{2N_c}\left\langle \mathrm{tr}(V_\mathbf{x}^\dagger V_\mathbf{y}) - \mathrm{tr}(V_\mathbf{y}^\dagger V_\mathbf{x})\right\rangle_\tau. \tag{2.3}$$

This allows us to identify the operator for C-odd exchanges in the dipole-CGC scattering ("the dipole odderon operator") as

$$O(\mathbf{x},\mathbf{y}) \equiv \frac{1}{2iN_c}\mathrm{tr}(V_\mathbf{x}^\dagger V_\mathbf{y} - V_\mathbf{y}^\dagger V_\mathbf{x}) = -O(\mathbf{y},\mathbf{x}). \tag{2.4}$$

One can check that the operator (2.4) is indeed C-odd by using the transformation property of the gauge fields, $CA_\mu C^{-1} = -(A_\mu)^T$, or, $CVC^{-1} = (V^\dagger)^T$ for a generic

Wilson line V built with A_μ. Note that the C-odd contribution (2.3) is the imaginary part of the S-matrix element:

$$\langle O(\mathbf{x},\mathbf{y})\rangle_\tau = \Im m\, S_\tau(\mathbf{x},\mathbf{y}). \tag{2.5}$$

Correspondingly, the C-even, pomeron exchange, amplitude, that we shall denote as $N(\mathbf{x},\mathbf{y})$, is identified with the real part of the S-matrix:

$$N(\mathbf{x},\mathbf{y}) \equiv 1 - \frac{1}{2N_c}\text{tr}(V_\mathbf{x}^\dagger V_\mathbf{y} + V_\mathbf{y}^\dagger V_\mathbf{x}), \tag{2.6}$$

$$\langle N(\mathbf{x},\mathbf{y})\rangle_\tau = 1 - \Re e\, S_\tau(\mathbf{x},\mathbf{y}). \tag{2.7}$$

From perturbative QCD, we expect that the lowest order contribution to the odderon exchange is of the form $d^{abc}A_\mu^a(\mathbf{x})A_\nu^b(\mathbf{y})A_\rho^c(\mathbf{z})$ with $d^{abc} = 2\text{tr}(\{t^a,t^b\}t^c)$ being a totally symmetric tensor. A similar structure indeed emerges from the CGC operator (2.4) in the weak-field limit. By expanding the Wilson lines (2.1) up to *cubic* order in the field α in the exponent

$$\begin{aligned}V_\mathbf{x}^\dagger[\alpha] &\approx 1 + ig\int dx^-\alpha^a(x^-,\mathbf{x})t^a \\ &- \frac{g^2}{2}\int dx^-\int dy^-\alpha^a(x^-,\mathbf{x})\alpha^b(y^-,\mathbf{x})[\theta(x^- - y^-)t^a t^b + \theta(y^- - x^-)t^b t^a] \\ &+ \{\text{cubic term in } \alpha\},\end{aligned} \tag{2.8}$$

one finds the lowest non-trivial contribution to Eq. (2.4):

$$O(\mathbf{x},\mathbf{y}) \simeq \frac{-g^3}{24N_c}d^{abc}\left\{3(\alpha_\mathbf{x}^a\alpha_\mathbf{y}^b\alpha_\mathbf{y}^c - \alpha_\mathbf{x}^a\alpha_\mathbf{x}^b\alpha_\mathbf{y}^c) + (\alpha_\mathbf{x}^a\alpha_\mathbf{x}^b\alpha_\mathbf{x}^c - \alpha_\mathbf{y}^a\alpha_\mathbf{y}^b\alpha_\mathbf{y}^c)\right\}, \tag{2.9}$$

where $\alpha_\mathbf{x}^a = \int dx^-\alpha^a(x^-,\mathbf{x})$. As expected, this expression is cubic in α^a with the color indices contracted symmetrically by the d–symbol. Note that this combination of trilinear field operators is gauge invariant by construction.

2.2. The 3-quark–CGC scattering

We now turn to the 3-quark–CGC scattering at high energies. The 3-quark colorless state may be given by the "baryonic" operator $\epsilon^{ijk}\psi^i(\mathbf{x})\psi^j(\mathbf{y})\psi^k(\mathbf{z})$, where ϵ^{ijk} is the complete antisymmetric symbol ($i,j,k=1,2,3$). For our present purpose, flavor dependences are irrelevant. By using the same eikonal approximation as for the dipole-CGC scattering, one obtains the following S-matrix:

$$S_\tau(\mathbf{x},\mathbf{y},\mathbf{z}) = \frac{1}{3!}\epsilon^{ijk}\epsilon^{lmn}\left\langle V_{il}^\dagger(\mathbf{x})V_{jm}^\dagger(\mathbf{y})V_{kn}^\dagger(\mathbf{z})\right\rangle_\tau, \tag{2.10}$$

where $\mathbf{x}, \mathbf{y}$ and $\mathbf{z}$ are transverse positions of the three quarks. The odderon contribution is given again by the imaginary part of the S-matrix:

$$\langle O(\mathbf{x},\mathbf{y},\mathbf{z})\rangle_\tau = \Im m\, S_\tau(\mathbf{x},\mathbf{y},\mathbf{z}), \tag{2.11}$$

where the "3-quark odderon operator" $O(\mathbf{x},\mathbf{y},\mathbf{z})$ has been introduced as

$$O(\mathbf{x},\mathbf{y},\mathbf{z}) = \frac{1}{3!2i}\left(\epsilon^{ijk}\epsilon^{lmn}V_{il}^\dagger(\mathbf{x})V_{jm}^\dagger(\mathbf{y})V_{kn}^\dagger(\mathbf{z}) - \text{c.c.}\right). \tag{2.12}$$

Gauge invariance of this operator becomes manifest if one rewrites this as

$$\begin{aligned}O(\mathbf{x},\mathbf{y},\mathbf{z}) &= \frac{1}{3!2i}\Big[\text{tr}(V_\mathbf{x}^\dagger V_\mathbf{w})\text{tr}(V_\mathbf{y}^\dagger V_\mathbf{w})\text{tr}(V_\mathbf{z}^\dagger V_\mathbf{w}) - \text{tr}(V_\mathbf{x}^\dagger V_\mathbf{w})\text{tr}(V_\mathbf{y}^\dagger V_\mathbf{w}V_\mathbf{z}^\dagger V_\mathbf{w}) \\ &\quad - \text{tr}(V_\mathbf{y}^\dagger V_\mathbf{w})\text{tr}(V_\mathbf{x}^\dagger V_\mathbf{w}V_\mathbf{z}^\dagger V_\mathbf{w}) - \text{tr}(V_\mathbf{z}^\dagger V_\mathbf{w})\text{tr}(V_\mathbf{x}^\dagger V_\mathbf{w}V_\mathbf{y}^\dagger V_\mathbf{w}) \\ &\quad + \text{tr}(V_\mathbf{x}^\dagger V_\mathbf{w}V_\mathbf{y}^\dagger V_\mathbf{w}V_\mathbf{z}^\dagger V_\mathbf{w}) + \text{tr}(V_\mathbf{x}^\dagger V_\mathbf{w}V_\mathbf{z}^\dagger V_\mathbf{w}V_\mathbf{y}^\dagger V_\mathbf{w}) - \text{c.c.}\Big],\end{aligned} \tag{2.13}$$

which is easily done with the help of the identity related to the definition of the determinant for SU(3) matrices: $\epsilon^{ijk}\epsilon^{lmn}V_{il}(\mathbf{w})V_{jm}(\mathbf{w})V_{kn}(\mathbf{w}) = 3!\det V(\mathbf{w}) = 3!$ ($\mathbf{w}$ is an arbitrary transverse coordinate). By construction, this expression is independent of $\mathbf{w}$ when $N_c = 3$. Note that it can be simplified by choosing $\mathbf{w}$ to be one of the quark coordinates, say $\mathbf{w} = \mathbf{z}$:

$$O(\mathbf{x},\mathbf{y},\mathbf{z}) = \frac{1}{3!2i}\Big[\text{tr}(V_\mathbf{x}^\dagger V_\mathbf{z})\text{tr}(V_\mathbf{y}^\dagger V_\mathbf{z}) - \text{tr}(V_\mathbf{x}^\dagger V_\mathbf{z}V_\mathbf{y}^\dagger V_\mathbf{z}) - \text{c.c.}\Big]. \tag{2.14}$$

Furthermore, when two of the coordinates are the same, the 3-quark odderon operator reduces to the dipole odderon operator, Eq. (2.4):

$$O(\mathbf{x},\mathbf{z},\mathbf{z}) = O(\mathbf{x},\mathbf{z}) = -O(\mathbf{x},\mathbf{x},\mathbf{z}). \qquad (N_c = 3) \tag{2.15}$$

This is physically reasonable because the diquark state is equivalent to an antiquark as far as color degrees of freedom are concerned.

In the weak-field approximation, as obtained after expanding to lowest non-trivial order (i.e., to cubic order in α) the Wilson lines, one finds again a gauge invariant linear combination of trilinear field operators with the d-symbol:

$$O(\mathbf{x},\mathbf{y},\mathbf{z}) \simeq \frac{g^3}{144}d^{abc} \tag{2.16}$$
$$\times \left\{(\alpha_\mathbf{x}^a - \alpha_\mathbf{z}^a) + (\alpha_\mathbf{y}^a - \alpha_\mathbf{z}^a)\right\}\left\{(\alpha_\mathbf{y}^b - \alpha_\mathbf{x}^b) + (\alpha_\mathbf{z}^b - \alpha_\mathbf{x}^b)\right\}\left\{(\alpha_\mathbf{z}^c - \alpha_\mathbf{y}^c) + (\alpha_\mathbf{x}^c - \alpha_\mathbf{y}^c)\right\}.$$

Lastly, one can similarly introduce the "3-quark *pomeron* operator" $N(\mathbf{x},\mathbf{y},\mathbf{z})$ by the real part of the scattering matrix:

$$\langle N(\mathbf{x},\mathbf{y},\mathbf{z})\rangle_\tau = 1 - \Re\mathfrak{e}\, S_\tau(\mathbf{x},\mathbf{y},\mathbf{z}), \tag{2.17}$$

$$N(\mathbf{x},\mathbf{y},\mathbf{z}) = 1 - \frac{1}{3!2}\left(\epsilon^{ijk}\epsilon^{lmn}V_{il}^\dagger(\mathbf{x})V_{jm}^\dagger(\mathbf{y})V_{kn}^\dagger(\mathbf{z}) + \text{c.c.}\right). \tag{2.18}$$

This operator should be important in describing the high-energy behavior of the proton-nucleus collisions, which is, however, out of the scope of the present paper.

3. ODDERON EVOLUTION

Once we know the relevant operators for the C-odd scattering amplitudes, we can apply the JIMWLK equation (or its simplified version proposed in Ref. [3]) to the operators to derive the evolution equations for them. We consider the two cases which were discussed in the previous section: the dipole-CGC scattering and the 3-quark–CGC scattering, and then discuss the relation between our result and the BKP equation.

3.1. The dipole–CGC scattering

For the dipole-CGC scattering, the evolution equations obeyed by the average amplitudes $\langle N(\mathbf{x},\mathbf{y}) \rangle_\tau$ and $\langle O(\mathbf{x},\mathbf{y}) \rangle_\tau$ can be easily derived from the first Balitsky equation because the operators $N(\mathbf{x},\mathbf{y})$ and $O(\mathbf{x},\mathbf{y})$ are, respectively, the real part and the imaginary part of the dipole-CGC scattering operator $(1/N_c)\mathrm{tr}(V_\mathbf{x}^\dagger V_\mathbf{y})$ which satisfies the Balitsky equation. Therefore, the respective equations can be simply obtained by separating the real part and the imaginary part in the Balitsky equation. The result is

$$\frac{\partial}{\partial \tau} \langle O(\mathbf{x},\mathbf{y}) \rangle_\tau = \frac{\bar{\alpha}_s}{2\pi} \int d^2\mathbf{z}\, \mathcal{M}_{\mathbf{xyz}} \Big\langle O(\mathbf{x},\mathbf{z}) + O(\mathbf{z},\mathbf{y}) - O(\mathbf{x},\mathbf{y})$$
$$- O(\mathbf{x},\mathbf{z})N(\mathbf{z},\mathbf{y}) - N(\mathbf{x},\mathbf{z})O(\mathbf{z},\mathbf{y}) \Big\rangle_\tau, \quad (3.1)$$

$$\frac{\partial}{\partial \tau} \langle N(\mathbf{x},\mathbf{y}) \rangle_\tau = \frac{\bar{\alpha}_s}{2\pi} \int d^2\mathbf{z}\, \mathcal{M}_{\mathbf{xyz}} \Big\langle N(\mathbf{x},\mathbf{z}) + N(\mathbf{z},\mathbf{y}) - N(\mathbf{x},\mathbf{y})$$
$$- N(\mathbf{x},\mathbf{z})N(\mathbf{z},\mathbf{y}) + O(\mathbf{x},\mathbf{z})O(\mathbf{z},\mathbf{y}) \Big\rangle_\tau, \quad (3.2)$$

where we have defined the dipole kernel

$$\mathcal{M}_{\mathbf{xyz}} = \frac{(\mathbf{x}-\mathbf{y})^2}{(\mathbf{x}-\mathbf{z})^2(\mathbf{z}-\mathbf{y})^2}. \quad (3.3)$$

Several comments are in order about these equations:

- As is the case with the Balitsky equations, the equations above do not close by themselves (since they contain both two-point and four-point functions, as mentioned before), but rather belong to an infinite hierarchy.

- In the weak-field limit, both of the evolution equations reduce to the (linear) BFKL equation. However, the BFKL equation for the odderon exchange must be solved with the antisymmetric condition (2.4). Therefore, even if the evolution equations are the same, the respective solutions behave differently. In particular, it is known that the highest intercept of the BFKL solution with C being odd is given by 1 which is smaller than the (hard) pomeron intercept [6].

- The non-linear terms in these equations couple the evolution of C-odd and C-even operators. For instance, the last term, quadratic in O, in the r.h.s. of Eq. (3.2) for $\langle N \rangle_\tau$ describes the merging of two odderons into one pomeron ('merging' from the target point of view). This has not been discussed before in connection with the Balitsky hierarchy.

- In the mean-field approximation, Eqs. (3.1)–(3.2) reduce to a closed system of coupled, non-linear, equations for $\langle N \rangle_\tau$ and $\langle O \rangle_\tau$:

$$\frac{\partial}{\partial \tau} \langle O(\mathbf{x},\mathbf{y}) \rangle_\tau = \frac{\bar{\alpha}_s}{2\pi} \int d^2\mathbf{z}\, \mathcal{M}_{\mathbf{xyz}} \Big[\langle O(\mathbf{x},\mathbf{z}) \rangle_\tau + \langle O(\mathbf{z},\mathbf{y}) \rangle_\tau - \langle O(\mathbf{x},\mathbf{y}) \rangle_\tau \quad (3.4)$$
$$- \langle O(\mathbf{x},\mathbf{z}) \rangle_\tau \langle N(\mathbf{z},\mathbf{y}) \rangle_\tau - \langle N(\mathbf{x},\mathbf{z}) \rangle_\tau \langle O(\mathbf{z},\mathbf{y}) \rangle_\tau \Big],$$

$$\frac{\partial}{\partial \tau} \langle N(\mathbf{x},\mathbf{y}) \rangle_\tau = \frac{\bar{\alpha}_s}{2\pi} \int d^2\mathbf{z}\, \mathcal{M}_{\mathbf{xyz}} \Big[\langle N(\mathbf{x},\mathbf{z}) \rangle_\tau + \langle N(\mathbf{z},\mathbf{y}) \rangle_\tau - \langle N(\mathbf{x},\mathbf{y}) \rangle_\tau \quad (3.5)$$
$$- \langle N(\mathbf{x},\mathbf{z}) \rangle_\tau \langle N(\mathbf{z},\mathbf{y}) \rangle_\tau + \langle O(\mathbf{x},\mathbf{z}) \rangle_\tau \langle O(\mathbf{z},\mathbf{y}) \rangle_\tau \Big].$$

The first of these equations has been already proposed in Ref. [6], as a plausible non-linear generalization of the BFKL equation in the C-odd channel. As for Eq. (3.5), this is the BK equation supplemented by a new term describing the merging of two odderons.

- One of the significant consequences of the nonlinear effects in the factorized evolution equation (3.4) is that the odderon amplitude $\langle O \rangle_\tau$ will decay into zero with increasing energy. This is most easily seen by noting that when the pomeron amplitude $\langle N \rangle_\tau$ is close to 1 (deeply in saturation regime which is expected to realize at high energy), the nonlinear terms in Eq. (3.4) cancel the first two terms on the r.h.s. and the resulting equation for $\langle O \rangle_\tau$ simply implies decrease of the solution. Therefore, as one goes to higher energies, the odderon contribution becomes less and less important.

3.2. The 3-quark–CGC scattering

Since we know the full non-linear expression of the relevant operator (2.12) for the 3-quark–CGC scattering, there is no difficulty at conceptual level in deriving the evolution equation. A straightforward application of the JIMWLK equation (or its simplified version) to this operator automatically leads to the result. However, the resulting equation turned out to be complicated and not very illuminating: Through their non-linear terms, they couple the 3-quark odderon operator to other types of operators with different color structures. Therefore, in this talk, we rather show the evolution equation for the weak-field version of the 3-quark odderon operator, Eq. (2.16). This is indeed sufficient to discuss the correspondence with the BKP equation [4, 5], which is a *linear* evolution equation for the odderon exchange.

After a straightforward but lengthy calculation, the following linear evolution equation for $\langle O_{xyz}\rangle_\tau \equiv \langle O(\mathbf{x},\mathbf{y},\mathbf{z})\rangle_\tau$ is obtained

$$\frac{\partial}{\partial \tau}\langle O_{xyz}\rangle_\tau = \frac{3\alpha_s}{4\pi^2}\int d^2\mathbf{w}\, \mathcal{M}_{xyw}\Big(\langle O_{xwz}\rangle_\tau + \langle O_{wyz}\rangle_\tau - \langle O_{xyz}\rangle_\tau$$
$$-\langle O_{wwz}\rangle_\tau - \langle O_{xxw}\rangle_\tau - \langle O_{yyw}\rangle_\tau - \langle O_{xyw}\rangle_\tau\Big)$$
$$+\big\{2\text{ cyclic permutations}\big\}. \tag{3.6}$$

Note that this is a *closed* equation for $\langle O_{xyz}\rangle_\tau$, which was expected from the viewpoint of gauge invariance: the only gauge invariant C-odd operators available are O_{xyz} and $O(\mathbf{x},\mathbf{y}) = O_{xyy}$ (cf. Eq. (2.15)). The linear combination of O's in the integrand vanishes at the points $\mathbf{w} = \mathbf{x}$ and $\mathbf{w} = \mathbf{y}$ where lie the poles of the dipole kernel $\mathcal{M}_{xyw}$, so the poles are harmless. Also, one can easily check that the above equation is consistent with the relation (2.15) between the dipole and the 3-quark odderon amplitudes: if one sets $\mathbf{z} = \mathbf{y}$, Eq. (3.6) reduces indeed to the BFKL equation which is the evolution equation of the dipole-CGC scattering in the weak-field regime.

Now we come to the final point: the comparison of our result (3.6) with the BKP equation. First of all, our result (3.6) does not look equivalent to the BKP equation. In fact, within our framework, the BKP equation rather appears as the evolution equation for the 3-point Green's function defined by

$$f_\tau(\mathbf{x},\mathbf{y},\mathbf{z}) \equiv d^{abc}\langle \alpha_\mathbf{x}^a \alpha_\mathbf{y}^b \alpha_\mathbf{z}^c\rangle_\tau. \tag{3.7}$$

Indeed, the evolution equation for this Green's function reads

$$\frac{\partial}{\partial \tau}f_\tau(\mathbf{x},\mathbf{y},\mathbf{z}) = \frac{\bar{\alpha}_s}{4\pi}\int d^2\mathbf{w}\,\mathcal{M}_{xyw}\Big(f_\tau(\mathbf{x},\mathbf{w},\mathbf{z}) + f_\tau(\mathbf{w},\mathbf{y},\mathbf{z}) - f_\tau(\mathbf{x},\mathbf{y},\mathbf{z}) - f_\tau(\mathbf{w},\mathbf{w},\mathbf{z})\Big)$$
$$+\big\{2\text{ cyclic permutations}\big\}. \tag{3.8}$$

Notice that this equation is nothing but the Fourier transform of the BKP equation which is usually written in the momentum space. Since the 3-quark odderon operator Eq. (2.16) can be represented as a linear combination of the 3-point Green's functions, the equivalence between our result (3.6) and the BKP equation is essentially established. However, there is a caveat when we write Eq. (3.8). In fact, since the Green function $f_\tau(\mathbf{x},\mathbf{y},\mathbf{z})$ is not gauge invariant as it is, if one applied the original JIMWLK equation to this operator, one would obtain a result which is different from Eq. (3.8) and is even ill-defined due to infra-red divergences (the evolution equation (3.8) is finite and well-defined). Instead of doing this, we have derived Eq. (3.8) from the simplified version of the JIMWLK equation which is free of any infra-red divergences and is justified for gauge invariant operators. This means that we can use the simplified version of the JIMWLK equation to gauge *variant* operators as far as we finally consider gauge invariant quantities (infra-red divergences are canceled among themselves in the final result). In other words, the use of the simplified JIMWLK equation for the Green function corresponds to a kind of regularization of the resulting evolution equation.

ACKNOWLEDGMENTS

The author is grateful to Yoshitaka Hatta, Edmond Iancu and Larry McLerran, with whom the results presented in this talk were obtained [3].

REFERENCES

1. E. Iancu and R. Venugopalan, in *Quark-Gluon Plasma 3*, Eds. R. C. Hwa and X.-N. Wang, World Scientific (2003), hep-ph/0303204.
2. C. Ewerz, " *The odderon in quantum chromodynamics*", hep-ph/0306137.
3. Y. Hatta, E. Iancu, K. Itakura, and L. McLerran," *Odderon in the Color Glass Condensate*", hep-ph/0501171.
4. J. Bartels, *Nucl. Phys.* B **175** (1980) 365.
5. J. Kwiecinski and M. Praszalowicz, *Phys. Lett.* B **94** (1980) 413.
6. Y. V. Kovchegov, L. Szymanowski and S. Wallon, *Phys. Lett.* B **586** (2004) 267.

Hard diffractive production of vector mesons

R. Kirschner

Inst. Theor. Physik, Universität Leipzig, Augustusplatz 10, D-04109 Leipzig, Germany
e-mail: roland.kirschner@itp.uni-leipzig.de

Abstract. The large Q^2 behaviour of diffractive electroproduction and the large t behaviour of diffractive photoproduction are considered relying on models of the meson and real photon light-cone wave functions. The large size quark-antiquark dipole contribution to the impact factors of all helicity configurations are discussed.

Keywords: Hard diffraction
PACS: 13.60.Le,12.38.Bx,13.88,+e

1. INTRODUCTION

A good starting point for analyzing high energy diffractive processes is the impact factor representation of the amplitude written in terms of partial waves,

$$M^{\lambda_i \lambda_f}(s,Q,q) = s \int_{-i\infty}^{i\infty} \frac{d\omega}{2\pi i} F^{\lambda_i \lambda_f}(\omega,Q,q) \left[\left(\frac{s}{M^2(Q,m,q)} \right)^\omega + \left(\frac{-s}{M^2(Q,m,q)} \right)^\omega \right],$$

$$F^{\lambda_i \lambda_f}(\omega,Q,q) = \int d^2\kappa d^2\kappa' \Phi^{\lambda_i \lambda_f}(\kappa,Q,q) \, \mathcal{G}(\kappa,\kappa',q,\omega) \, \Phi^P(\kappa',q) \qquad (1)$$

This form is typically obtained in perturbative calculations, however it relies on more general arguments of impulse approximation [1]. The field mediating the exchange interaction acts on the scattering particles in a short time compared to the typical time of their binding interaction. The field sees just an intermediate fluctuation of the particle states. The diffractive exchange $\mathcal{G}$ is called Pomeron.

In the case of hard diffraction the intermediate fluctuation is squeezed into a narrow space-time region in the vicinity of the light cone. Only short-distance modes in $\mathcal{G}$ can interact with this fluctuation and therefore the coupling between scattering particle and exchange field can be described perturbatively. Moreover, the fluctuations involving a small number of partons dominate.

In the diffractive vector meson production by virtual photons the factorization in terms of a $q\bar{q}$ fluctuation coupling to the pomeron by two gluons has been proven on a rigorous level [2] in the case where the short distance scale is provided by the longitudinally polarized virtual photon momentum squared. As an obstacle to factorization in electroproduction with transverse polarization one encounters large contributions by $q\bar{q}$ fluctuations with one of the two quarks carrying small longitudinal momentum fraction. We have shown that those end-point contributions can be factorized in terms of $q\bar{q}$ exchange instead of two-gluon exchange [3].

Diffractive photoproduction at relatively large momentum transfer, $s \gg -t \gg m_V^2$, is another example of hard diffraction. Small-size $q\bar{q}$ fluctuations of the photon give an important contribution. However the squeezing by large t is not effective in configurations with small longitudinal momentum fractions of one of the quarks (end-point contributions, like in electroproduction) and with the large momentum transfer shared by the exchanged gluons in a portion determined by the longitudinal momentum fraction (Landshoff mechanism [4]).

The small dipole contributions to the impact factors of all helicities have been calculated in [5] where the non-perturbative states of the photon and the vector meson have been represented in terms of distribution amplitudes. In a recent paper we have studied the role of the large-size $q\bar{q}$ dipole contributions to the impact factor [6]. There is a recent detailed review of diffractive vector meson production [7].

2. IMPACT FACTOR

The impact factor of the $q\bar{q}$ contribution can be written as a convolution of the dipole impact factor with the light cone wave functions of the incoming photon and the outgoing vector meson,

$$\Phi^{\lambda_i \lambda_f}(\kappa_1, \kappa_2) = \int d^2\ell_1 d^2\ell_2 dz \psi_i^{\lambda_i}(\ell_1, z) \phi^{dipole}(\ell_1, \ell_2, \kappa_1, \kappa_2) \psi_f^{\lambda_f *}(\ell_2 - zq, z),$$
$$\phi^{dipole}(\ell_1, \ell_2, \kappa_1, \kappa_2) = \alpha_s [\delta^{(2)}(\ell_2 - \ell_1) + \delta^{(2)}(\ell_2 - \ell_1 + \kappa_1 + \kappa_2) - \delta^{(2)}(\ell_2 - \ell_1 + \kappa_1) - \delta^{(2)}(\ell_2 - \ell_1 + \kappa_2)] \quad (2)$$

The first argument in the light-cone wave functions $\psi_{i/f}$ is the transverse momentum relative to the momentum direction of corresponding particle. κ_i are the transverse momenta components of the exchange gluons, $\kappa_1 + \kappa_2 = -q$ is the transverse part of the momentum transfer.

In the case of the virtual photon the wave function results from perturbative calculation,

$$\Psi^{(\gamma)\lambda}(\ell, z, Q) = e \frac{V^\lambda(\ell, z, Q)}{[Q^2 + \frac{|\ell|^2 + m_q^2}{z\bar{z}}]}, \quad V^{(0)} = Q, \; V^{(+1)} = \frac{\ell^*}{z}, \; V^{(-1)} = \frac{\ell}{\bar{z}}. \quad (3)$$

We describe the real photon by a wave function obtained from the above by substituting $Q = 0$.

As a model for the light cone wave function of the vector meson we assume

$$\Psi^{V\lambda}(\ell, z) = f_V \frac{V^\lambda(\ell, z, m_V)}{m_V^2} \exp\left[-\frac{|\ell|^2 + m_q^2}{z\bar{z} m_V^2}\right]. \quad (4)$$

The form is motivated by QCD sum rules, it is formally obtained by Borel transformation of the propagator factor in (3) with respect to Q^2 and by the substitution of the Borel variable by m_V^2, where m_V is the meson mass. This wave function, being close to the one of γ^*, is a particular realization of the phenomenologically successful concept of vector dominance. Actually the explicit form of Ψ^V involves more information than necessary for the asymptotic estimate at large Q. The essential point in changing from Ψ^γ to Ψ^V is removing the hard (singular in impact parameter, the Fourier conjugate to ℓ) component while keeping the helicity structure.

In the case of virtual photons the transverse momentum integral in (2) is dominated by $\ell \sim \sqrt{z\bar{z}}Q$ and the dipole size $r \sim \frac{1}{\ell}$ is small outside the end-point regions.

In the case of photoproduction at large t the wave functions do not suppress the contributions from large dipole sizes. Such a suppression can result rather from the dipole impact factor involving the large momentum transfer. Transforming the dipole impact factor to coordinate representation we have

$$\int e^{i(\ell_1 r_1 - \ell_2 r_2)} d\ell_1 d\ell_2 \phi^{dip}(\ell_1, \ell_2, \kappa_1, \kappa_2)$$

$$= e^{izqr} \left(e^{-i\kappa_1 r} + e^{-i\kappa_2 r} - 1 - e^{i(\kappa_1+\kappa_2)r} \right) \delta^2(r_1 - r_2). \tag{5}$$

The large momentum transfer q leads to the dominance of small dipole sizes, $r_1 = r_2 = \mathcal{O}(q^{-1})$ for generic values of the momenta κ_1, κ_2 ($\kappa_1 + \kappa_2 = -q$) of the exchanged gluons. This hard contribution to the photon-meson impact factor can be constructed with distribution amplitudes of both the photon and the vector meson. In [5] only this contribution has been considered. A particular feature is that one exchanged gluon carries large and the other relatively small momentum, $\kappa_1 \ll q$ or $\kappa_2 \ll q$. Eq(5) shows that we have two further regions, where the dominant dipole size is not small. The first one is the vicinity of the end points, $z = 0, z = 1$. The second one corresponds to small values of $\kappa_1 + zq$ or $\kappa_1 + \bar{z}q$. This means that here the large momentum transfer is shared by the two gluons. In this respect it is reminiscent of the Landshoff mechanism proposed for pp elastic scattering at large t [4]. It turns out that the photon-meson impact factor has extra terms which contribute to this region but are exponentially small outside of it.

The effect of the BFKL leading $\ln s$ corrections to the two-gluon exchange has been studied in [8].

There is no divergence in the impact factor expressions, in particular not in the integration over the longitudinal momentum fraction z. Spurious end-point singularities appear if one substitutes the integrand by its twist expansion, i.e. by the power expansion in Q^2 or t valid for $z = \mathcal{O}(1)$. The end-point regions contribute in dependence on the helicities λ_i, λ_f by an logarithm of Q^2 or t. This logarithmic contribution can be identified as the one of a generalized parton evolution leading from exchanged gluons to exchanged $q\bar{q}$. In this case the hard sub-process is of Compton type rather than of dipole scattering type.

3. ELECTROPRODUCTION HELICITY AMPLITUDES

In the case of factorization via two exchanged partons one can write

$$\mathcal{G}(k,k',q,\omega) = \frac{1}{|k|^2|k+q|^2}\tilde{\mathcal{G}}(k,k',q,\omega). \tag{6}$$

In the considered case of electroproduction we have in eq. (1) $M^2(Q^2, m, q) \approx Q^2$. We pick up the leading contribution in the ω integral by writing

$$M^{\lambda_i \lambda_j} = \sum_p \int d^2\kappa \Phi_p(\kappa, q, Q) \frac{1}{|k|^2|\kappa+q|^2} G'_p(x_1, x_2, q, \kappa). \tag{7}$$

G'_p stands for the unintegrated generalized parton distribution of the proton resulting from the convolution of the proton impact factor Φ^P with the projection of the exchange $\mathcal{G}$ that couples by two exchange partons (p) in the small Bjorken variable limit, $x_1 = \frac{Q^2}{s}$, $x_2 = \frac{m_V^2}{s}$. The skewedness $\xi = \frac{x_1 - x_2}{2}$ is small; we have non-vanishing transverse momentum transfer q.

As a simplification we replace G'_p by the derivative of the gluon distribution at small x

$$G'_p = \frac{\partial}{\partial \ln |\kappa|^2}[G_p(x_1, x_2, |q|^2, |\kappa|^2)\, T(\kappa^2, Q^2))]. \tag{8}$$

Here $T(\kappa^2, Q^2)$ is the parton Sudakov form factor; it is equal to 1 at $\kappa^2 = Q^2$ and small for $\kappa^2 \ll Q^2$.

The decomposition of the z range in the impact factor into $0 < z < z_0, z = \mathcal{O}(1), 1 - z_0 < z < 1$ results in the amplitude as a sum of corresponding 3 terms, $M^{\lambda_i, \lambda_f} = M_1^{\lambda_i, \lambda_f} + M_{z_0}^{\lambda_i, \lambda_f} + M_{\bar{z}_0}^{\lambda_i, \lambda_f}$. The leading contribution of $z = \mathcal{O}(1)$ can be expressed in terms of the generalized gluon distribution. For example, in the case $\lambda_i = \lambda_f = 1$ we have

$$M_1^{11} = \left(\int_{z_0}^{\bar{z}_0} (\frac{1}{z\bar{z}} - 2)\, dz\right) \frac{m_V^2}{(Q^2 + m_V^2)^2} \cdot G_g(x_1, x_2, q; Q^2). \tag{9}$$

The end point contributions are not small with z_0 for the terms involving $\frac{1}{z\bar{z}}$. The logarithmic z integral is a generalized parton evolution term resulting in the effective quark distribution

$$\tilde{G}_q(x_1, x_2, q, z_0 Q^2) = \int_{m_V^2}^{z_0 Q^2} \frac{d|\ell|^2}{|\ell|^2} \frac{\alpha_S(|\ell|^2)}{\alpha_S(Q^2)} G_g(x_1, x_2, q, |\ell|^2). \tag{10}$$

In the example $\lambda_i = \lambda_f = 1$ this leads to

$$M_{z_0}^{11} = M_{\bar{z}_0}^{11} = \frac{m_V^2}{(Q^2 + m_V^2)^2} \cdot \tilde{G}_q(x_1, x_2, q, z_0 Q^2). \tag{11}$$

With the helicity amplitudes we calculate the ratio of longitudinal to transverse virtual photon diffractive cross section,

$$R(Q^2) = \frac{\sigma_L(Q^2)}{\sigma_T(Q^2)}, \qquad (12)$$

where

$$\sigma_L(Q^2) = \int dt (|M^{00}|^2 + 2|M^{01}|^2),$$
$$\sigma_T(Q^2) = \int dt (|M^{11}|^2 + |M^{10}|^2 + |M^{1-1}|^2). \qquad (13)$$

The coefficients in the angular-decay distribution are more sensitive to the helicity dependence, because in some of them the smaller flip amplitudes enter in the first power,

$$r_{00}^{04} \propto \frac{1}{N}(|M^{00}|^2 + |M^{10}|^2)$$
$$r_{00}^{5} \propto \frac{1}{N} Re(M^{00*} M^{10})$$
$$r_{11}^{5} \propto (Re(M^{01} M^{*11}) - Re(M^{01} M^{*1-1}))$$
$$r_{00}^{1} \propto -\frac{1}{N}|M^{10}|^2$$
$$r_{11}^{1} \propto \frac{1}{N}(M^{1-1} M^{*11} + M^{11} M^{*1-1}),$$
$$N = |M^{00}|^2 + |M^{10}|^2 + 2|M^{01}|^2 + |M^{11}|^2 + |M^{1-1}|^2. \qquad (14)$$

For a detailed discussion of the results in comparison with HERA data we refer to our publication [3].

4. PHOTOPRODUCTION HELICITY AMPLITUDES

At large t the diffractive exchange interacts with a single quark in the disintegrating proton. We write down the helicity amplitudes of diffractive scattering on a quark. In this case the proton impact factor reduces to a constant and the coupling of the exchange to the disintegrating proton does not influence the t-dependence.

We obtain the diffractive γV amplitudes in terms of a sum of the hard and Landshoff-type contributions, each of them has the form

$$M^{\lambda_i \lambda_f} = is \int d^2\kappa dz z \bar{z} \varphi_4^{\lambda_i \lambda_f}(\kappa, q, z) \frac{\mathcal{G}(s, q, \kappa, \bar{\kappa})}{|\kappa|^2 |\kappa - q|^2} \qquad (15)$$

Here $\mathcal{G}$ represents the effect of the interaction of the exchanged gluons; in the case of simple two-gluon exchange we put $\mathcal{G} = 1$. The hard contribution to the cases $\lambda_f = 0$, and $\lambda_i = -\lambda_f$ can be calculated in this case without further approximation with the results

$$M_{1,h}^{1,0} = -2C m_V \pi^2 (4 - \frac{\pi^2}{3}) \frac{q}{t^2}, \qquad (16)$$

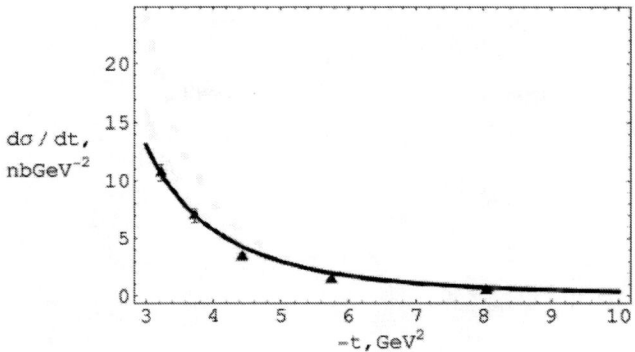

FIGURE 1. t-dependence of the diffractive photoproduction cross section.

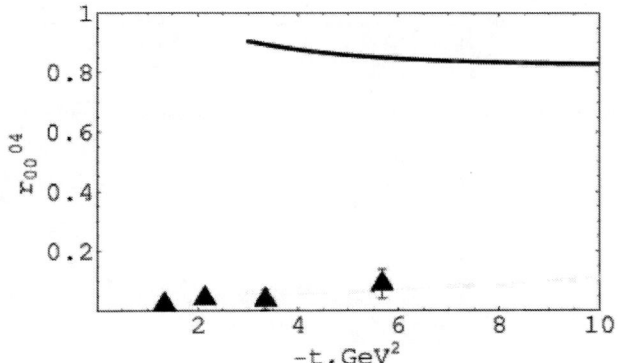

FIGURE 2. The angular-decay coefficient r_{00}^{04}

$$M_{1,h}^{1,-1} = C\frac{2\pi^2}{t^2}(\frac{2\pi^2}{3} - 8).$$

The factor C denotes $C = is\frac{2}{3}\alpha_s^2 eQ_q f_V \Phi_P$.
The Landshoff-type contributions for 2-gluon case are

$$M_{1,L}^{1,1} = C\frac{2\pi^2}{|q|^4}\int_{m_V/q}^{1-m_V/q}(\frac{1}{z\bar{z}} - 2)dz. \qquad (17)$$

The helicity amplitudes calculated above allow to evaluate the t dependence of the vector meson production cross section and of the angular-decay coefficients.

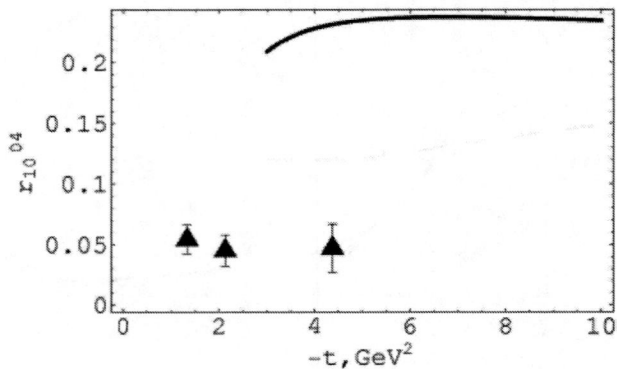

FIGURE 3. The angular-decay coefficient r_{10}^{04}

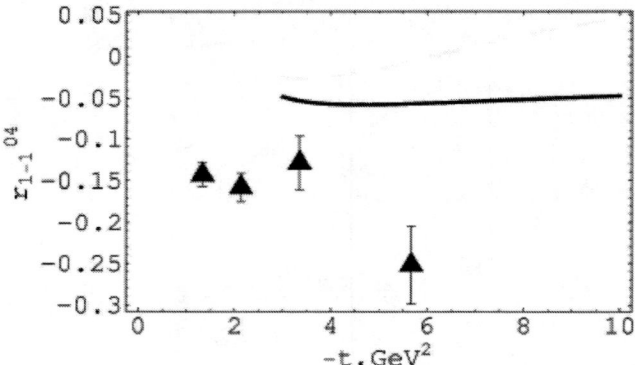

FIGURE 4. The angular-decay coefficient r_{1-1}^{04}

For the latter we use the relations,

$$r_{00}^{04} \propto \frac{1}{N}|M^{10}|^2$$

$$r_{10}^{04} \propto \frac{1}{2N}(M^{10*}M^{11} + M^{10*}M^{1-1}) \qquad (18)$$

$$r_{1-1}^{04} \propto \frac{1}{N}(M^{1-1*}M^{11})$$

$$N = |M^{10}|^2 + |M^{11}|^2 + |M^{1-1}|^2$$

Our estimates are done for large t, therefore it makes sense to extract results for values of $|t|$ above 3 GeV^2. Without fitting parameters we obtain a reasonable comparison with the cross section data. There appears a clear discrepancy with the data on r_{00}^{04}, [9].

ACKNOWLEDGMENTS

I am grateful to J.R. Cudell, J. Cugnon and J.P. Lansberg for hospitality in Liège and in Spa. The work has been supported in part by the Joint Research Activity"Generalized Parton Distributions" in the Integrated Infrastructure Initiative "Hadron Physics" of the European Union, contract No. RII3-CT-2004-506078.

REFERENCES

1. V. N. Gribov, "Space-Time Description Of Hadron Interactions At High Energies," In *Moscow 1 ITEP school, v.1 'Elementary particles'*, 65, 1973, [arXiv:hep-ph/0006158].
2. J. C. Collins, L. Frankfurt and M. Strikman, Phys. Rev. D **56** (1997) 2982 [arXiv:hep-ph/9611433], arXiv:hep-ph/9709336.
3. A. Ivanov and R. Kirschner, Eur. Phys. J. C **29** (2003) 353 [arXiv:hep-ph/0301182].
4. P. V. Landshoff, Phys. Rev. D **10** (1974) 1024. A. Donnachie and P. V. Landshoff, Z. Phys. C **2** (1979) 55 [Erratum-ibid. C **2** (1979) 372], Phys. Lett. B **123** (1983) 345.
5. D. Yu. Ivanov, R. Kirschner, A. Schafer, and L. Szymanowski, Phys. Lett. B **578** (2000) 101-113 [arXiv:hep-ph/0001255].
6. A. Ivanov and R. Kirschner, Eur. Phys. J. C **36** (2004) 43 [arXiv:hep-ph/0311077].
7. I. P. Ivanov, N. N. Nikolaev and A. A. Savin, arXiv:hep-ph/0501034.
8. R. Enberg, J. R. Forshaw, L. Motyka and G. Poludniowski, JHEP **0309** (2003) 008 [arXiv:hep-ph/0306232]; G. G. Poludniowski, R. Enberg, J. R. Forshaw and L. Motyka, JHEP **0312** (2003) 002 [arXiv:hep-ph/0311017].
9. S. Chekanov et al. [ZEUS Collaboration], Eur. Phys. J. C **26** (2003) 389 [arXiv:hep-ex/0205081].

HIGH TEMPERATURE AND DENSITY

Heavy-quarkonium interaction and the temperature dependence of the QCD static potential

François Arleo

LPTHE, 4, place Jussieu, 75252 Paris cedex 05, France
Email: arleo@lpthe.jussieu.fr

Abstract. We explore the temperature dependence of the heavy-quarkonium interaction based on the Bhanot - Peskin leading order perturbative QCD analysis. The Wilson coefficients are computed solving the Schrödinger equation in a screened Coulomb heavy-quark potential. The inverse Mellin transform of the Wilson coefficients then allows for the computation of the 1S and 2S heavy-quarkonium gluon and pion total cross section at finite screening.

Keywords: Heavy-quarkonium, QCD calculation, Debye screening
PACS: 14.40.Gx, 12.38.Bx, 25.75.Nq

1. MOTIVATIONS

Heavy quark bound states may no longer exist well above the deconfinement critical temperature due to the Debye screening between two opposite color charges seen in the QCD static potential computed at finite temperature T on the lattice [1]. The immediate consequence is the disappearance – or at least a significant suppression – of heavy-quarkonium production in high energy heavy-ion collisions where quark-gluon plasma is expected to be formed [2].

In order to interpret the so-called J/ψ "anomalous" suppression measured in the most central lead-lead collisions ($\sqrt{s} \simeq 17$ GeV) by the NA50 experiment at the CERN SPS [3], one has first to understand all nuclear effects that could affect its production in heavy-ion reactions. While the effects of nuclear shadowing or parton energy loss should remain reasonably small at these energies [4], the inelastic interaction of heavy quark bound states with gluons or pions produced in the nuclear collision is expected to be large. Several approaches have been suggested in the past to determine heavy-quarkonium total cross sections, from meson exchange [5] or constituent quark models [6] to the perturbative framework developed by Bhanot and Peskin [7, 8] upon which the present paper relies.

Although derived from first principles in QCD perturbation theory, the Bhanot - Peskin result describes the interaction of Coulombic bound states, that is for which the heavy-quark potential is well approximated by the perturbative one-gluon exchange potential. As noted in [8], this stringent condition should be fulfilled provided the heavy quark mass to be larger than 25 GeV ! In this talk, I would like to discuss how to go beyond this assumption and in particular how to explore possible medium effects to the heavy-quarkonium interaction at finite screening.

The perturbative QCD framework as well as our model are first described in Section 2. Numerical results are presented in Section 3 and discussed in Section 4. Finally, Section 5 is devoted to a summary of what has been carried out.

2. MODEL

At twist two, the forward heavy-quarkonium (Φ) - hadron (h) scattering amplitude $\mathcal{M}_{\Phi h}$ is an operator product expansion of perturbative Wilson coefficients d_{2k} evaluated in the heavy-quarkonium state times soft matrix elements A_{2k} in the hadron state. It reads [7]

$$\mathcal{M}_{\Phi h}(\lambda) = \left(\frac{g^2 N_c}{16\pi}\right) a_0^2 \epsilon \sum_{k \geq 1} d_{2k} A_{2k} (\lambda/\epsilon)^{2k}. \tag{1}$$

where a_0 and ϵ stand, respectively, for the Bohr radius and the binding energy for the Φ system, g the QCD coupling, N_c the number of colors and λ the hadron energy in the Φ rest frame. Expressing the Wilson coefficients in terms of their Mellin moments,

$$d_{2k} = \int_0^1 \frac{dx}{x} x^{2k} \tilde{d}(x),$$

the power series (1) can be continued analytically throughout the whole complex plane of energies [9]. One then obtains the total heavy-quarkonium cross section

$$\sigma_{\Phi h}(\lambda) = \int_0^1 dx\, G(x)\, \sigma_{\Phi g}(x\lambda), \tag{3}$$

where the heavy-quarkonium gluon cross section is defined as

$$\sigma_{\Phi g}(\omega) = \left(\frac{g^2 N_c}{32}\right) a_0^2 \frac{\epsilon}{\omega} \tilde{d}\left(\frac{\epsilon}{\omega}\right) \tag{4}$$

and the gluon energy $\omega = \lambda x$ in the Φ rest frame.

Resuming all diagrams contributing to leading order in g^2 to the $\Phi - h$ interaction, Peskin made explicit the heavy-quarkonium Wilson coefficients [7]. They are given by

$$d_{2k} = \frac{16\pi}{N_c^2 a_0^2} \epsilon^{2k-1} \langle \phi | r^i \frac{1}{(H_a + \epsilon)^{2k-1}} r^j | \phi \rangle, \tag{5}$$

where

- $|\phi\rangle$ denotes the heavy-quarkonium wavefunction;
- $H_{s,a}$ the internal Hamiltonian in the singlet and adjoint representation

$$H_{s,a} = \frac{k^2}{m_Q} + V_{s,a}(r); \tag{6}$$

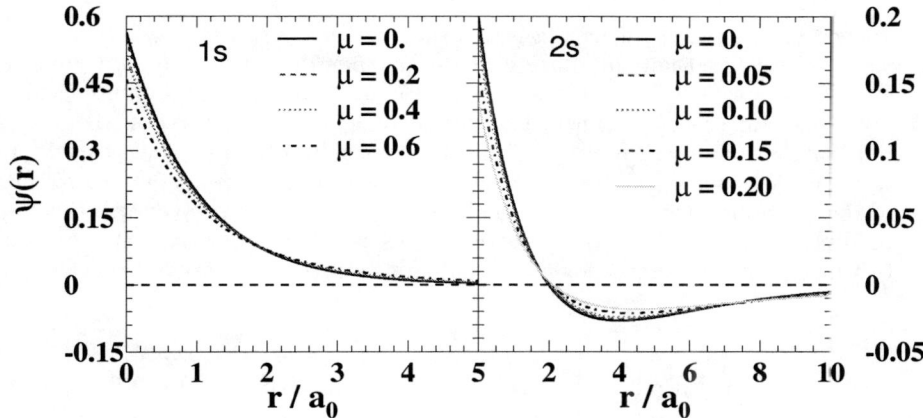

FIGURE 1. 1S (*left*) and 2S (*right*) heavy-quarkonium radial wavefunctions as a function of the quark-antiquark distance for several values of the screening parameter μ.

- the binding energy ϵ is given by the Schrödinger equation,

$$H_s |\phi\rangle = -\epsilon |\phi\rangle. \qquad (7)$$

In order to discuss possible medium modifications to the heavy-quarkonium interaction, we introduced a simple model in Ref. [10] based on one single assumption formulated by Bhanot and Peskin [8]: as long as heavy-quarkonium states are small, the dipole coupling of gluons (hidden in the Wilson coefficients) identified in the Coulombic regime may survive perturbative modifications to the potential. We shall therefore compute the Wilson coefficients Eq. (5) using a screened Coulomb heavy-quark potential

$$V_s = -\frac{g^2 N_c}{8\pi r} \exp(-\mu r/a_0),$$
$$V_a = 0, \qquad (8)$$

characterized by the dimensionless screening parameter μ. In the $\mu \to 0$ limit, one should then recover the one-gluon exchange potential (at large N_c) and hence the original Bhanot-Peskin result.

3. RESULTS

The first step is to determine both the heavy-quarkonium wavefunction $|\phi\rangle$ as well as the binding energy ϵ. For the illustration, Figure 1 displays the 1S and 2S radial wavefunctions plotted as a function of the quark-antiquark distance for several values of the screening parameter μ. Once these wavefunctions are known, the Wilson coefficients d_{2k} (and their Mellin transform $\tilde{d}(x)$) are determined which allows for the computation of the partonic heavy-quarkonium gluon total cross sections (4). These are shown in Figure 2 in the charmonium (*left*) and bottomonium (*right*) channel. The dominant effect of the screened heavy quark potential is the decrease of the 1S heavy-quarkonium binding energy from ϵ_0 to ϵ which leads to a lower threshold for the inelastic process. The medium modifications of the Φ – gluon total cross sections are nevertheless not only due to the smaller binding energy, yet the characteristic shapes of the cross sections are reminiscent to what is already known for pure Coulombic states, $\mu = 0$ (Figure 2, *solid*).

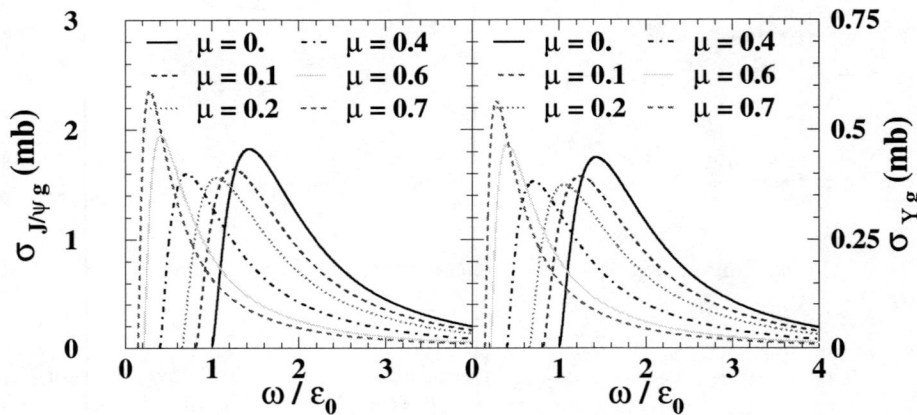

FIGURE 2. 1S charmonium (*left*) and bottomonium (*right*) gluon total cross section as a function of the gluon energy ω for various values of the screening parameter μ.

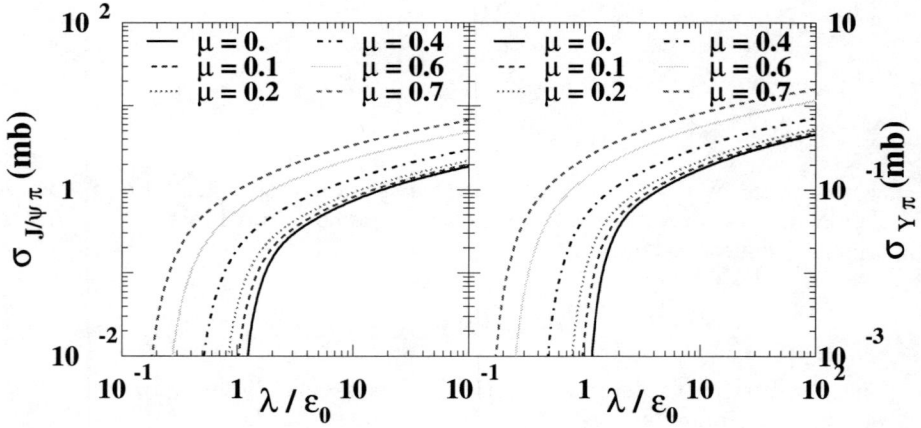

FIGURE 3. $J/\psi - \pi$ (*left*) and $\Upsilon - \pi$ (*right*) total cross section as a function of the pion energy λ for various values of the screening parameter μ.

Let us now discuss the heavy-quarkonium hadron cross section. Since heavy-quarkonia plunged into the hot medium are most likely to interact with pions, we shall only consider the $\Phi - \pi$ channel and choose the GRV LO parameterization for the gluon distribution [11]. The $J/\psi - \pi$ and $\Upsilon - \pi$ cross sections are computed in Figure 3 as a function of the pion energy λ. Again, the threshold for the process, located at $\lambda = \epsilon$, gets shifted to lower values leading to a strong modification of the heavy-quarkonium pion interaction in this region. At high energy, small $x = \mathcal{O}(\epsilon/\lambda)$ gluons dissociate heavy-quarkonia, thereby increasing the $\Phi - \pi$ cross section by a factor $(\epsilon_0/\epsilon)^\delta$ where $\delta \simeq 0.3$ governs the rise of the gluon distribution at small x, $xG(x) \propto x^{-\delta}$ [12].

4. LIMITATIONS

The starting point of the calculation is the forward scattering amplitude $\mathcal{M}_{\Phi h}$ originally derived for Coulomb bound states. To go beyond this one-gluon exchange picture would require to include light quark loops in the diagrammatics, to which the soft gluon source may couple, that we have not attempted. However, as conjectured in [8], it is appealing to guess that the generic dipole coupling appearing to leading order in g^2 in the heavy-quarkonium Wilson coefficients (5) survives perturbative and non-perturbative modifications of the $Q\bar{Q}$ binding potential. Therefore we believe that taking the literal expression for the Coulomb states Wilson coefficients and compute them in a Coulomb screened potential appears sensible, at least as long as the screening remains reasonable, $m_D a_0 \ll 1$. This is certainly the case when the temperature is kept small as compared to the heavy quark mass. Typical space time scale becomes increasingly larger with the temperature, thus strongly limiting our confidence in the high temperature regime. Finally, one should keep a clear factorization between the gluon source and the heavy-quarkonium swimming in the gluon bath. Such a separation should be achieved as long as the Debye mass is small as compared to the bound state Rydberg energy, that is for temperatures $T \lesssim 350$ MeV.

5. SUMMARY

A numerical calculation of the heavy-quarkonium cross section with gluons and pions has been presented, taking into account the possible medium-modifications of the heavy-quark potential at finite temperature. Such a work can therefore be useful to estimate heavy-quarkonium production in high energy heavy-ion collisions. In particular, we feel it would be interesting to explore the phenomenological consequences of such corrections comparing them to present calculations based on the vacuum heavy-quarkonium interaction.

ACKNOWLEDGMENTS

This work has been done in collaboration with J. Cugnon and Y. Kalinovsky. I warmly thank the organizers of this stimulating conference.

REFERENCES

1. F. Karsch, H.W. Wyld, Phys. Lett. **B213**, 505 (1988).
2. T. Matsui, H. Satz, Phys. Lett. **B178**, 416 (1986).
3. NA50 collaboration, M. C. Abreu *et al.*, Phys. Lett. **B477**, 28 (2000).
4. F. Arleo, Phys. Lett. **B532** (2002) 231.
5. S.G. Matinyan, B. Müller, Phys. Rev. **C58**, 2994 (1998);
   Y. Oh, T. Song, S.H. Lee, Phys. Rev. **C63**, 034901 (2001);
   L. Maiani, F. Piccinini, A.D. Polosa, V. Riquer, Nucl. Phys. **A741**, 273 (2004).
6. K. Martins, D. Blaschke, E. Quack, Phys. Rev. **C51**, 2723 (1995);
   C.-Y. Wong, E.S. Swanson, T. Barnes, Phys. Rev. **C62**, 045201 (2000).
7. M.E. Peskin, Nucl. Phys. **B156**, 365 (1979).
8. G. Bhanot, M.E. Peskin, Nucl. Phys. **B156**, 391 (1979).
9. F. Arleo, P.-B. Gossiaux, T. Gousset, J. Aichelin, Phys. Rev. **D65**, 014005 (2002).
10. F. Arleo, J. Cugnon, Y. Kalinovsky, hep-ph/0410295.
11. M. Glück, E. Reya, A. Vogt, Z. Phys. **C53**, 651 (1992).
12. H1 collaboration, S. Aid *et al.*, Nucl. Phys. **B470**, 3 (1996).

Light front approach to hot and dense quark matter[1]

M. Beyer[*,†], S. Mattiello[*], S. Strauß[*], T. Frederico[**] and H.J. Weber[‡]

[*]*Institute of Physics, University of Rostock, D-18051 Rostock, Germany*
[†]*E-mail: michael.beyer@uni-rostock.de*
[**]*Dep. de Física, Instituto Tecnológico de Aeronáutica, Centro Técnico Aeroespacial, 12.228-900 São José dos Campos, São Paulo, Brazil*
[‡]*Dept. of Physics, University of Virginia, Charlottesville VA, 22901 U.S.A.*

Abstract. In recent years we have extended light-front quantization to systems of finite temperature and density, hence providing a novel approach towards nuclear matter under extreme conditions. Light-front field theory is particularly suited, since boost invariance is essential, e.g., for a fireball created in a heavy ion collision. The approach is based on a relativistic Hamiltonian and allows to treat the perturbative as well as the non-perturbative regime of QCD. Presently, we are in particular interested in few-quark correlations, i.e. quark-antiquark and three-quark correlations. The later ones have hardly been considered in this context, although they should be important as quark matter hadronizes below the critical temperature of the chiral phase transition to form nucleons. To this end we have derived relativistic Faddeev type in-medium equations and investigated the stability of three-quark bound states (nucleons) in hot and dense quark matter. We have used an effective zero-range interaction, and compared our results with the more traditional instant form approach, where applicable.

Keywords: light cone quantization, finite-temperature field theory, quark-gluon plasma, few-body physics, relativistic quantum field theory
PACS: 11.10.Wx 12.38.Mh 25.75.Nq

1. INTRODUCTION

Light front quantization recognized by Dirac [1] provides a framework to describe the perturbative and the nonperturbative regime of quantum chromodynamics (QCD), see e.g. [2, 3]. A description involving both regimes is necessary, if one is interested in the dynamics close to the chiral and the confinement-deconfinement transition that is suggested by lattice QCD [4, 5], model analyses [6, 7, 8, 9], and through recent interpretations of certain RHIC results [10, 11, 12, 13, 14, 15]. It seems that the quark gluon plasma does not appear as a weakly interacting gas of quarks and gluons but rather as a strongly correlated system.

Here we focus on three-quark correlations that have hardly been considered in this context. They should be important (at least at finite chemical potentials) as quark matter hadronizes to nucleons that are the relevant degrees of freedom to

---

[1] presented by M. Beyer.

form nuclear matter below the critical temperature. We treat quark matter in the framework of many-body Green functions defined in standard textbooks of many-body physics, see e.g. Refs. [16, 17]. The inclusion of correlations is provided by the Dyson equation approach developed in the context of nonrelativistic nuclear physics, reviewed in [18], and generalized here to the light front to comply with the requirements of special relativity. The front form provides a rigorous relativistic many-body theory that in the context of relativistic heavy ion collisions has to be generalized to systems of finite temperature. This has been accomplished in Ref. [20] where distribution functions, Green functions and Matsubara frequencies have been given for light front quantization. Further developments have been carried out and more examples given in Refs. [21, 22, 23, 24, 25, 26, 27, 28, 29, 30, 31, 32, 33, 34, 35, 36, 37]. Todate many aspects of light front field theory at finite temperature have been touched upon including spontaneous symmetry breaking and restoration. Light front field theory seems to work also at finite temperature although it requires generalization of some widespread intuitive viewpoints.

2. LIGHT-FRONT STATISTICAL PHYSICS

A formal framework of covariant calculations at finite temperatures in instant form has been given in Refs. [38, 39, 40] in a different context. The grand canonical partition function is given by

$$Z_G = \mathrm{Tr}\ \exp\{-(u \cdot P - \mu N)/T\}, \tag{1}$$

where T is the temperature and μ the chemical potential. The velocity of the medium is given by the normalized time-like vector $u_\nu u^\nu = 1$ [38]. In light front quantization and following notation of Ref. [2] the four-momentum operator P^μ is given by

$$P^\mu = \int d\omega_+\ T^{+\mu}(x) \equiv \int dx_- d^2 x_\perp\ T^{+\mu}(x), \tag{2}$$

where $T^{\nu\mu}(x)$ denotes the energy momentum tensor, $x^\pm = x^0 \pm x^3$, $\vec{x}_\perp = (x^1, x^2)$. The number operator is

$$N = \int d\omega_+\ J^+(x), \tag{3}$$

where $J^\nu(x)$ is the conserved current. With the choice $u^\nu = (u^+, u^-, \vec{u}^\perp) = (1, 1, 0, 0)$ the partition function can then be written as

$$Z_G = \mathrm{Tr}\ \exp\left\{-\frac{1}{T}(P_+ + P_- - \mu N)\right\}. \tag{4}$$

and the statistical operator is defined accordingly.

The light-cone time-ordered (causal) Green function for a given number of particles is

$$\mathcal{G}_{\alpha\beta}(x-y) = \theta(x^+ - y^+) \mathcal{G}^{>}_{\alpha\beta}(x-y) + \theta(y^+ - x^+) \mathcal{G}^{<}_{\alpha\beta}(x-y), \tag{5}$$

with the correlation functions

$$\mathcal{G}^>_{\alpha\beta}(x-y) = -i\langle\Psi_\alpha(x)\bar{\Psi}_\beta(y)\rangle \qquad (6)$$
$$\mathcal{G}^<_{\alpha\beta}(x-y) = \mp(-i)\langle\bar{\Psi}_\beta(y)\Psi_\alpha(x)\rangle. \qquad (7)$$

Here and in the following the upper (lower) sign is for fermions (bosons). The anti-causal Green function can be defined correspondingly. In the generalized Heisenberg picture the light-cone time dependence of operators is given by

$$\mathcal{O}(x^+) = e^{iH_{\text{eff}}x^+}\mathcal{O}e^{-iH_{\text{eff}}x^+}, \qquad H_{\text{eff}} = P_+ + P_- - \mu N. \qquad (8)$$

This differs from the true Heisenberg picture for the isolated case by the thermodynamic constraints. Since u is a time like four vector also P_- appears here in addition to the usual particle number N, since there is no frame were the velocity of the system can be given such that $u \cdot P$ equals to the "pure" Hamiltonian P_+. Evaluating Eq. (5) for the vacuum, i.e. $\langle\ldots\rangle = \langle 0|\ldots|0\rangle$ (isolated case), the Fourier transform of the Green function becomes

$$G(k) = \frac{\gamma k_{\text{on}} + m}{\frac{1}{2}k^- - \frac{1}{2}k^-_{\text{on}} + i\varepsilon}\frac{\theta(k^+)}{2k^+} + \frac{\gamma k_{\text{on}} + m}{\frac{1}{2}k^- - \frac{1}{2}k^-_{\text{on}} - i\varepsilon}\frac{\theta(-k^+)}{2k^+} = S_F(k) - \frac{\gamma^+}{2k^+} \qquad (9)$$

where $S_F(k)$ is the usual Feynman propagator. We note here that for fermions the causal Green functions (5) and (9) differ from the Feynman propagator S_F in the front form by a contact term and hence (9) coincides with the light-front propagator given in Ref. [41].

Only two Green functions (of the four possible) are independent. In equilibrium the average is taken over the (equilibrium) grand canonical statistical operator ρ_G, viz. $\langle\cdots\rangle = \text{tr}\{\rho_G\ldots\}$ and in addition $\mathcal{G}^<_{\alpha\beta}$ and $\mathcal{G}^>_{\alpha\beta}$ are related by (anti) periodic boundary conditions,

$$\mathcal{G}^<_{\alpha\beta}(x^+) = \pm\mathcal{G}^>_{\alpha\beta}(x^+ - i\beta). \qquad (10)$$

Hence, in equilibrium, only one Green function needs to be considered, and, alternatively to (5), we may introduce the thermodynamic Green function [17]

$$\mathcal{G}^{\tau-\tau'}_{\alpha\beta} = -\langle T_\tau \Psi_\alpha(\tau)\bar{\Psi}_\beta(\tau')\rangle. \qquad (11)$$

Setting $x^+ = -i\tau$ ("imaginary time") in (8) we achieve the thermodynamic Heisenberg picture. The Green functions (5) and (11) are related by their spectral representation utilizing analytic continuation of the respective spectral functions just as in the instant form case [17]. To be more specific, in momentum space we introduce the spectral function $A(k)$

$$G^<(k) = f(k)A(k) \qquad (12)$$
$$G^>(k) = (1-f(k))A(k) \qquad (13)$$

where $A(k) = G^>(k) + G^<(k)$. Using (5) along with (12) and (13) the causal Green function can be represented as

$$G(k) = \int_{-\infty}^{\infty}\frac{d\tilde{k}^-}{2(2\pi)}\left[\frac{f(\tilde{k})A(\tilde{k})}{\frac{1}{2}k^- - \frac{1}{2}\tilde{k}^- - i\varepsilon} + \frac{(1-f(\tilde{k}))A(\tilde{k})}{\frac{1}{2}k^- - \frac{1}{2}\tilde{k}^- + i\varepsilon}\right] \qquad (14)$$

where $\tilde{k} = (\tilde{k}^-, k^+, \vec{k}_\perp)$. A similar equation holds for (11), where the $i\varepsilon$ term can be dropped

$$G(k_n) = \int_{-\infty}^{\infty} \frac{d\tilde{k}^-}{2(2\pi)} \frac{A(\tilde{k})}{\frac{1}{2}k_n^- - \frac{1}{2}\tilde{k}^-} \qquad (15)$$

with $k_n = (k_n^-, k^+, \vec{k}_\perp)$ and the Matsubara frequencies are given by [20]

$$\frac{1}{2}k_n^- = \begin{cases} i(2n+1)\pi T + \mu - \frac{1}{2}k^+ & \text{fermions}, \\ i2n\pi T + \mu - \frac{1}{2}k^+ & \text{bosons}. \end{cases} \qquad (16)$$

In this case the spectral function is given by

$$A(k) = i \lim_{\varepsilon \to 0} \left(G(k_n)|_{k_n^- = k^- + i\varepsilon} - G(k_n)|_{k_n^- = k^- - i\varepsilon} \right) \qquad (17)$$

where the Matsubara-Fourier representation of the thermodynamic single particle Green function is

$$G(k_n) = \frac{\gamma k_{\text{on}} + m}{k_n^2 - m^2}, \qquad (18)$$

which has been given in similar form in [26, 32]. For an ideal gas (and also in Hartree-Fock approximation utilized later on) the imaginary part of the spectral function vanishes and therefore (18) can be used to achieve the spectral function.

$$A(k) = 2\pi \frac{\gamma k + m}{2k^+} \delta(\tfrac{1}{2}k^- - \tfrac{1}{2}k_{\text{on}}^-) \qquad (19)$$

Inserting (19) into (14) eventually leads to

$$\begin{aligned}
G(k) &= \frac{\gamma k_{\text{on}} + m}{\frac{1}{2}k^- - \frac{1}{2}k_{\text{on}}^- + i\varepsilon} \frac{\theta(k^+)}{2k^+} (1 - f^+(k^+)) + \frac{\gamma k_{\text{on}} + m}{\frac{1}{2}k^- - \frac{1}{2}k_{\text{on}}^- - i\varepsilon} \frac{\theta(k^+)}{2k^+} f^+(k^+) \\
&+ \frac{\gamma k_{\text{on}} + m}{\frac{1}{2}k^- - \frac{1}{2}k_{\text{on}}^- + i\varepsilon} \frac{\theta(-k^+)}{2k^+} f^-(-k^+) + \frac{\gamma k_{\text{on}} + m}{\frac{1}{2}k^- - \frac{1}{2}k_{\text{on}}^- - i\varepsilon} \frac{\theta(-k^+)}{2k^+} (1 - f^-(-k^+)).
\end{aligned} \qquad (20)$$

For a grand canonical ensemble in equilibrium the Fermi distribution functions of particles $f^+ \equiv f$ and antiparticles f^- are given by

$$f^\pm(k^+, \vec{k}_\perp) = \left[\exp\left\{ \frac{1}{T} \left(\frac{1}{2}k_{\text{on}}^- + \frac{1}{2}k^+ \mp \mu \right) \right\} + 1 \right]^{-1} \qquad (21)$$

and $k_{\text{on}}^- = (\vec{k}_\perp^2 + m^2)/k^+$. The particle propagator ($f^- = 0$) has been given previously [20, 23].

3. LIGHT FRONT CLUSTER GREEN FUNCTIONS

We generalize the definitions of Green functions given in the previous section to include multi-particle correlations. We follow the Dyson equation approach to correlations functions developed originally in the context of nonrelativistic many-body physics, reviewed in [18]. The causal $\mathcal{G}_{\alpha\beta}$ and anti-causal $\mathcal{G}^a_{\alpha\beta}$ many-body Green function along with the correlation functions $\mathcal{G}^<_{\alpha\beta}$ and $\mathcal{G}^>_{\alpha\beta}$ can be defined as follows,

$$\mathcal{G}^>_{\alpha\beta}(x^+ - y^+) = -i\langle A_\alpha(x^+) A^\dagger_\beta(y^+)\rangle \tag{22}$$

$$\mathcal{G}^<_{\alpha\beta}(x^+ - y^+) = \mp(-i)\langle A^\dagger_\beta(y^+) A_\alpha(x^+)\rangle \tag{23}$$

$$\mathcal{G}_{\alpha\beta}(x^+ - y^+) = \theta(x^+ - y^+)\mathcal{G}^>_{\alpha\beta}(x^+ - y^+) + \theta(y^+ - x^+)\mathcal{G}^<_{\alpha\beta}(x^+ - y^+) \tag{24}$$

$$\mathcal{G}^a_{\alpha\beta}(x^+ - y^+) = \theta(y^+ - x^+)\mathcal{G}^>_{\alpha\beta}(x^+ - y^+) + \theta(x^+ - y^+)\mathcal{G}^<_{\alpha\beta}(x^+ - y^+) \tag{25}$$

The average $\langle \cdots \rangle$ is taken over the exact ground state. The Heisenberg operators $A_\alpha(x^+)$, see (8), could be build out of any number of field operators (fermions and/or bosons). For quark-antiquark correlations (e.g. pions) one may choose $A_\alpha(x^+) = \Psi_{\alpha_1}(x^+)\Psi^c_{\alpha_2}(x^+)$ with the anti-particle field Ψ^c. For the three-quark system considered here, $A_\alpha(x^+) = \Psi_{\alpha_1}(x^+)\Psi_{\alpha_2}(x^+)\Psi_{\alpha_2}(x^+)$. Note that all field operators are taken at equal (global) time, which sets the basis for the Dyson expansion also for multi-particle systems. It is understood that α are multi-indices labeling Dirac indices and additive quantum numbers. As mentioned before, in equilibrium, only one Green function is independent and one may alternatively use the thermodynamic Green function [17]

$$\mathcal{G}_{\alpha\beta}(\tau - \tau') = -\langle T_\tau A_\alpha(\tau) A^\dagger_\beta(\tau')\rangle. \tag{26}$$

where the time ordering is according to the value of τ, with the smallest at the right and the imaginary unit is dropped compared to (24). For $x^+ = -i\tau$ ("imaginary time") in (8) we achieve the thermodynamic Heisenberg picture. Dyson equations can be established for both forms [18]. In the light-front formalism the Dyson equations are given by [20]

$$i\frac{\partial}{\partial x^+}\mathcal{G}_{\alpha\beta}(x^+ - y^+) = \delta(x^+ - y^+)\langle[A_\alpha, A^\dagger_\beta]_\pm(x^+)\rangle$$
$$+ \sum_\gamma \int d\bar{x}^+ \mathcal{M}_{\alpha\gamma}(x^+ - \bar{x}^+) \mathcal{G}_{\gamma\beta}(\bar{x}^+ - y^+). \tag{27}$$

The mass matrix that appears in (27) is given by

$$\mathcal{M}_{\alpha\beta}(x^+ - y^+) = \delta(x^+ - y^+)\mathcal{M}_{0,\alpha\beta}(x^+) + \mathcal{M}_{r,\alpha\beta}(x^+ - y^+) \tag{28}$$

$$(\mathcal{M}_0\mathcal{N})_{\alpha\beta}(x^+) = \langle[[A_\alpha, H](x^+), A^\dagger_\beta(x^+)]_\pm\rangle \tag{29}$$

$$(\mathcal{M}_r\mathcal{N})_{\alpha\beta}(x^+ - y^+) = \sum_\gamma \langle T_{x^+}[A_\alpha, H](x^+), [A^\dagger_\beta, H](y^+)]\rangle_{\text{irreducible}} \tag{30}$$

where $\mathcal{N}_{\alpha\beta}(x^+) = \langle[A_\alpha, A_\beta^\dagger]_\pm(x^+)\rangle$. The first term in (28) is instantaneous and related to the mean field approximation, the second term is the retardation or memory term. This form suggests to first solve the mean field problem (neglecting memory) and evaluate higher order contributions involving $\mathcal{M}_{r,\alpha\beta}(x^+ - y^+)$ in the mean field basis. In this approximation (27) can be written in the following form

$$\left(i\frac{\partial}{\partial x^+} - \mathcal{M}_0(x^+)\right)\mathcal{G}(x^+ - y^+) = \delta(x^+ - y^+)\mathcal{N}(x^+), \qquad (31)$$

where $\mathcal{M}, \mathcal{G}$ etc. are understood as matrices acting in the index space of $\alpha, \beta \ldots$ As an example we give results for the ideal gas (of particles or quasiparticles). Following (almost) the notation of [2] the Fock space representation of H is given by

$$H = \sum \int dp^+ d^2 p_\perp \left[(u \cdot p - \mu)b^\dagger b + (u \cdot p + \mu)d^\dagger d\right], \qquad (32)$$

where the sum is over all quantum numbers relevant for the particles b or antiparticles d. It is now straight forward to evaluate $\mathcal{N}$ and $\mathcal{M}_0$ using free fermion field operators in (31), $A(x^+) = \Psi(x^+)$ and for fermions $A^\dagger(x^+) = \bar{\Psi}(x^+)$ [2]. The Fourier transform of the resulting single-particle light-cone time ordered Green function coincides with (20). The ensemble averages needed are

$$\langle b^\dagger b \rangle = f^+, \quad \langle bb^\dagger \rangle = 1 - f^+, \quad \langle d^\dagger d \rangle = f^-, \quad \langle dd^\dagger \rangle = 1 - f^-. \qquad (33)$$

4. TWO- AND THREE-BODY CORRELATIONS

To treat particle correlations we use Hartree-Fock approximation, i.e. the traces appearing in (31) are evaluated for a medium of independent particles. For particles that are part of a larger n-body correlation (presently $n = 2, 3$) it is convenient to introduce fractions $x = k^+/P_n^+$, where P_n^+ is the plus component of the cluster's c.m. momentum, $P_n^2 = M_n^2$. For a cluster at rest $P_n^+ = M_n$, where M_n is the mass of the cluster.

We use a zero range interaction [42, 43], which can be considered as the lowest order effective interaction appearing in a $1/N_c$ expansion of QCD. The Nambu Jona-Lasinio model is a chiral realization of such an approach. The gap equation, quark-antiquark and two-quark correlations have been considered elsewhere during this meeting [44]. In view of the technical difficulties related to the spin (see e.g. Ref. [45]), we presently used spin-averaged quarks to calculate the correlations. This leads to an effective bosonic Faddeev equation. Despite of this approximation the statistical properties of quarks are properly treated as fermionic. Because of the zero-range interaction the two-body t-matrix simplifies and summation of the diagrams leads to a geometrical series that can be represented by the following reduced t-matrix $t(M_2)$

$$t(M_2) = \left(i\lambda^{-1} - B(M_2)\right)^{-1}. \qquad (34)$$

The expression for $B(M_2)$ is represented by a loop, which in the rest system of the two-body system is given by

$$B(M_2) = -\frac{i}{(2\pi)^3} \int_{LB} \frac{dx d^2 k_\perp}{x(1-x)} \frac{1 - f^+(x, \vec{k}_\perp^2) - f^+(1-x, \vec{k}_\perp^2)}{M_2^2 - M_{20}^2}, \qquad (35)$$

where the virtual two-body mass is $M_{20}^2 = (\vec{k}_\perp^2 + m^2)/x(1-x)$ and $x = k^+/P_2^+$. Note that M_{20} is invariant for boosts along z-direction. The integral has a logarithmic singularity that is regularized using $M_{20}^2 < \Lambda^2$, where Λ is chosen as a multiple of the constituent quark mass. A bound state (if it appears) is given by the pole condition

$$i\lambda^{-1} = B(M_{2B}). \qquad (36)$$

The quark mass $m(\mu, T)$ as well as the two-body masses $M_2(\mu, T)$, $M_{2B}(\mu, T)$, and $M_{20}(\mu, T)$ depend on the chemical potential and the temperature of the medium. Also there is an additional dependence on the cluster's c.m. momentum related to the momentum dependence of the blocking factors. Hence several changes of kinematical variables are necessary in the two-body amplitude before eq. (34) can be used in a three-body cluster. They are discussed in the following for the three-body system at rest. The two-body mass is

$$M_2^2 = (P_3 - q)^2 = (M_3 - q^+)\left(M_3 - \frac{\vec{q}_\perp^2 + m^2}{q^+}\right) - \vec{q}_\perp^2, \qquad (37)$$

where q denotes the momentum of the third particle. Also the blocking factors depend on the cluster momentum and hence M_2 is different for a moving system not only because of eq. (37), but also due to medium effects that depend on the momentum. As a consequence the arguments of the blocking factors of eq. (35) have to be properly replaced by

$$2f^+(M_{20}) \to f^+(x, \vec{k}_\perp^2) + f^+(1 - x - y, (\vec{k} + \vec{q})_\perp^2). \qquad (38)$$

where $x = k^+/M_3$ and $y = q^+/M_3$ in the three-body rest system. The two-body t-matrix is determined with the above mentioned changes of arguments and implemented into the three-body equation.

The relativistic Faddeev-type equation for the vertex functions Γ on the light front including the effects of finite temperature and density can then be written as (compare Ref. [42, 43] for the isolated case)

$$\Gamma(y, \vec{q}_\perp) = \frac{i}{(2\pi)^3} t(M_2) \int_{LB} \frac{dx d^2 k_\perp}{x(1-y-x)}$$
$$\times \frac{1 - f^+(x, \vec{k}_\perp^2) + f^+(1-x-y, (\vec{k}+\vec{q})_\perp^2)}{M_3^2 - M_{30}^2} \Gamma(x, \vec{k}_\perp), \qquad (39)$$

The mass of the virtual three-particle state (in the rest system) is

$$M_{30}^2 = \frac{\vec{k}_\perp^2 + m^2}{x} + \frac{\vec{q}_\perp^2 + m^2}{y} + \frac{(\vec{k}+\vec{q})_\perp^2 + m^2}{1-x-y}, \qquad (40)$$

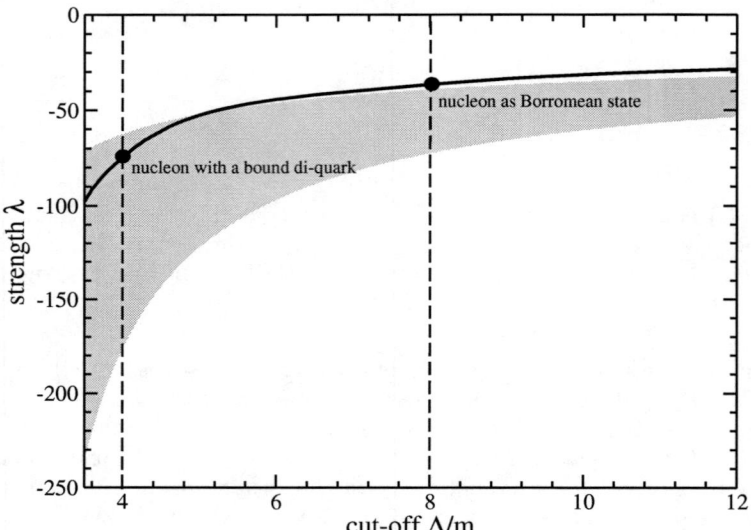

FIGURE 1. Solid line shows the dependence of coupling strength Λ on the cut-off parameter Λ/m_Q, $m = 336$ MeV for the isolated three-body system. The shaded area indicates parameter values, where a bound di-quark state exists, upper bound $M_2 = 2m$ (continuum) lower bound $M_2 = 0$. Vertical dashed lines indicate values chosen for $\Lambda = 4m$, nucleon includes a bound di-quark, $\Lambda = 8m$ without bound di-quark, i.e. the nucleon is a Borromean state.

which is the sum of the on-shell minus-components of the three particles. For consistency we use $M_{30}^2 < \Lambda^2$.

For the in-medium quark mass we use values of $m(\mu, T)$ given elsewhere during this meeting [44]. They are consistent with the instant form values provided earlier by a Nambu-Jona-Lasinio model [46]. That model's approximations are close to those used in the present solution of the relativistic three-body equation with a simple zero range interaction. We take $m = 336$ MeV for the isolated constituent mass and determine the coupling strength λ as a function of Λ to reproduce the nucleon mass $m_N = 938$ MeV. The resulting line is shown in Figure 1.

The two-body and three-body binding energies, B_2 and B_3 are shown in Figure 2 for a temperature of $T = 50$ MeV. The binding energies are defined by

$$B_3(\mu, T) = m(\mu, T) + M_{2B}(\mu, T) - M_{3B}(\mu, T) \qquad (41)$$
$$B_2(\mu, T) = 2m(\mu, T) - M_{2B}(\mu, T). \qquad (42)$$

For the parameter values chosen the nucleon is more stable than the di-quark. For $\Lambda = 8m$, no di-quark exists in this model even for the isolated case, compare Figure 1.

The three-body mass $M_3(\mu, T)$ as a function of the chemical potential for different temperatures and a cut-off parameter $\Lambda = 8m$ are shown in Figure 3. The chemical potential where the solid lines intersects the continuum define the Mott

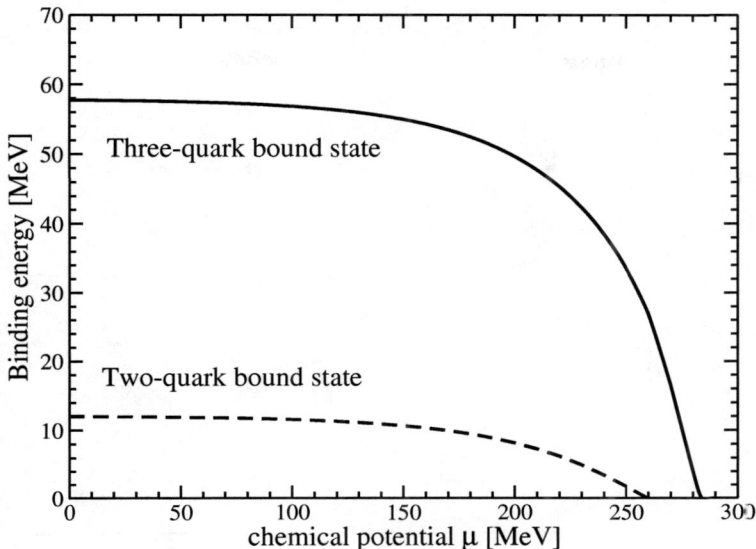

FIGURE 2. Binding energy of two-quark (dashed) and three-quark (solid) bound states for a temperature $T = 50$ MeV and $\Lambda = 4m$. The di-quark is weaker bound and vanishes at its Mott density before the nucleon is dissolved.

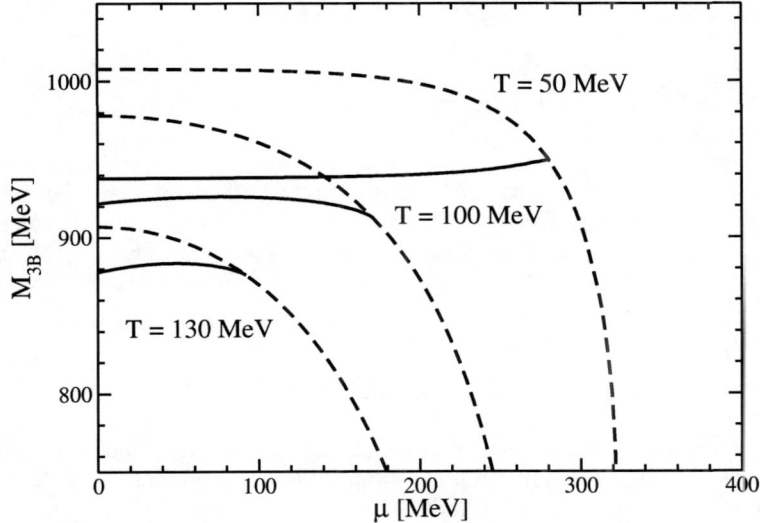

FIGURE 3. Binding energy of the three-quark bound state (solid) for different temperatures as a function of the chemical potential. The respective continuum lines (3m) are indicated as dashed line, $\Lambda = 8m$.

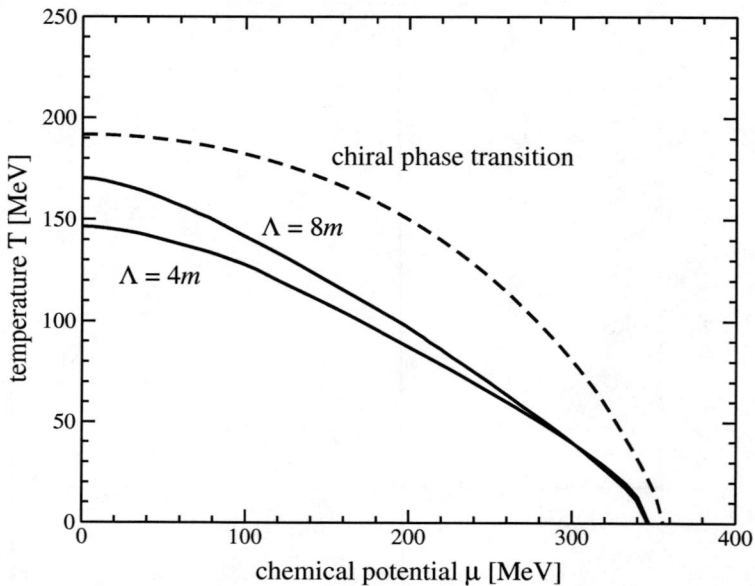

FIGURE 4. Mott line for the three-body system at rest in the medium for $\Lambda = 4m$ (with di-quark substate) and $\Lambda = 8m$ (Borromean nucleon) as extreme cases. For values of T and μ below the Mott lines three-body bound states can be formed. The chiral phase transition from the NJL model is indicated as dashed line, compare Ref [44]

transition. The condition is given by $M_3(\mu_{\text{Mott}}, T_{\text{Mott}}) = 3m(\mu_{\text{Mott}}, T_{\text{Mott}})$ for the three-quark continuum and $M_3(\mu_{\text{Mott}}, T_{\text{Mott}}) = m(\mu_{\text{Mott}}, T_{\text{Mott}}) + M_2(\mu_{\text{Mott}}, T_{\text{Mott}})$ for the quark-diquark continuum. Note that the three-quark bound state is at rest and the influence of the medium is expected to become smaller for a moving cluster, compare, e.g., Ref. [47] and refs. therein, where the α-particle has been investigated in the Dyson equation approach. The corresponding Mott lines are shown in Figure 4 along with the chiral phase transition line calculated in the NJL model [44].

5. CONCLUSION

We have shown that three-quark correlations can be investigated in a relativistic framework using in-medium Faddeev equations that include the dominant medium effects, such as Pauli blocking and self energy corrections (gap-equation). Several approximations have been done during the course of investigation, the most serious one is averaging the quark spin to achieve simpler boson-type Faddeev equations (than the more complicated fermionic Faddeev equation). The inclusion of spin is currently on it way. One important result is, that light front quantization is applicable to finite temperature systems and leads to meaningful results. Several lines have to be followed for future investigation.

- Include three-quark correlations in the distribution function, and the thermodynamical potentials to see how quark matter actually changes to nucleonic matter when the temperature is lowered.
- The parameter values chosen are fixed to the nucleon mass which eventually should give the strength for the two-quark correlation that is important in the studies of color superconductivity.
- In fact, influence of three-quark correlations has so far not been investigated in the context of color superconductivity. The Dyson approach provides the necessary tools to do so.
- Besides the correlation issues mentioned in this contribution, light front quantization may allow to use light front QCD in the context of finite temperature systems, such as the quark gluon plasma, which is a challenging and exciting possibility. See also contribution of J. Raufeisen to this meeting [37].

ACKNOWLEDGMENTS

It is a great pleasure to thank the organizers for the fruitful and exciting meeting. Work is supported by the Deutsche Forschungsgemeinschaft (DFG) and the HWP program of the Bildungsministerium des Landes Mecklenburg-Vorpommern.

REFERENCES

1. P.A.M. Dirac, Rev. Mod. Phys. **21** (1949) 392
2. S. J. Brodsky, H. C. Pauli and S. S. Pinsky, Phys. Rept. **301**, 299 (1998), and refs. therein.
3. 17th International Conference On Ultra Relativistic Nucleus-Nucleus Collisions (Quark Matter 2004) 11-17 Jan 2004, Oakland, California, proceedings, eds. Hans Georg Ritter and Xin-Nian Wang published: J. Phys. G: Nucl. Part. Phys. **30** S633-S1429.
4. F. Karsch, Nucl. Phys. A **698** (2002) 199 [arXiv:hep-ph/0103314].
5. F. Karsch, Lect. Notes Phys. **583** (2002) 209 [arXiv:hep-lat/0106019].
6. M. G. Alford, K. Rajagopal and F. Wilczek, Phys. Lett. B **422** (1998) 247.
7. M. G. Alford, K. Rajagopal and F. Wilczek, Nucl. Phys. B **537** (1999) 443.
8. R. Rapp, T. Schafer, E. V. Shuryak and M. Velkovsky, Phys. Rev. Lett. **81** (1998) 53.
9. K. Rajagopal and F. Wilczek, Shifman, M. (ed.): At the frontier of particle physics, vol. 3* 2061, hep-ph/0011333.
10. M. Gyulassy and L. McLerran, arXiv:nucl-th/0405013.
11. M. Gyulassy, arXiv:nucl-th/0403032.
12. H. Stoecker, arXiv:nucl-th/0406018.
13. X. N. Wang, arXiv:nucl-th/0405017.
14. E. V. Shuryak, arXiv:nucl-ph/0405066.
15. U. W. Heinz, AIP Conf. Proc. **739** (2005) 163 [arXiv:nucl-th/0407067].
16. Kadanov L P and Baym G 1962, *Quantum Statistical Mechanics* (Mc Graw-Hill, New York).
17. Fetter A L and Walecka J D 1971, *Quantum Theory of Many-Particle Systems* (McGraw-Hill, New York), and Dover reprints 2003.
18. Dukelsky J., Röpke G. and Schuck P. (1998), Nucl. Phys. **A 628** 17-40.
19. S. A. Sofianos and M. Beyer, arXiv:nucl-th/0408073.
20. M. Beyer, S. Mattiello, T. Frederico and H. J. Weber, Phys. Lett. B **521** (2001) 33.
21. S. Elser and A. C. Kalloniatis, Phys. Lett. B **375** (1996) 285 [arXiv:hep-th/9601045].

22. S. J. Brodsky, Fortsch. Phys. **50**, 503 (2002).
23. S. Mattiello, M. Beyer, T. Frederico and H. J. Weber, Few Body Syst. **31**, 159 (2002).
24. S. J. Brodsky, Nucl. Phys. Proc. Suppl. **108**, 327 (2002).
25. S. Mattiello, M. Beyer, T. Frederico and H. J. Weber, Few Body Syst. Suppl. **14** (2002) 379.
26. V. S. Alves, A. Das and S. Perez, Phys. Rev. D **66**, 125008 (2002).
27. M. Beyer, S. Mattiello, T. Frederico and H. J. Weber, Hirschegg Proceedings, arXiv:nucl-th/0202032.
28. M. Beyer, in Bled Workshops in Physics, eds. B. Golli, M. Rosina, S. Sirca, (Ljublana 2003) p.11, arXiv:hep-ph/0310324.
29. M. Beyer, S. Mattiello, T. Frederico and H. J. Weber, arXiv:hep-ph/0310222.
30. H. A. Weldon, Phys. Rev. D **67**, 085027 (2003).
31. H. A. Weldon, Phys. Rev. D **67**, 128701 (2003).
32. A. N. Kvinikhidze and B. Blankleider, hep-th/0305115.
33. A. Das and X. x. Zhou, Phys. Rev. D **68**, 065017 (2003).
34. J. Raufeisen and S. J. Brodsky, Phys. Rev. D **70** (2004) 085017 [arXiv:hep-th/0408108].
35. J. Raufeisen and S. J. Brodsky, arXiv:hep-th/0409157.
36. A. Das, J. Frenkel and S. Perez, arXiv:hep-th/0502243.
37. J. Raufeisen, these proceedings.
38. W. Israel, Annals Phys. **100**, 310 (1976).
39. W. Israel, Physics **106A**, 204 (1981).
40. H. A. Weldon, Phys. Rev. D **26**, 1394 (1982).
41. S.J. Chang, R.G. Root and T.-M. Yan, Phys. Rev. **D7**, (1973) 1133; S.J. Chang and T.-M. Yan, Phys. Rev. **D7** (1973) 1147.
42. T. Frederico, Phys. Lett. B **282**, 409 (1992).
43. W. R. de Araújo, J. P. de Melo and T. Frederico, Phys. Rev. C **52**, 2733 (1995).
44. S. Strauss, M. Beyer, S. Mattiello, these proceedings.
45. M. Beyer, C. Kuhrts and H. J. Weber, Annals Phys. **269** (1998) 129.
46. S. P. Klevansky , Rev. Mod. Phys. **64** (1992) 649.
47. M. Beyer, S. Strauss, P. Schuck and S. A. Sofianos, Eur. Phys. J. A **22** (2004) 261.

Hadronic resonances above the QCD phase transition

D. Blaschke

Bogoliubov Laboratory for Theoretical Physics, JINR Dubna, RU-141980 Dubna, Russia,
Present address: GSI Darmstadt, Planckstr. 1, D-64291 Darmstadt, Germany
E-mail: Blaschke@theory.gsi.de

Abstract. We describe matter formed in ultrarelativistic heavy ion collisions within a generalized Hagedorn resonance gas model where hadrons have a vanishing width below the Hagedorn temperature T_H and a Hagedorn spectrum-like width above T_H. Such an approach not only eliminates the divergence of the thermodynamic functions above T_H, but it is able to successfully describe the lattice quantum chromodynamics (QCD) data on the energy density. It also allows to explain the absence of heavy resonance contributions in the fit of the experimentally measured particle ratios at SPS and RHIC energies. We present an application of the approach to the description of the NA50 experiment which suggests that the anomalous suppression of J/ψ production may be explained by the increase of the effective number of degrees of freedom at the Hagedorn temperature. We estimate the shear viscosity of the generalized Hagedorn resonance gas model and find remarkable agreement with the perfect liquid bound $\eta/s \geq 1/(4\pi)$.

Keywords: QCD thermodynamics, Mott transition, in-medium resonance broadening, J/ψ suppression, viscosity bound
PACS: 12.38.Mh, 13.85.Ni, 14.40.Gx, 25.75.Dw

1. INTRODUCTION

The lattice QCD simulations not only provide the strongest theoretical support of the quark gluon plasma (QGP) existence, but they also give detailed information on the properties of strongly interacting matter over a wide range of temperatures. A recent analysis [1] of the lattice energy density showed that a resonance gas model can perfectly explain the steep rise in the number of degrees of freedom at $T \approx T_c$. On the other hand, lattice QCD has also revealed that hadronic correlations persist for $T > T_c$ [2]. The question arises whether it is more appropriate to describe hot QCD matter in terms of hadronic correlations rather than in terms of quarks and gluons. Therefore, in the present contribution, we would like to discuss a generalization of the Hagedorn resonance gas (statistical bootstrap) model which allows for the extension of a hadronic description above T_c.

The statistical bootstrap model (SBM) [3] is based on the hypothesis that hadrons are made of hadrons, with constituent and compound hadrons being treated on the same footing. This implies an exponentially growing form of the hadronic mass spectrum $\rho_H(m) \approx C_H m^{-a} \exp[m/T_H]$ for $m \to \infty$. The parameter T_H, the Hagedorn temperature, was interpreted as a limiting temperature reached at infinite energy density. The extensive investigation of the SBM has led to a formulation of both the important physical ideas and the mathematical methods for modern statistical mechanics of strongly interacting matter [4].

However, up to now the formulation of the SBM had some severe problems. The first one is the absence of a width for the heavy resonances. From the Particle Data Group [5] we know that heavy resonances with masses $m \geq 3.5$ GeV may have widths comparable to their masses. Taking the widths into account will effectively reduce the statistical weight of the resonance. The second problem arises while discussing the results of the hadron gas (HG) model [6, 7]. The HG model accounts for all strong decays of resonances according their partial width given in [5], and, hence, it describes remarkably well the light hadron multiplicities measured in nucleus-nucleus collisions at CERN SPS [6] and BNL RHIC [7] energies. This model is nothing else than the SBM of light hadrons which accounts for the proper volume of hadrons with masses below 2.5 GeV, but neglects the contribution of the exponentially growing mass spectrum.

Thus, one immediately faces a severe problem: "Why do the heavy resonances with masses above 2.5 GeV predicted by the SBM not contribute in the particle spectra measured in heavy-ion collisions at SPS and RHIC energies?" Note that the absence of heavy resonance contributions in the particle ratios cannot be due to the statistical suppression of the Hagedorn mass spectrum because the latter should not be strong in the quark-hadron phase transition region, where those ratios are believed to be formed [6, 7].

In the present contribution we suggest that the introduction of a finite width of the resonances can solve the above problems of the SBM. In the next section we formulate a simple statistical model that incorporates besides of the Hagedorn mass spectrum also medium dependent resonance widths due to the hadronic Mott effect, and analyze its mathematical structure. In Section 3 we discuss a model fit to recent lattice data of QCD thermodynamics [1] and some possible consequences for heavy-ion physics, whereas Section 4 is devoted to the discussion of the J/ψ anomalous suppression within the model developed here.

2. RESONANCE WIDTH MODEL: MOTT TRANSITION

According to QCD, hadrons are not elementary, point-like objects but rather color singlet bound states of quarks and gluons with a finite spatial extension of their wave function. While at low densities a hadron gas description can be sufficient, at high densities and temperatures, when hadronic wave functions overlap, nonvanishing quark exchange matrix elements between hadrons occur in order to fulfill the Pauli principle. This leads to a Mott-Anderson type delocalization transition with frequent rearrangement processes of color strings (string-flip [8]) so that hadronic resonances become off-shell with a finite, medium-dependent width. Such a Mott transition has been thoroughly discussed for light hadron systems in [9] and has been named *soft deconfinement*. The Mott transition for heavy mesons may serve as the physical mechanism behind the anomalous J/ψ suppression phenomenon [10].

We introduce the width Γ of a resonance in the statistical model with the Hagedorn mass spectrum through the spectral function

$$A(s,m) = N_s \frac{\Gamma\, m}{(s-m^2)^2 + \Gamma^2\, m^2}, \qquad (1)$$

a Breit-Wigner distribution of virtual masses with a maximum at $\sqrt{s} = m$ and the normalization factor

$$N_s = \left[\int_{m_B^2}^{\infty} ds \frac{\Gamma m}{(s-m^2)^2 + \Gamma^2 m^2}\right]^{-1} = \frac{1}{\frac{\pi}{2} + \arctan\left(\frac{m^2 - m_B^2}{\Gamma m}\right)}. \qquad (2)$$

The energy density of this model with zero resonance proper volume for given temperature T and baryonic chemical potential μ can be cast in the form

$$\begin{aligned}\varepsilon(T,\mu) &= \sum_{i=\pi,\rho,\ldots} g_i\, \varepsilon_M(T,\mu_i; m_i) \\ &+ \sum_{A=M,B} \int_{m_A}^{\infty} dm \int_{m_B^2}^{\infty} ds\, \rho_H(m)\, A(s,m)\, \varepsilon_A(T,\mu_A;\sqrt{s}),\end{aligned} \qquad (3)$$

where the energy density per degree of freedom with a mass m is

$$\varepsilon_A(T,\mu_A;m) = \int \frac{d^3 k}{(2\pi)^3} \frac{\sqrt{k^2+m^2}}{\exp\left(\frac{\sqrt{k^2+m^2}-\mu_A}{T}\right) + \delta_A}, \qquad (4)$$

with the degeneracy g_A and the baryonic chemical potential μ_A of hadron A. For mesons, $\delta_M = -1$, $\mu_M = 0$ and for baryons $\delta_B = 1$ and $\mu_B = \mu$, respectively. According to Eq. (3) the energy density of hadrons consists of the contribution of light hadrons for $m_i < m_A$ and the contribution of the Hagedorn mass spectrum $\rho_H(m)$ for $m \geq m_A$.

A new element of Eq. (3) in comparison to the SBM is the presence of the $\sqrt{s}$-dependent spectral function. The analysis shows that, depending on the behavior of the resonance width Γ in the limit $m \to \infty$, there are the following possibilities:

- For vanishing resonance width, $\Gamma = 0$, Eq. (3) evidently reproduces the usual SBM.
- For final values of the resonance width, $\Gamma = $ const, Eq. (3) diverges for all temperatures T because, in contrast to the SBM, the statistical factor in Eq. (3) behaves as $\{\exp[(m_B - \mu_A)/T] + \delta_A\}^{-1}$ so that it cannot suppress the exponential divergence of the Hagedorn mass spectrum $\rho_H(m)$.
- For a resonance width growing with mass like the Hagedorn spectrum $\Gamma \sim C_\Gamma \exp\left[\frac{m}{T_H}\right]$ or faster, Eq. (3) converges again.

Indeed, in the latter case the Breit-Wigner spectral function behaves as

$$N_s \frac{\Gamma m}{(s-m^2)^2 + \Gamma^2 m^2}\bigg|_{m\to\infty} \to \frac{2}{\pi\,\Gamma} \sim \exp\left(-\frac{m}{T_H}\right) \qquad (5)$$

and cancels the exponential divergence of the Hagedorn mass spectrum. Hence, the energy density remains finite. Note that both the analytical properties of model (3) and the right hand side of Eq. (5) remain the same, if a Gaussian shape of the spectral function is chosen instead of the Breit-Wigner one.

It can be shown that the behavior of the width at finite resonance masses is not essential for the convergence of the energy density (3). In other words, for a convergent energy density (3) above T_H it is sufficient to have a very small probability density (5) (or smaller) for a resonance of mass m to be found in the state with the virtual mass $\sqrt{s}$. Since there is no principal difference between the high and low mass resonances, we can use the same functional dependence of the width Γ for all masses. Thus, for the following model ansatz

$$\Gamma(T) = \begin{cases} 0, & \text{for } T \leq T_H, \\ C_\Gamma \left(\frac{m}{T_H}\right)^{N_m} \left(\frac{T}{T_H}\right)^{N_T} \exp\left(\frac{m}{T}\right), & \text{for } T > T_H, \end{cases} \quad (6)$$

the energy density (3) is finite for all temperatures and the divergence of the SBM is removed. At $T = T_H$, depending on choice of parameters, it may have either a discontinuity or its partial T derivative may be discontinuous. As discussed above, for $T \leq T_H$ such a model corresponds to the usual SBM, but for high temperatures $T > T_H$ it remains finite for a wide choice of powers N_m.

Note that for heavy resonances having the widths (6) the resulting mass distribution will be a power law which is seen both in hadron-hadron reactions [11] and nucleus-nucleus reactions [12] at high energies.

3. APPLICATIONS FOR LATTICE QCD AND HEAVY-ION COLLISIONS

3.1. Lattice QCD energy density

As one can see from Fig. 1 the Hagedorn gas model correctly reproduces the lattice QCD results below the critical temperature T_c and just in a vicinity above T_c, but not for large temperatures. Fig. 1 shows a comparison of the same lattice QCD data [1] with the Mott-Hagedorn gas (6) where the parameters of the spectral function are $N_T = 2.325$, $N_m = 2.5$ and $T_H = 165$ MeV and $m_A = m_B = 1$ GeV. The successful description of the lattice energy density [1] indicates that above T_c the strongly interacting matter may be well described in terms of strongly correlated hadronic degrees of freedom. This result is based on the concept of soft deconfinement and provides an alternative to the conventional explanation of the deconfinement transition as the emergence of quasifree quarks and gluons.

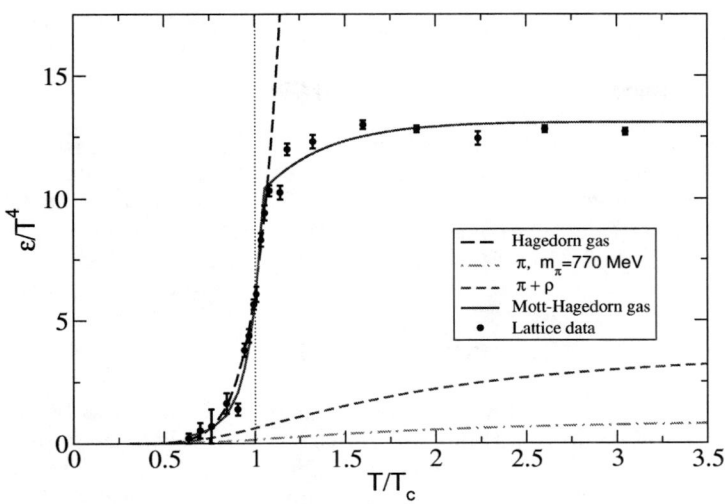

FIGURE 1. Fit of the lattice QCD data [1] with the Mott-Hagedorn resonance gas model (6). For details see text.

3.2. Fast equilibration and spectra in heavy-ion collisions

Another interesting feature of the model (6) is that it allows to explain naturally the absence of heavy resonance contributions to the particle yields measured at highest SPS and all RHIC energies, where QGP conditions are expected [6, 7]. In order to find out whether a given resonance has a chance to survive till the freeze-out it is necessary to compare its lifetime with the typical timescale in the system. There are two typical timescales usually discussed in nucleus-nucleus collisions, the equilibration time τ_{eq} and the formation time τ_f. The equilibration time tells when the matter created in collision process reaches a thermal equilibrium which allows to use the hydrodynamic and thermodynamic descriptions. For Au + Au collisions at RHIC energies it was estimated to be about $\tau_{eq} \approx 0.5$ fm [13]. On the other hand in transport calculations the formation time is used: the time for constituent quarks to form a hadron. The formation time depends on the momentum and energy of the created hadron, but is of the same order $\tau_f \approx 1-2$ fm [14] as the equilibration time.

Since within our model the QGP is equivalent to a resonance gas with medium dependent widths, all hadronic resonances with life time $\Gamma^{-1}(m)$ shorter than $\max\{\tau_f, \tau_{eq}\}$ will have no chance to be formed in the system. Therefore, the upper limit of the the integrals over the resonance mass m and over the virtual mass $\sqrt{s}$

in Eq. (3) should be reduced to a resonance mass defined by

$$\Gamma(m)^{-1} = \max\{\tau_f, \tau_{eq}\}. \tag{7}$$

This reduction may essentially weaken the energy density gap at the transition temperature or even make it vanish. Thus, the explicit time dependence should be introduced into the resonance width model (3) while applying it to nuclear collisions, and this finite time (size) effect, as we discussed, may change essentially the thermodynamics of the hadron resonances formed in the nucleus-nucleus collisions.

3.3. Anomalous J/ψ Suppression

The phenomenon of anomalous J/ψ suppression [15] as observed by the NA50 collaboration at CERN-SPS with ultrarelativistic Pb-Pb collisions at 158 GeV/A did not yet find a satisfactory explanation. It has been demonstrated that the extrapolation from results with pA collisions and light ion beams (O, S) fails to describe the E_T dependence of the J/ψ survival probability (production cross section) above $E_T \sim 40$ GeV Successful fits to the data are obtained with models assuming a critical phenomenon where the conditions for the onset of the new phase are fulfilled at the above E_T value. Among these models are in particular:

- percolation (Satz)
- J/ψ Mott transition (Blaizot/Ollitrault)
- D-meson Mott transition (Blaschke/Burau)

Here we would like to generalize the kinetic theory approach to J/ψ suppression as it was formulated in Ref. [10] and study charmonium dissociation in a Mott-Hagedorn resonance gas. Besides the impact by off-shell π and ρ mesons which has been studied in [10], the resonace gas (3)-(6) consists of more massive resonances so that no reaction threshold occurs for the breakup into open-charm hadrons (D-mesons, Σ_c, Λ_c, excited states).

The key quantities for the solution of a kinetic equation of the charmonium distribution function are the rate coefficients which can be estimated within the binary collision approximation. This approximation shall be applicable since the first collision is expected to destroy the charmonium and thus determines the lifetime as long as the charm density is too low for the reverse process of charmonium gain (fusion of $D\bar{D}$, $\Lambda_c \bar{D}$, ...).

We estimate the charmonium lifetime to be

$$\tau_\psi^{-1}(T) = \sigma_{\psi R \to X} v n_R(T) \tag{8}$$

under the assumption that the charmonium breakup cross section is rather universal and may be estimated by the value for ψ absorption measured in pA collision experiments of charm hadroproduction. In pA and S-U collisions an absorption cross section σ_{abs} of 4.4 ± 0.5 mb has been extracted from the data by using a Glauber model analysis [15]. The velocity of a typical hadron resonance

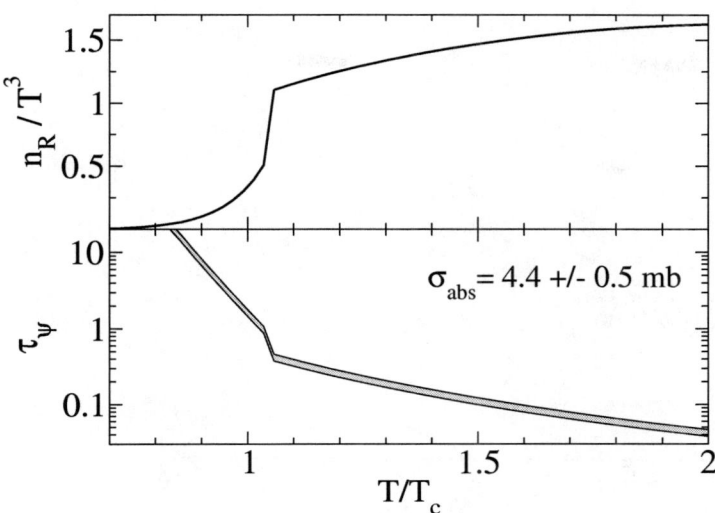

FIGURE 2. Density of resonances (upper panel) and mean J/ψ lifetime (lower panel) in a Mott-Hagedorn resonance gas.

is of the order of 0.5 c and the number density of resonances at a given temperature is given by

$$n_R(T) = \sum_{i=\pi,\rho,..} g_i\, n_M(T,\mu;m_i)$$
$$+ \sum_{A=M,B} \int_{m_A}^{\infty} dm \int_{m_B^2}^{\infty} ds\, \rho_H(m)\, A(s,m)\, n_A(T,\mu_A;\sqrt{s})\,, \qquad (9)$$

where the number density per degree of freedom is as follows

$$n_A(T,\mu_A;m) = \int \frac{d^3k}{(2\pi)^3} \frac{1}{e^{(\sqrt{k^2+m^2}-\mu_A)/T} + \delta_A}\,, \qquad (10)$$

see Fig. 2 (upper panel). The result for the charmonium lifetime is given in Fig. 2 (lower panel) and exhibits a sharp drop at the critical temperature for the Mott transition in the resonance gas. This behavior is then reflected in a threshold-like behavior of the charmonium survival probability

$$S(T_0) = \exp\left(-\int_{\tau_0}^{\infty} d\tau\, \tau_\psi^{-1}(T(\tau))\right)\,, \qquad (11)$$

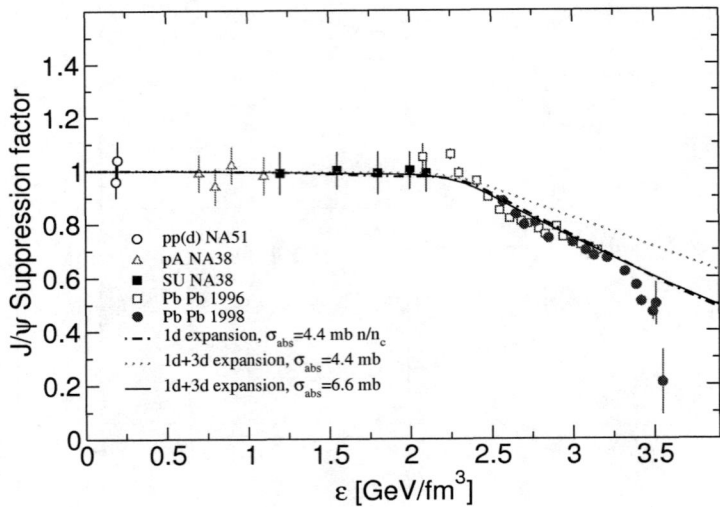

FIGURE 3. J/ψ survival probability in a Mott-Hagedorn resonance gas as a function of the energy density. Data are from the NA50 collaboration [15].

estimated within the one-dimensional hydrodynamic expansion (Bjorken scenario) with the following dependence $T^3(\tau)\tau = T^3(\tau_0)\tau_0$ of the temperature of the system on the proper time τ.

The result is shown in Fig. 3. As soon as the initial temperature $T_0 = T(\tau_0)$ is above the critical one, the survival probability drops according to the increase in the effective density of degrees of freedom measured by $n_R(T)$, see Fig. 2. Therefore, we suggest that the step-like drop in the J/ψ production cross section in Pb-Pb collisions at SPS energies can be explained by the dramatic increase in the number of hadronic resonances at the Hagedorn temperature in the resonance gas, see also [23].

3.4. Ideal fluid properties

The radial and elliptic flow observed in ultrarelativistic heavy ion collisions at RHIC can be well described by ideal hydrodynamics with a very small viscosity, $\eta/s \approx 0.1 - 0.2$ [24, 25, 26] which is very similar to the KSS bound [27] for this number, $1/(4\pi)$, obtained in CFT treatment of N=4 supersymmetric Yang-Mills theory. Such a small number for the viscosity to entropy ratio has recently been found in lattice simulations [28], see Fig. 4. While in Refs. [24, 25] this deviation from the behavior of a gas of weakly interacting quarks and gluons is attributed to the occurence of weakly bound states, we want to point out that this behavior finds a natural explanation within the Mott-Hagedorn resonance gas model. For temperatures just above the critical one, the resonant massive correlations with short lifetime lead to a liquid-like behavior of the hot and dense matter. Such a behavior has been described long ago within the string-flip model of quark matter [8], where concepts from liquid theory have been applied. Similar to Ref. [24] we use the Einstein formula

$$D = \frac{k_B T}{6\pi r_0 \eta} \quad (12)$$

for a spherical particle with radius r_0 in a suspension to derive for the viscosity-to-entropy ratio

$$\frac{\eta}{s} = \frac{\lambda \sigma}{2\pi <v>} \frac{n/T^3}{\varepsilon/T^4} . \quad (13)$$

We have used that the diffusion constant is $D = <v> l_{\text{mfp}}/3 = <v>/(n\sigma)$, where the mean free path l_{mfp} has been estimated using a typical hadron resonance cross section σ and the resonance density n, given in Fig. 2, while $s = k_B n$ is the entropy density. Motivated by a string-type linear confining potential we assume that the resonance radius scales with the mass, $r_0 = m/\lambda = \varepsilon/(n\lambda)$, with the string tension $\lambda \sim 1$ GeV/fm. Using the numerical results $\varepsilon/(nT) \leq 10$, $<v> \sim <nv>/<n> \leq 1$ obtained for the Mott-Hagedorn resonance gas in the temperature domain $1 < T/T_c < 2$, we find astonishing correspondence to the previous estimates for the viscosity-to-entropy ratio, see Fig. 4,

$$\frac{\eta}{s} \geq \frac{1}{4\pi} \left[\frac{\sigma}{10 \text{mb}} \right] , \quad (14)$$

in remarkable agreement with the KSS bound $\eta/s \geq 1/(4\pi)$.

4. CONCLUSIONS AND REMARKS

The statistical bootstrap model allows one to interpret the QGP as the hadron resonance gas dominated by the state of infinite mass (and infinite volume). As we show, it is necessary to include the resonance width into the SBM in order to avoid the contradiction with experimental data on hadron spectroscopy. We found that the simple model (3)-(6) may not only eliminate the divergence of the thermodynamic functions above T_H, but it is also able to successfully describe the lattice QCD data [1] for energy density. Such a model also explains the absence of heavy resonance contributions in the fit of the experimentally measured particle ratios at SPS and RHIC energies.

However, such a modification of the SBM requires an essential change in our view on QGP: it is conceivable that hadrons of very large masses which should be associated with a QGP cannot be formed in nucleus-nucleus collisions because of their very short lifetime.

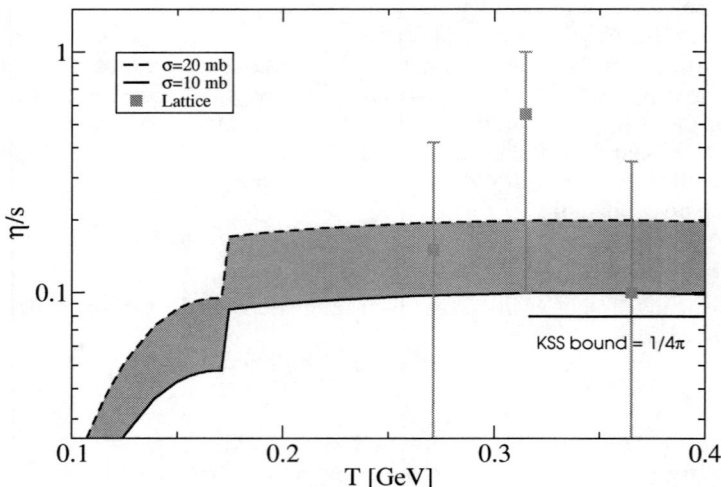

FIGURE 4. The ratio of shear viscosity and entropy in a Mott-Hagedorn resonance gas according to Eq. (13) for two choices of a phenomenological hadron-hadron cross section: $\sigma = 10$ mb (solid), $\sigma = 10$ mb (dashed). Also shown are the KSS bound [27] and Lattice data from Ref. [28] for $16^3 \times 8$ lattices.

We also showed that the dramatic increase in the number of hadronic resonances at the Hagedorn temperature of this model can explain the step-like drop in the J/ψ production cross section in Pb-Pb collisions at SPS energies [23].

It is also necessary to mention that the presented model should be applied to experimental data with care: it can be successfully applied to describe either the quantities associated with the chemical freeze-out, i.e. particle ratios or spectra of Ω hyperons, ϕ, J/ψ and ψ' mesons that are freezing out at hadronization [16, 17, 18, 19]. But as discussed in Refs. [20, 21, 22] the model presented here should not be used for the post freeze-out momentum spectra of other hadrons produced in the nucleus-nucleus collisions. Perhaps only such weakly interacting hadrons like Ω, ϕ, J/ψ and ψ' will allow us to test the model presented here.

ACKNOWLEDGMENTS

The work reported here is the result of a collaboration with Kyrill Bugaev, which I have the pleasure to acknowledge. I am also grateful to the many discussions about this model with colleagues at Bielefeld University, in particular Frithjof Karsch, Krzysztof Redlich, Abdelnasser Tawfik, and at the meetings of the Virtual

Institute of the Helmholtz Association "Dense hadronic matter and QCD phase transition", VH-VI-041.

REFERENCES

1. F. Karsch, K. Redlich and A. Tawfik, *Eur. Phys. J.*, **C 29**, 549 (2003).
2. I. Wetzorke et al., *Nucl. Phys. Proc. Suppl.*, **106**, 510 (2002);
   M. Asakawa, T. Hatsuda, Y. Nakahara, *Nucl. Phys.*, **A 715**, 863 (2003).
3. R. Hagedorn, *Suppl. Nuovo Cimento*, **3**, 147 (1965);
   R. Hagedorn and J. Ranft, *Suppl. Nuovo Cimento*, **6**, 169 (1968).
4. more references can be found in J. Letessier, J. Rafelski and A. Tounsi, *Phys. Lett.*, **B 328**, 499 (1994); and D. B. Blaschke and K. A. Bugaev, *Fizika*, **B 13**, 491 (2004).
5. Particle Data Group, *Phys. Rev.*, **D 66** (2002).
6. P. Braun–Munzinger, I. Heppe and J. Stachel, *Phys. Lett.*, **B 465**, 15 (1999);
   G. D. Yen and M. I. Gorenstein, *Phys. Rev.*, **C 59**, 2788 (1999);
   F. Becattini et al., *Phys. Rev.*, **C 64**, 024901 (2001).
7. P. Braun–Munzinger et al., *Phys. Lett.*, **B 518**, 41 (2001);
   N. Xu and M. Kaneta, *Nucl. Phys.*, **A 698**, 306 (2002).
8. D. Blaschke et al., *Phys. Lett.*, **151 B**, 439 (1985);
   G. Röpke, D. Blaschke and H. Schulz, *Phys. Rev.*, **D 34**, 3499 (1986).
9. J. Hüfner, S.P. Klevansky and P. Rehberg, *Nucl. Phys.*, **A 606**, 260 (1996).
10. G. Burau, D. Blaschke and Yu. Kalinovsky, *Phys. Lett.*, **B 506**, 297 (2001).
11. M. Gazdzicki and M. I. Gorenstein, *Phys. Lett.*, **B 517**, 250 (2001).
12. J. Schaffner-Bielich et al., *Nucl. Phys.*, **A 705**, 494 (2002).
13. X.-N. Wang, M. Gyulassy and M. Plümer, Phys. Rev. **D 51** (1995) 3436;
    R. Baier et al., *Phys. Lett.*, **B 345**, 277 (1995).
14. see, for instance, S. A. Bass et al., *Prog. Part. Nucl. Phys.*, **41**, 225 (1998).
15. M. C. Abreu et al. [NA50 Collaboration], *Phys. Lett. B*, **477**, 28 (2000).
16. more references can be found in K. A. Bugaev, M. Gazdzicki, M.I. Gorenstein, *Phys. Lett.*, **B 523**, 255 (2001); *Phys. Lett.*, **B 544**, 127 (2002); and *Phys. Rev.*, **C 68**, 017901 (2003).
17. M. I. Gorenstein, K. A. Bugaev and M. Gazdzicki, *Phys. Rev. Lett.*, **88**, 132301 (2002).
18. S. Bass and A. Dumitru, *Phys. Rev.*, **C 61**, 064909 (2000).
19. D. Teaney, J. Lauret and E. V. Shuryak, *Phys. Rev. Lett.*, **86**, 4783 (2001); and nucl-th/0110037 (2001).
20. K. A. Bugaev, *Nucl. Phys.*, **A 606**, 559 (1996); *J. Phys.*, **G 28** 1981 (2002); *Phys. Rev. Lett.*, **90**, 252301 (2003).
21. K. A. Bugaev and M. I. Gorenstein, nucl-th/9903072.
22. K. A. Bugaev, M. I. Gorenstein and W. Greiner, *J. Phys.*, **G 25**, 2147 (1999); *Heavy Ion Phys.*, **10**, 333 (1999).
23. D. B. Blaschke and K. A. Bugaev, *Prog. Part. Nucl. Phys.*, **53**, 197 (2004).
24. E. Shuryak, Prog. Part. Nucl. Phys. **53** (2004) 273 [arXiv:hep-ph/0312227].
25. E. V. Shuryak and I. Zahed, Phys. Rev. C **70** (2004) 021901 [arXiv:hep-ph/0307267].
26. D. A. Teaney, *J. Phys. G*, **30**, S1247 (2004).
27. G. Policastro, D.T. Son and A.O. Starinets, *Phys. Rev. Lett.*, **87**, 081601 (2001); P. Kovtun, D.T. Son and A.O. Starinets, *JHEP*, **0310**, 064 (2003).
28. A. Nakamura and S. Sakai, *Phys. Rev. Lett.*, **94**, 072305 (2005).

RG flow of the Polyakov-loop potential
- first status report -

J. Braun*, H. Gies* and H.-J. Pirner*

Institute for Theoretical Physics, University of Heidelberg, 69120 Heidelberg

Abstract. We study SU(2) Yang-Mills theory at finite temperature in the framework of the functional renormalization group. We concentrate on the effective potential for the Polyakov loop which serves as an order parameter for confinement. In this first status report, we focus on the behaviour of the effective Polyakov-loop potential at high temperatures. In addition to the standard perturbative result, our findings provide information about the "RG improved" backreactions of Polyakov-loop fluctuations on the potential. We demonstrate that these fluctuations establish the convexity of the effective potential.

Keywords: renormalization group, quark deconfinement, gauge theories
PACS: 64.60.Ak, 25.75.Nq, 11.15.-q

1. INTRODUCTION

An understanding of strongly interacting matter at finite temperature is a prominent problem of contemporary physics that deserves to be analyzed with great effort in view of the current and future experiments at heavy ion colliders. Since the forces between quarks as elementary constituents are governed by a non-Abelian gauge theory, already the understanding of gauge boson dynamics is an important challenge. In this latter case of pure gluodynamics, the expected transition to a deconfined phase can be studied with the aid of the Polyakov loop [1], being the order parameter for this transition:

$$\mathcal{P}(\vec{x}) = \frac{1}{N} \text{Tr}_F P \exp\left(i\bar{g} \int_0^\beta A_0^a(\vec{x},t) t^a dt\right). \tag{1}$$

Here t^a are the generators of $SU(N)$ and β denotes the inverse temperature. The subscript F alludes to the fundamental representation. The negative logarithm of the Polyakov-loop expectation value can be interpreted as the free energy of a single static fundamental color source [2]. In this sense, an infinite free energy associated with confinement is indicated as $\langle \mathcal{P} \rangle \to 0$, whereas $\langle \mathcal{P} \rangle \neq 0$ signals deconfinement.

Moreover, $\langle \mathcal{P} \rangle$ measures whether center symmetry, a discrete symmetry of Yang-Mills theory, is realized by the thermodynamic ensemble [2, 3]. Gauge transformations which differ at Euclidean times $x_0 = 0$ and $x_0 = \beta$ by a center element of the gauge group change $\mathcal{P}$ by a phase $e^{2\pi i k/N}$, k integer, but leave the action and the functional integration measure invariant. This implies that a center-symmetric ground state automatically ensures $\langle \mathcal{P} \rangle = 0$, whereas deconfinement $\langle \mathcal{P} \rangle \neq 0$ is related to the breaking of this symmetry.

In fact, lattice simulations have not only collected strong evidence for a second order phase transition in $SU(2)$ Yang-Mills theory [4, 5], but reveal moreover that the critical exponents agree with those of a 3D $Z(2)$ Ising model [6]. The latter corresponds exactly to the conjectured universality class obtained from the Polyakov-loop criterion [7].

In recent years, an effective theory consisting of gauge-invariant powers of the Polyakov loop has been developed based on mean-field arguments, see [8] for an overview. Such considerations show a good agreement with lattice data [9]. Moreover, inverse Monte-Carlo techniques have recently facilitated a precise lattice determination of the Polyakov-loop effective action for SU(2) [10].

A perturbative calculation of the effective potential V for the order parameter was first performed by Weiss [11]. For this, it is convenient to work with the "Polyakov gauge", which rotates the zeroth component A_0 of the gauge field into the Cartan subalgebra of SU(N); furthermore, the condition $\partial_0 A_0 = 0$ is imposed. Focussing solely on the Polyakov loop, it suffices to consider a single scalar degree of freedom $\phi(\vec{x})$, defined by

$$A(\vec{x})^a_\mu = n^a \delta_{\mu 0} \phi(\vec{x}). \tag{2}$$

Here n^a denotes a constant unit vector in color space, e.g. $n^a = \delta^{a3}$. Using this gauge, the order parameter is fully determined by ϕ,

$$\mathcal{P}(\vec{x}) = \cos\left(\frac{\beta \bar{g} \phi(\vec{x})}{2}\right), \tag{3}$$

and, consequently, the effective potential can purely be expressed in terms of ϕ, being a compact variable $\beta \bar{g} \phi \in [0, 2\pi]$. To leading order in a derivative expansion, the effective potential yields [11]

$$\beta^4 V(\beta \bar{g} \phi) = -\sum_{n=1}^{\infty} \frac{4}{n^4 \pi^2} \cos(n \beta \bar{g} \phi), \tag{4}$$

displayed here as a Fourier-cosine series, and depicted in Fig. 1. The potential has minima at $\beta \bar{g} \phi = 2\pi n$ and is $Z(2)$-symmetric, i.e., invariant under $\beta \bar{g} \phi \to 2\pi - \beta \bar{g} \phi$. Furthermore the order parameter is finite for $\beta \bar{g} \phi = 2\pi n$ and therefore the $Z(2)$-symmetry is spontaneously broken and the system is in the deconfined phase. This perturbative result agrees with the expectation that perturbation theory holds for high temperature where the coupling is small. It fails to describe the confinement phase.

In the confined phase, the potential should have its minimum at $\beta \bar{g} \phi = \pi$, implying a vanishing order parameter. A sketch of a possible form of the potential with a finite IR regulator (to circumvent the convexity obstruction, see below) is shown in Fig. 2 for illustration.

Various generalizations to Weiss's result have been worked out within perturbation theory, for instance, the inclusion of a magnetic background field [12] or higher-order derivative expansions [13] to name a few. However, in order to investigate the transition to the confinement phase, reliable nonperturbative tools are

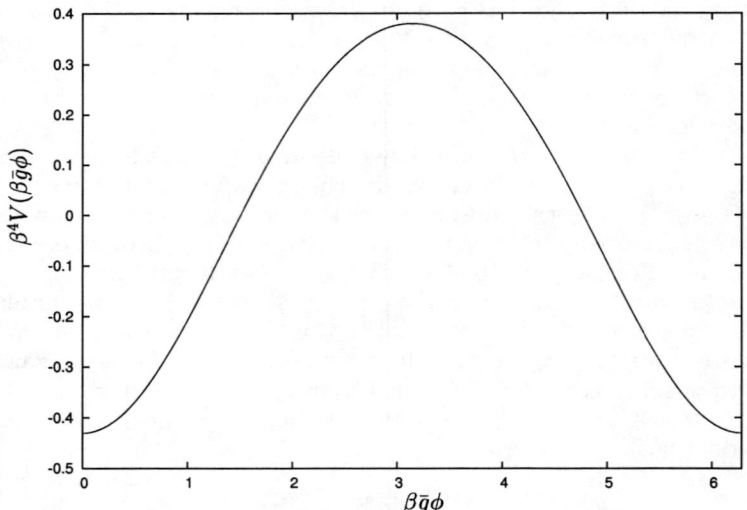

FIGURE 1. Perturbative effective Polyakov-loop potential Eq. (4) [11].

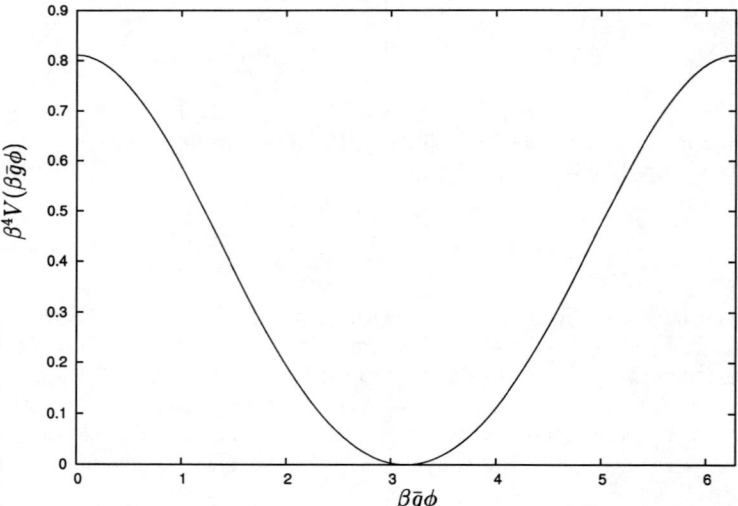

FIGURE 2. A sketch of a possible form of the effective potential in the confining phase.

required. We will base our study on the functional (or "Exact") renormalization group [14] formulated in terms of a flow equation for the effective action [15].

The paper is organized as follows: In Sect. 2 we briefly review RG flow equations for Yang-Mills theories and present the flow equation for the effective potential of the order parameter in a propertime approximation. The flow of the potential is analyzed in Sect. 3. Conclusions and future directions are discussed in Sect. 4.

2. FLOW EQUATION FOR THE POLYAKOV-LOOP POTENTIAL

2.1. Exact Renormalization Group

In the flow equation approach, we consider the effective average action Γ_k which includes all quantum fluctuations with momenta $|p| > k$, with the scale k serving as an infrared (IR) regularization. The boundary condition for the flow equation is fixed at an ultraviolet (UV) scale Λ in terms of the bare action Γ_Λ to be quantized. Quantum fluctuations are successively integrated out by lowering the scale k. In the limit $k \to 0$, the full quantum effective action $\Gamma_{k \to 0}$, i.e., the generating functional of the 1PI Green's functions, is obtained. The flow of Γ_k, i.e., the RG trajectory from the UV scale Λ to the deep IR, is obtained from a functional differential equation [15].

Flow equations for gauge theories require a careful control of gauge invariance, because the IR regulator scale k introduces sources of gauge-symmetry breaking in addition to standard gauge-fixing terms. Nevertheless, standard gauge symmetry can be obtained in the physical limit of vanishing regulator scale $k \to 0$ by controlling the symmetry constraints with the aid of regulator-modified Ward-Takahashi identities [16]. In this paper, we employ the flow equation with the background-field method [17, 18] for a simplified control of gauge invariance within our approximation. We work along the lines of [17, 19] where this technique has been used for zero-temperature gluodynamics. The flow equation for the effective average action reads [17]

$$k \partial_k \Gamma_k[A, \bar{A}] \equiv \partial_t \Gamma_k[A, \bar{A}] = \frac{1}{2} \text{Tr} \frac{\partial_t R_k(\Gamma_k^{(2)}[\bar{A}, \bar{A}])}{\Gamma_k^{(2)}[A, \bar{A}] + R_k(\Gamma_k^{(2)}[\bar{A}, \bar{A}])}. \tag{5}$$

Here the trace runs over all internal indices including momenta. The classical gauge field is given by A, whereas the background-field is denoted by $\bar{A}$ (the ghost fields are not displayed for brevity). $\Gamma_k^{(2)}$ denotes the second functional derivative with respect to the fluctuating fields. The regulator function R_k implements the IR regularization at the scale k, see below. Inserting the background-field dependent $\Gamma^{(2)}$ into the regulator leads to an adjustment of the regularization to the spectral flow of the fluctuations as discussed in [19], implying a potential improvement when it comes to approximations.

The boundary condition for the effective action at the UV scale Λ consists of:

$$\Gamma_\Lambda[A, \bar{A}] = \Gamma^{cl}[A, \bar{A}] + \Gamma_\Lambda^{gf}[A, \bar{A}] + \Gamma_\Lambda^{gh}[A, \bar{A}], \tag{6}$$

with the bare Yang-Mills action Γ^{cl}. The gauge-fixing and ghost terms are given by

$$\Gamma_\Lambda^{gf}[A,\bar{A}] = \frac{1}{2\xi_\Lambda}\int d^d x (D_\mu[\bar{A}]a_\mu)^2, \quad \Gamma_\Lambda^{gh}[A,\bar{A}] = -\int d^d x \bar{c} D_\mu[\bar{A}]D_\mu[A]c, \quad (7)$$

respectively, with the quantum fluctuations a_μ defined by $a = A - \bar{A}$. The gauge parameter is denoted by ξ_k.

Let us briefly summarize a few properties of the regulator function R_k that can conveniently be written as

$$R_k(x) = xr(y), \quad y := \frac{x}{\mathcal{Z}_k k^2}, \quad (8)$$

with $r(y)$ being a dimensionless regulator shape function of dimensionless argument. Here $\mathcal{Z}_k$ denotes a wave-function renormalization. Note that both R_k and $\mathcal{Z}_k$ are matrix-valued in field space.

The IR regularization is implemented by the property

$$\lim_{x/k^2 \to 0} R_k(x) = \mathcal{Z}_k k^2 \quad \Leftrightarrow \quad r(y) \xrightarrow{y \to 0} \frac{1}{y}. \quad (9)$$

The second and third properties of the regulator read

$$\lim_{k^2/x \to 0} R_k(x) = 0 \quad \text{and} \quad \lim_{k \to \Lambda} R_k(x) \to \infty, \quad (10)$$

and ensure that the regulator is removed in the limit $k \to 0$ and that the initial bare action is approached at the UV scale Λ. Moreover, the second property guarantees that the regulator vanishes for modes with $|p| \gg k$, i.e., the theory is not affected by the regulator for large momenta.

The flow equation (5) can be mapped onto a generalized propertime representation, once the background field is identified with the full quantum field[1] [21, 19]:

$$\partial_t \Gamma_k[A=\bar{A},\bar{A}] = \frac{1}{2}\mathrm{STr}\frac{\partial_t R_k(\Gamma_k^{(2)})}{\Gamma_k^{(2)} + R_k(\Gamma_k^{(2)})} = \frac{1}{2}\int_0^\infty ds\,\mathrm{STr}\hat{f}(s,\eta)\exp\left(\frac{s}{k^2}\Gamma_k^{(2)}\right), \quad (11)$$

where the operator $\hat{f}(s,\eta)$ is given by[2]

$$\hat{f}(s,\eta) = \tilde{h}(s)(2-\eta) - (\tilde{h}(s)-\tilde{g}(s))(2-\eta) + (\tilde{H}(s)-\tilde{G}(s))\frac{1}{s}\partial_t. \quad (12)$$

---

[1] This identification itself involves an approximation for $\Gamma^{(2)}$ in the denominator of the flow equation, as discussed in more detail in [17, 21, 19].

[2] Terms proportional to $(\tilde{h}(s)-\tilde{g}(s))$ and $(\tilde{H}(s)-\tilde{G}(s))$ in $\hat{f}(s,\eta)$ arise due to the use of the chain rule for $\partial_t R_k$ in the flow equation (5) and manifestly represent terms arising from the spectral adjustment of the flow $\sim \partial_t \Gamma^{(2)}$.

In addition, we have introduced the (matrix-valued) anomalous dimension

$$\eta := -\partial_t \ln \mathcal{Z}_k = -\frac{1}{\mathcal{Z}_k} \partial_t \mathcal{Z}_k. \tag{13}$$

The connection between the function $\hat{f}(s,\eta)$ and the regulator function R_k is given by the functions $\tilde{g}(s)$, $\tilde{G}(s)$ and $\tilde{H}(s)$ which are defined as:

$$h(y) = \frac{-yr'(y)}{1+r(y)}, \quad h(y) = \int_0^\infty ds\, \tilde{h}(s) e^{-ys}, \quad \frac{d}{ds}\tilde{H}(s) = \tilde{h}(s), \quad \tilde{H}(0) = 0, \tag{14}$$

$$g(y) = \frac{r(y)}{1+r(y)}, \quad g(y) = \int_0^\infty ds\, \tilde{g}(s) e^{-ys}, \quad \frac{d}{ds}\tilde{G}(s) = \tilde{g}(s), \quad \tilde{G}(0) = 0. \tag{15}$$

The generalized propertime representation (11) of the flow equation has the advantage that the evaluation of the trace becomes considerably simplified.

2.2. Truncated Flow Equations

Even in the propertime form of Eq. (11), the flow equation cannot be solved in closed form, which necessitates further approximations. For this, we truncate the space of action functionals down to a set of operators that are considered to represent the relevant degrees of freedom for the system or at least for a particular parametric regime of the system.

For obtaining a first glance at finite-temperature gluodynamics, we concentrate on the Polyakov-loop potential, employing the simple ansatz:

$$\Gamma_k[A,\bar{A}] = \int d^d x \left\{ \frac{Z_k}{4} F^a_{\mu\nu} F^a_{\mu\nu} + V_k\big((v_\mu n^a A^a_\mu)^2\big) \right\} + \Gamma^{gf}_k[A,\bar{A}] + \Gamma^{gh}_k[A,\bar{A}]. \tag{16}$$

Here v_μ denotes the heat-bath velocity, for which we choose $v_\mu = \delta_{\mu 0}$. In the following, we neglect any running in the ghost- and gauge-fixing sectors, maintaining the form of Eq. (7), $\Gamma^{gf,gh}_k = \Gamma^{gf,gh}_\Lambda$. Furthermore, we neglect any other gauge-field operators except for the classical action with a wave function renormalization Z_k, and, of course, the Polyakov-loop potential $V\big((v_\mu A_\mu)^2\big)$. In order to derive the flow of the potential, it suffices to evaluate the flow for a trial background field of the simple form

$$A^a_\mu = (n^a \phi, 0)^T, \tag{17}$$

where n^a denotes a constant unit vector in color space. Furthermore, we exploit the freedom to choose suitable wave function renormalizations in the regulator function for an optimal adjustment of the regulator, cf. Eq. (8),

$$\mathcal{Z}_k = \left\{ Z^{gh}_k = 1, \quad Z^L_k = \frac{1}{\xi}, \quad Z^T_k \equiv Z_k \right\} \tag{18}$$

for the corresponding ghost, longitudinal and transversal degrees of freedom with respect to the background field. In particular, we set the transversal wave-function renormalization equal to the background-field wave-function renormalization. The choice for Z_k^L renders the truncated flow independent of the gauge-fixing parameter ξ, so that we can implicitly choose the Landau gauge $\xi_k \equiv 0$ which is known to be an RG fixed point [22].

It is convenient to express the flow equation in dimensionless renormalized quantities,

$$g_k^2 = k^{d-4} Z_k^{-1} \bar{g}^2, \quad \varphi = \beta \bar{g} \phi, \quad v_k = g^2 k^{-d} V_k, \qquad (19)$$

where $\varphi \in [0, 2\pi]$. Within our truncation, the flow equation in $d = 4$ dimensions reads [3]

$$\partial_t v_k(\varphi) = -(4 - \eta_k) v_k(\varphi) + \frac{\alpha_k}{4\pi} \sum_{n=-\infty}^{\infty} \left\{ (4 - 3\eta_k) \exp\left[-\frac{1}{4} n^2 \left(\frac{k}{T}\right)^2 \right] \overbrace{\cos(n\varphi)}^{\text{perturbation theory}} \right.$$

$$\left. + 4(2 - \eta_k) \left(\frac{T}{k}\right) \int_0^\infty dx\, x^2 \exp\left[-\tilde{\omega}_n^2 - x^2 - \left(1 - \frac{\tilde{\omega}_n^2}{\tilde{\omega}_n^2 + x^2}\right) \left(\frac{k}{T}\right)^2 \partial_\varphi^2 v_k(\varphi) \right] \right\}. (20)$$

Here we have used the abbreviations

$$\alpha_k = \frac{g_k^2}{4\pi}, \quad \eta_k = -\partial_t \ln Z_k, \quad \tilde{\omega}_n = 2\pi \frac{T}{k} n. \qquad (21)$$

Moreover, we have not displayed terms arising from $\propto \partial_t \Gamma^{(2)}$, e.g. terms $\propto \partial_t \partial_\varphi^2 v_k$, on the right-hand side of the flow equation. We have dropped these terms in the following preliminary numerical investigation for reasons of simplicity. As a consequence, the result of Eq. (20) corresponds to a standard propertime flow [20]. In future work, these terms arising within the Exact RG flow will be included to facilitate a quantitative study of the differences between the Exact and the standard propertime RG in the present case. In deriving Eq. (20), we have furthermore employed a regulator which yields a particulary simple representation in propertime (or Laplace) space,

$$\tilde{h}(s) = \delta(s - 1). \qquad (22)$$

At this point, it is useful to study the overlap of the present result with perturbation theory. In fact, we rediscover Weiss's result of Eq. (4) if we (i) hold the coupling fixed, $\alpha_k = $const., (ii) set the anomalous dimension to zero, $\eta_k = 0$, and (iii) drop the complete second line of Eq. (20). The resulting simplified equation is an ordinary differential equation, that can immediately be integrated from $k = \Lambda$ to $k = 0$, leading us to the perturbative one-loop result of Eq. (4).

---

[3] Details of the calculation will be presented in forthcoming publication [23].

Now, this observation helps estimating the nonperturbative content of the full Eq. (20): the occurrence of the running gauge coupling α_k and the k-dependent anomalous dimension on the right-hand side signal the "RG improvement", i.e., a resummation of an infinite set of Feynman diagrams, performed by the flow equation. Finally, the second line of Eq. (20) depends on (derivatives of) the potential itself. This term is truly nonperturbative, being induced by fluctuations of the Polyakov-loop variable on top of its own potential minimum. Since fluctuations of the degrees of freedom associated with the order parameter become highly important near the phase transition, we expect that particularly terms of this kind can cover important aspects of the nonperturbative dynamics.

In principle, the present truncation also allows for a rough calculation of the running coupling g_k including finite-temperature effects. For instance, the β function for the running coupling using the background field gauge reads

$$\partial_t g_k^2 \equiv \beta_{g_k^2} = \eta_k g_k^2, \tag{23}$$

and is thus related solely to the background wave function renormalization that is part of the truncation. However, important features of the running coupling in the deep infrared require much larger truncations [19, 24], hence we decide to take the running coupling as an external input in this work.

3. RESULTS

Let us now discuss the flow of the Polyakov-loop potential from the perturbative UV regime to the IR on the basis of our minimal approximation given by Eq. (20). In order to solve this partial differential equation, we rewrite it as an infinite set of coupled first-order differential equations by projecting it on a Fourier cosine series; this projection is naturally suggested by the fact that the $Z(2)$ symmetry of the potential and its dependence on a compact variable allows for such a Fourier expansion of the potential itself,

$$v_k(\varphi) = \sum_{n=0}^{\infty} v_k^{(n)} \cos n\varphi. \tag{24}$$

With this procedure, we obtain flow equations for the dimensionless Fourier coefficients $v_k^{(n)}$ by Fourier transforming the right-hand side (RHS) of Eq. (20),

$$\partial_t v_k^{(n)} = \frac{1}{\pi} \int_0^{2\pi} d\varphi \left\{ \text{RHS-Eq.}(20)[\varphi, v_k(\varphi)] \right\}. \tag{25}$$

For a numerical evaluation of the Polyakov-loop potential, we have to truncate the infinite set at some $n = N_{max}$ and set all $v_k^{(n)}$ with $n > N_{max}$ equal to zero. The results, shown in this work, are calculated for $N_{max} = 2$ for simplicity. But we have confirmed that the inclusion of higher Fourier orders does not change our results

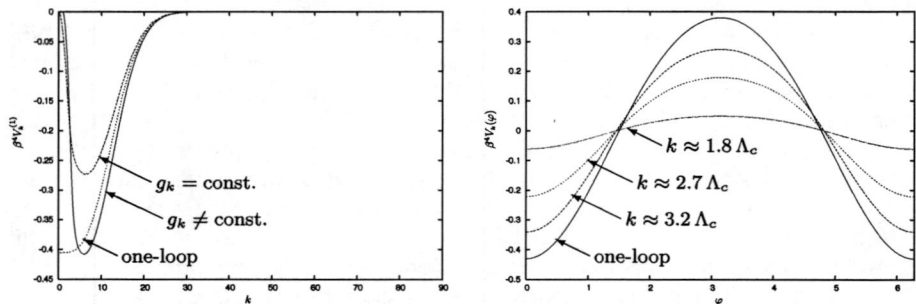

FIGURE 3. Left panel: flow of the first Fourier coefficient of $V_k = k^d g_k^{-2} v_k$ scaled by T^4. Right panel: effective potential V_k at different scales k for high temperature $T = 5\Lambda_c$ and g_k as in Eq. (26).

significantly. As the boundary conditions for the flow equation at the UV scale $\Lambda = 90$ GeV, we employ the result of the one-loop calculation Eq. (4). [4]

Finally, our simple choice for the running coupling g_k in this work is

$$g_k \propto \left[\ln\left(k/\Lambda_c\right)\right]^{-1}, \qquad (26)$$

where Λ_c denotes a "strong scale" for which we choose $\Lambda_c = 1$ GeV for simplicity.

In Fig. 3, we plot the results for the flow of the coefficient $v_k^{(1)}$ (left panel) and the Polyakov-loop potential at different values of k as a function of φ for $T = 5\Lambda_c$ (right panel). The flow of the first Fourier coefficient with the running coupling of Eq. (26) is compared to that for a constant coupling $g_k \equiv$ const. and the one-loop result of Eq. (4). We find that the three different calculations are in good agreement if the scale k is larger than the temperature. In this regime, the potential is built up by thermal fluctuations. At $k \approx T$, the coefficient of the nonperturbative flow-equation result develops a (negative) minimum and increases to zero for even smaller k. Since we observe a similar behaviour also for all other Fourier coefficients, the potential flattens for $k < T$,

$$v_k^{(n)} \xrightarrow{k \to 0} 0 \; (g_k \equiv \text{const.}) \quad \text{and} \quad v_k^{(n)} \xrightarrow{k \to \Lambda_c} 0 \; (g_k \text{ of Eq. (26)}). \qquad (27)$$

For the running coupling g_k, this behavior is depicted on the right panel of Fig. 3. By contrast, the coefficients of the one-loop result stay finite for $k \to 0$. This flatness of the potential is, in fact, nothing but the explicit manifestation of the convexity property that has to hold for the effective action in general, but is missed by perturbation theory.

Even though the observation of a convex potential is theoretically highly satisfactory, it does not tell us anything about the phase of the system. For this, the

---

[4] The integration from ∞ to Λ is controlled by perturbation theory, whereas the integration from Λ to the deep IR is controlled by the nonperturbative flow.

resulting expectation value of the Polyakov loop in the IR is relevant. At this point, we should stress that, even for a finally flat potential, the expectation value does not remain undetermined, but is well-defined by its $k \to 0^+$ limit (the effective potential is not flat for any nonzero k). However, as we can read off from Fig. 3, we do not observe a sign change of the Fourier coefficients under the flow; hence the minima of the effective potential are $Z(2)$ symmetry breaking and thus correspond to the deconfinement phase. In the present simple truncation, this holds true even for lower temperatures. Consequently, our truncation so far is only capable of describing deconfined dynamics.

4. CONCLUSIONS AND OUTLOOK

We have presented first steps towards a nonperturbative study of the Polyakov-loop potential based on an RG flow equation. Our intention so far mainly was to demonstrate the capability of our approach as a matter of principle by choosing a minimalistic approximation scheme. Already at this level, we observe important nonperturbative features, such as the backreactions of order-parameter fluctuations on top of the potential minimum and convexity of the effective potential. Confronting our first results with phenomena, we find that our truncation oversimplified the system, since we find deconfined dynamics on all scales. With hindsight, this is not too surprising, since, in addition to the Polyakov-loop dynamics, the gluon (and ghost) sector is basically approximated by its classical form, retaining "perturbative" gluon degrees of freedom all the way down to $k = 0$. From this point of view, our present minimalistic truncation leads to self-consistent results.

Future work will extend the present approach in various directions: on a technical level, the influence of the (so far neglected) terms from spectral adjustment $\propto \partial_t \Gamma_k^{(2)}$ has to be studied. This is an interesting task by its own, since it allows a quantitative analysis of the importance of these terms, distinguishing standard propertime flows from (background-field approximated) Exact RG flows. On a more conceptual level, the truncation has to be extended to include a larger class of gluonic (and/or ghost) degrees of freedom, covering important aspects of nonperturbative dynamics and allowing for an interplay with the Polyakov-loop sector. For instance, a gluonic potential $W_k(\frac{1}{4}F_{\mu\nu}F_{\mu\nu})$ replacing the present simple ansatz $\propto F_{\mu\nu}F_{\mu\nu}$ leaves room for the description of a nontrivial magnetic sector including gluon condensation as well as complex dynamics in combination with the Polyakov loop. Future work in this direction is in progress.

ACKNOWLEDGMENTS

We thank the organizers for the stimulating conference. We are grateful to J.M. Pawlowski for useful discussions. J.B. would like to thank the GSI Darmstadt for financial support. H.G. acknowledges financial support by the Deutsche Forschungsgemeinschaft (DFG) under contract Gi 328/1-2 (Emmy-Noether program).

REFERENCES

1. A. M. Polyakov, Phys. Lett. B **72** (1978) 477; L. Susskind, Phys. Rev. D **20** (1979) 2610.
2. B. Svetitsky, Phys. Rep. **132**, 1 (1986).
3. G. 't Hooft, Nucl. Phys. B **153** (1979) 141.
4. L. D. McLerran and B. Svetitsky, Phys. Rev. D **24** (1981) 450.
5. J. Engels, F. Karsch, H. Satz and I. Montvay, Nucl. Phys. B **205** (1982) 545.
6. S. Fortunato, F. Karsch, P. Petreczky and H. Satz, Phys. Lett. B **502**, 321 (2001); K. Langfeld, Phys. Rev. D **67**, 111501 (2003).
7. B. Svetitsky and L. G. Yaffe, Nucl. Phys. B **210**, 423 (1982).
8. R. D. Pisarski, arXiv:hep-ph/0203271
9. R. Fiore et al., arXiv:hep-lat/0409024.
10. L. Dittmann, T. Heinzl and A. Wipf, JHEP **0212**, 014 (2002).
11. N. Weiss, Phys. Rev. D **24** (1981) 475.
12. A.O. Starinets, A.S. Vshivtsev and V.Ch. Zhukovskii, Phys. Lett. B **322**, 403 (1994);P.N. Meisinger and M.C. Ogilvie, Phys. Lett. B **407**, 297 (1997).
13. M. Engelhardt and H. Reinhardt, Phys. Lett. B **430**, 161 (1998);H. Gies, Phys. Rev. D **63**, 025013 (2001); D. Diakonov and M. Oswald, Phys. Rev. D **68**, 025012 (2003).
14. F. Wegner, A. Houghton, Phys. Rev. **A 8** (1973) 401; K. G. Wilson and J. B. Kogut, Phys. Rept. **12** (1974) 75; J. Polchinski, Nucl. Phys. **B231** (1984) 269.
15. C. Wetterich, Phys. Lett. B **301** (1993) 90; M. Bonini, M. D'Attanasio and G. Marchesini, Nucl. Phys. B **409** (1993) 441; U. Ellwanger, Z. Phys. C **62** (1994) 503; T. R. Morris, Int. J. Mod. Phys. A **9** (1994) 2411.
16. U. Ellwanger, Phys. Lett. B **335** (1994) 364; M. Bonini, M. D'Attanasio and G. Marchesini, Nucl. Phys. B **437**, 163 (1995).
17. M. Reuter and C. Wetterich, Nucl. Phys. B **417** (1994) 181;Phys. Rev. D **56** (1997) 7893.
18. F. Freire, D. F. Litim and J. M. Pawlowski, Phys. Lett. B **495** (2000) 256.
19. H. Gies, Phys. Rev. D **66**, 025006 (2002).
20. R. Floreanini and R. Percacci, Phys. Lett. B **356**, 205 (1995); S. B. Liao, Phys. Rev. D **53**, 2020 (1996); B. J. Schaefer and H. J. Pirner, Nucl. Phys. A **660**, 439 (1999).
21. D. F. Litim and J. M. Pawlowski, Phys. Rev. D **66**, 025030 (2002).
22. U. Ellwanger, M. Hirsch and A. Weber, Z. Phys. C **69** (1996) 687; D. F. Litim and J. M. Pawlowski, Phys. Lett. B **435** (1998) 181.
23. J. Braun, H. Gies and H.-J. Pirner, forthcoming publication
24. J. M. Pawlowski, D. F. Litim, S. Nedelko and L. von Smekal, Phys. Rev. Lett. **93**, 152002 (2004); C. S. Fischer and H. Gies, JHEP **0410**, 048 (2004); B. Gruter, R. Alkofer, A. Maas and J. Wambach, arXiv:hep-ph/0408282.

Restoration of $U_A(1)$ symmetry and meson spectrum in hot or dense matter

P. Costa*, M. C. Ruivo*, C. A. de Sousa* and Yu.L. Kalinovsky[†,**]

*Centro de Física Teórica, Departamento de Física, Universidade, P3004-516 Coimbra, Portugal
[†]Physique théorique fondamentale, Département de Physique, Université de Liège, allée du 6 Août 17, bât. B5, B-4000 Liège 1, Belgium
[**]Laboratory of Information Technologies, Joint Institute for Nuclear Research, Dubna, Russia
E-mail: pcosta@fteor5.fis.uc.pt, maria@teor.fis.uc.pt, celia@teor.fis.uc.pt, kalinov@nusun.jinr.ru

Abstract. We explore the effects of breaking and restoration of chiral and axial symmetries using an extended three-flavor Nambu-Jona-Lasinio model that incorporates explicitly the axial anomaly through the 't Hooft interaction. We implement a temperature (density) dependence of the anomaly coefficient motivated by lattice results for the topological susceptibility. The spectrum of scalar and pseudoscalar mesons is analyzed bearing in mind the identification of chiral partners and the study of its convergence. We also concentrate on the behavior of the mixing angles that give us relevant information on the issue under discussion. The results suggest that the axial part of the symmetry is restored before the possible restoration of the full $U(3) \otimes U(3)$ chiral symmetry might occur.

Keywords: NJL model, $U_A(1)$ symmetry, Finite temperature and density
PACS: 11.10.Wx, 11.30.Rd, 14.40.Aq, 24.85.+p

1. INTRODUCTION

The study of properties of low-lying hadron spectrum from the viewpoint of QCD dynamics and symmetries has been an intrinsically interesting and relevant topic in physics of strong interactions. In this context, the explicit and spontaneous breaking of chiral symmetry, as well as the $U_A(1)$ anomaly, play a key role for the generation of the low-lying pseudoscalar meson masses. The importance for this process of the $U_A(1)$ symmetry breaking, and the consequent violation of the Okubo-Zweig-Iizuka (OZI) rule, has been stressed in many phenomenological investigations.

It is generally expected that ultra-relativistic heavy-ion experiments will provide the strong interaction conditions which will lead to new physics. Restoration of symmetries and deconfinement are expected to occur under those conditions, allowing for the search of signatures of quark gluon plasma. In particular, the order of the chiral phase transition and its consequences for aspects of the dynamical evolution of the system are important questions. It is also believed that at high temperatures the instanton effects are suppressed due to the Debye-type screening [1]. Then an effective restoration of $U_A(1)$ symmetry is expected to occur at high temperatures. In this context, it has been argued that the mass of the η' excitation in hot and dense matter should be small, being expected the return of this "prodigal Goldstone boson" [2].

There should be other indications of the restoration of the axial symmetry, like the vanishing of the topological susceptibility [3], χ, which, in pure color $SU(3)$ theory, can be linked to the η' mass through the Witten-Veneziano formula [4]. In addition, since the presence of the axial anomaly causes flavor mixing, with the consequent violation of the OZI rule, both for scalar and pseudoscalar mesons, restoration of axial symmetry should have relevant consequences for the phenomenology of meson mixing angles, leading to the recovering of ideal mixing.

We perform our calculations in the framework of an extended $SU(3)$ Nambu–Jona-Lasinio model Lagrangian density that includes the 't Hooft determinant:

$$\begin{aligned} \mathcal{L} &= \bar{q}(i\gamma.\partial - \hat{m})q + \frac{g_S}{2}\sum_{a=0}^{8}[(\bar{q}\lambda^a q)^2 + (\bar{q}i\gamma_5\lambda^a q)^2] \\ &+ g_D\{\det[\bar{q}(1+\gamma_5)q] + \det[\bar{q}(1-\gamma_5)q]\}. \end{aligned} \quad (1)$$

By using a standard hadronization procedure, an effective action is obtained, leading to gap equations for the constituent quark masses and to meson propagators from which several observables are calculated [5].

In what follows, we will concentrate first on the restoration of the symmetries at zero density and finite temperature, and the manifestations of the restoration on the scalar and pseudoscalar meson observables will be analyzed. In chiral models, when the coefficient of the anomaly term in the Lagrangian (g_D in the present case) is constant, it happens that, in spite of the decreasing of the $U_A(1)$ violating quantities, the axial symmetry is not restored due to the fact that the strange quark condensate does not decrease enough. However, an effective restoration may be achieved by assuming that the strength of the anomaly coefficient is a dropping function of the temperature [6, 7, 8]. The analysis of the temperature dependence of the mixing angles, allowing for the understanding of the evolution of meson quark content and, in particular, of the role of the strange order parameter in $SU(3)$ chiral partners like (σ, η) and (f_0, η'), can provide further indication of the restoration of the axial symmetry.

Preliminary results from calculations on lattice QCD at finite chemical potential [9] motivates also the study of the restoration of the $U_A(1)$ symmetry at finite density.

We study the effective restoration of chiral and axial symmetries with temperature and zero density in Sec. 2 and, with finite density and zero temperature in Sec. 3. Finally, we summarize our conclusions in Sect. 4.

2. FINITE TEMPERATURE AND ZERO DENSITY CASE

Restoration of the axial symmetry should be manifest when the effects of anomaly are no longer visible and, in the present work, we model this feature by means of a specific temperature dependence of g_D. So, following the methodology of [8], we

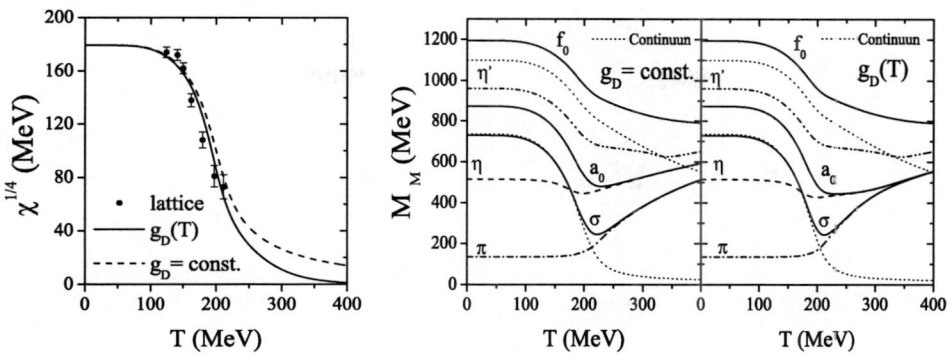

FIGURE 1. Topological susceptibility (left panel): from lattice data plotted with error bars [3]; the solid (dashed) line represents our fitting with constant (temperature dependent) g_D. Meson masses, as functions of the temperature, with g_D constant (middle panel) and $g_D(T)$ (right panel). The dotted lines indicate the continuum thresholds $2M_u$ and $2M_s$.

extract the temperature dependence of the anomaly coefficient g_D from the lattice results for the topological susceptibility [3] (see Fig. 1, left panel).

Our results, concerning the restoration of the chiral phase transition, at zero density and finite temperature, indicate, as usual, a smooth crossover: at temperatures around $T \approx 200$ MeV the mass of the light quarks drops to the current quark mass. The strange quark mass also starts to decrease significantly in this temperature range, however even at $T = 400$ MeV it is still 2 times the strange current quark mass. In fact, as $m_u = m_d < m_s$, the (sub)group SU(2)⊗SU(2) is a much better symmetry of the Lagrangian (1). So, the effective restoration of the above symmetry implies the degeneracy between the chiral partners (π, σ) and (a_0, η) around $T \approx 250$ MeV (see Fig. 1, right panel). For temperatures about $T \approx 350$ MeV, both a_0 and σ mesons become degenerate with the π and η mesons, showing an effective restoration of both chiral and axial symmetries. In fact, the $U_A(1)$ symmetry is effectively restored when the $U_A(1)$ violating quantities show a tendency to vanish, which means that the four meson masses degenerate and the topological susceptibility goes to zero. Without the restoration of $U_A(1)$ symmetry, the a_0 mass was moved upwards and never met the π mass as can be seen in Fig. 1, middle panel. The same argument is valid for σ meson comparatively with the η meson mass. We remember that the determinantal term acts in an opposite way for the scalar and pseudoscalar mesons. So, only after the effective restoration of $U_A(1)$ symmetry we can recover the SU(3) chiral partners (π, a_0) and (η, σ) which are now all degenerate. However, the η' and f_0 masses do not yet show a clear tendency to converge in the region of temperatures studied.

In order to understand this behavior, we also analyze the temperature dependence of the mixing angle (see Fig. 2): θ_S starts at 16° and goes, smoothly, to the

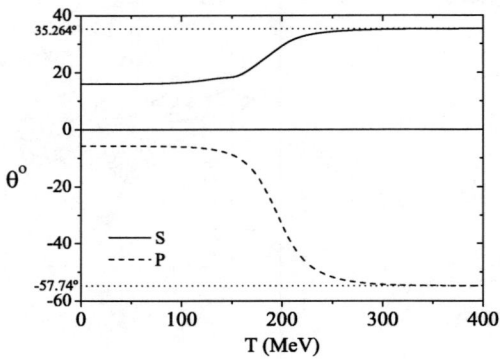

FIGURE 2. The pseudoscalar and scalar mixing angles as functions of the temperature.

ideal mixing angle 35.264° and θ_P starts at $-5.8°$ and goes to the ideal mixing angle $-54.7°$. This means that flavor mixing no more exists. In fact, analyzing the behavior of the SU(2) chiral partner (η, a_0) with the temperature (Fig. 1, right panel), we found that the a_0 meson is always a purely non-strange $q\bar{q}$ system while the η meson, at $T = 0$ MeV, has a strange component and becomes purely non-strange when θ_P goes to $-54.7°$ at $T \approx 250$ MeV. At this temperature they start to be degenerate. Concerning the SU(2) chiral partner (π, σ), at $T = 0$ MeV, π is always a light quark system and the σ meson has a strange component, but becomes purely non-strange when θ_S goes to 35.264° at $T \approx 250$ MeV. In summary, we see that the $U_A(1)$ symmetry is effectively restored at $T \approx 350$ MeV: π, σ, η and a_0 mesons become degenerate, the OZI rule is verified and χ goes asymptotically to zero [10]. We remember that π, σ, η_{ns} and a_0 form a complete representation of U(2)⊗U(2) symmetry, but the π and the a_0 do not belong to the same SU(2)⊗SU(2) multiplet. The partners (f_0, η'), which become purely strange at high temperatures, do not converge probably due to the fact that chiral symmetry is not restored in the strange sector.

3. FINITE DENSITY AND ZERO TEMPERATURE CASE

Recent calculations on lattice QCD at finite chemical potential motivates also the study of the restoration of the $U_A(1)$ symmetry at finite density. There are no firmly lattice results for the density dependence of χ, to be used as input, but the preliminary conclusions of [9] suggest a possible decrease of χ. So, it seems reasonable to model the density dependence of g_D extrapolating from our previous results for the finite temperature case and proceeding by analogy. Here we present an example (see Fig. 3, left panel) where we consider quark matter

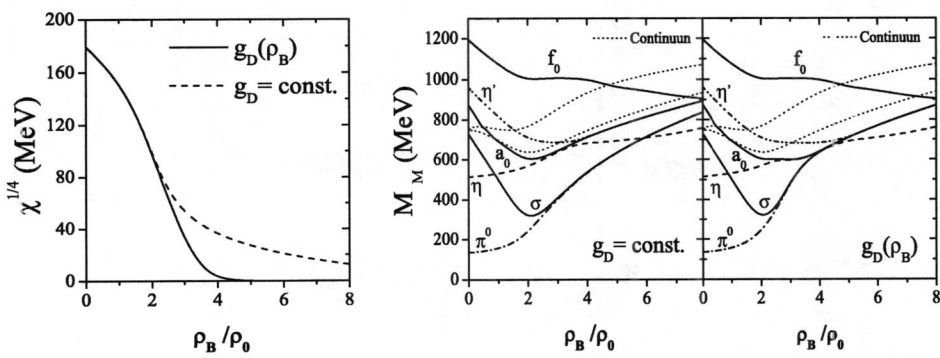

FIGURE 3. Topological susceptibility (left panel): the solid (dashed) line represents our fitting with constant (density dependent) g_D. Meson masses, as functions of density, with g_D constant (middle panel) and $g_D(\rho_B)$ (rigth panel). The dotted lines indicate the density depencence of the limits of the Dirac sea continua, defining $q\bar{q}$ thresholds for a_0 and η' mesons.

simulating "neutron" matter. This "neutron" matter is in β–equilibrium with charge neutrality, and undergoes a first order phase transition [5]. To begin with, we calculate the mixing angles for scalar and pseudoscalar mesons, θ_S and θ_P, respectively, that are plotted in Fig. 4. We observe that θ_S starts at 16° and increases up to the ideal mixing angle 35.264°. A different behavior is found for the angle θ_P that changes sign at $\rho_B \approx 4\rho_0$. In fact, it starts at $-5.8°$ and goes to the ideal mixing angle 35.264°, leading to a change of identity between η and η'. We think this result might be a useful contribution for the understanding of the somewhat controversial question: under extreme conditions will the pion degenerate with η or η'? The change of sign and the corresponding change of identity between η and η', effects that we do not observe in the finite temperature case, are due to differences in the behavior of the strange quark mass [5, 11].

The meson masses, as function of the density, are plotted in Fig. 3, right panel. The results for constant g_D are also presented (middle panel) for comparison purposes. The SU(2) chiral partners (π^0, σ) are bound states and become degenerate at $\rho_B = 3\rho_0$. With respect to the SU(2) chiral partners (η, a_0), the a_0 meson is always a purely non-strange quark system. For $\rho_B < 0.8\rho_0$ a_0 is above the continuum and, when $\rho_B \geq 0.8\rho_0$, a_0 becomes a bound state. At $\rho_B = 0$, the η has a strange component and, as the density increases, η becomes degenerate with a_0 at $4.0\rho_0 \leq \rho_B \leq 4.8\rho_0$ as expected. In this range of densities (η, a_0) and (π^0, σ) are all degenerate. Suddenly, the η mass separates from the others becoming a purely strange state. This is due to the behavior of θ_P that changes sign and goes to 35.264° at $\rho_B \approx 4.8\rho_0$. On the other hand, the η', that starts as an unbound state and becomes bounded at $\rho_B > 3.0\rho_0$, turns into a purely light quark system and degenerates with π^0, σ and a_0 mesons [10].

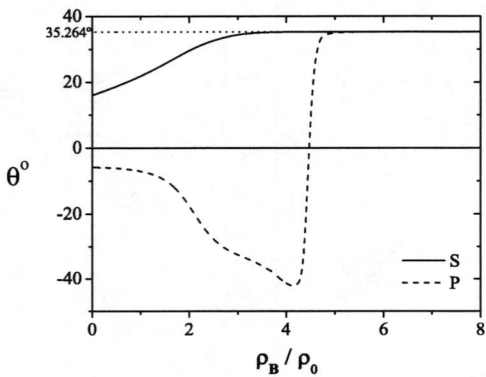

FIGURE 4. The pseudoscalar and scalar mixing angles as functions of the baryonic density.

Finally we analyze the behavior of charged mesons with density, plotted in Fig. 5. A glance at the figure immediately reveals that: the chiral partners (π^+, a_0^+) and (π^-, a_0^-), upper panel, become degenerate for $\rho_B \approx 4\rho_0$; the chiral partners $(K^+, \kappa^+), (K^-, \kappa^-)$, middle panel, and $(K^0, \kappa^0), (\bar{K}^0, \bar{\kappa}^0)$, lower panel, do not degenerate in the region of densities considered. We notice that, while the results for $(\pi^\pm, a_0^\pm)$ are affected by the dependence of g_D on density, we find no substantial differences for the kaonic mesons, whether g_D is constant or not. In order to understand this, let us remember that the effects of $U_A(1)$ symmetry breaking or restoration appear explicitly, in the gap equations and in the meson propagators, through the anomaly coupling, g_D, times a quark condensate. For the pion and the a_0 propagators, the dependence on the anomaly enters through $g_D < \bar{q}_s q_s >$ so, with g_D a decreasing function of the density, this term will affect less and less the mesons masses as the density increases. Then, the convergence of the mesons reflects the restoration of the $U_A(1)$ symmetry. Since for kaonic mesons the propagators depend on the anomaly through $g_D < \bar{q}_u q_u > (< \bar{q}_d q_d >)$, the anomaly has little effect on the kaonic masses, as the density increases, whether g_D is constant or not, due to the strong decrease of the mass of the non-strange quarks. The dominant factor for the calculation of the masses of those mesons is the mass of the strange quark, which, although decreasing, remains always very high. We can say that the restoration of the axial anomaly does not influence the behavior of kaons and of its chiral partners. In addition, we remark that M_u and M_d have different behaviors in neutron matter [5], and the chiral asymmetry $\chi_A = |M_u - M_d|/(M_u + M_d)$ is always different from zero, even for high densities.

We also analyze the low-lying solutions, which we denoted by the subscript S. Below the lower limit (ω_{low}, in Fig. 5) of the Fermi sea continuum of particle-hole excitations there are low bound states with quantum numbers of $K^-, \bar{K}^0$ and π^+. These collective particle-hole excitations of the Fermi sea, associated to a first-order

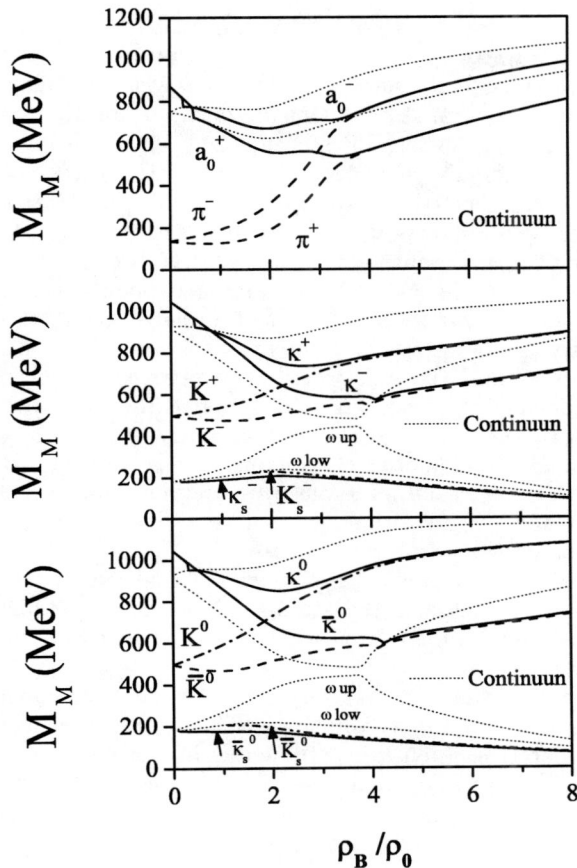

FIGURE 5. Meson masses, as functions of density, with $g_D(\rho_B)$. The dotted lines indicate the density dependence of the limits of the Dirac and the Fermi (ω_{low} and ω_{up}) sea continua, defining $q\bar{q}$ thresholds for the mesons. The low-lying mesonic solutions (denoted by the subscript S) are also included.

phase transition, are manifestations of the violation of the isospin symmetry by the environment conditions. It is interesting to look for the respective chiral partners and observe the behavior of the mass splitting between scalar and pseudoscalar mesons as the densities increases. The decreasing of this splitting for K^- and $\bar{K}^0$ mesons can be seen in Fig. 5.

We notice that the convergence between the different chiral partners always occurs at densities where the mesons are bound states (see Fig. 5), i.e., they are collective excitations defined below the respective $q\bar{q}$ threshold.

4. CONCLUSIONS

In this work we considered, via the $U_A(1)$ anomaly, an explicit breaking of the axial symmetry in the vacuum state. The subsequent restoration of axial symmetry at non-zero temperature (density) has been then discussed using a temperature (density) dependent anomaly coefficient. The case with $g_D = $ cte for all temperatures (densities) is also discussed. We verified that in this case there is always an amount of $U_A(1)$ symmetry breaking in the particle spectrum even when chiral symmetry restoration in the non-strange sector occurs at high temperature (density).

Since in all cases the chiral symmetry is explicitly broken by the presence of non-zero current quark mass terms, the chiral symmetry has been realized through parity doubling rather than by massless quarks. So, the identification of chiral partners and the study of its convergence is the criteria to study the effective restoration of chiral and axial symmetries. An important information is also provided by the mixing angles and we verify that, in the scenario of effective restoration of axial symmetry, the mixing angles converge to situations of ideal flavor mixing: (i) the σ and η mesons are pure non-strange $q\bar{q}$ states, while f_0 and η' are pure strange $s\bar{s}$ excitations for non-zero temperature case; (ii) the η and η' change identities for neutron matter case.

In the conditions of explicit breaking of chiral symmetry we worked, SU(3) symmetry is not exact and the strange sector does contribute with significant effects, even at high temperature (density) as it is visible in the behavior of f_0 and η (η') mesons.

We can conclude that the $U_A(1)$ symmetry is, at least partially, restored above the critical transition temperature of the SU(2) chiral phase transition. But, in the region of temperatures (densities) studied we do not observe signs indicating a full restoration of U(3)⊗U(3) symmetry. In fact, as we work in a real world scenario ($m_u = m_d << m_s$), we only observe the return to symmetries of the classical QCD Lagrangian in the non-strange sector.

ACKNOWLEDGMENTS

Work supported by grant SFRH/BD/3296/2000 (P. Costa), CFT and by FEDER/FCT under project POCTI/FIS/451/94. The work of Yu.L.K. was partly supported by the FNRS, Belgium.

REFERENCES

1. D. J. Gross, R. D. Pisarski and L. Yaffe, *Rev. Mod. Phys.* **53**, 43 (1981).
2. J. Kapusta, D. Kharzeev and L. McLerran, *Phys. Rev.* D **53**, 5028 (1996).
3. B. Allés, M. D'Elia and A. Di Giacomo, *Nucl. Phys.* B **494**, 281 (1997).
4. E. Witten, *Nucl. Phys.* B **156**, 269 (1979); G. Veneziano, *Nucl. Phys.* B **159**, 213 (1979).
5. P. Costa and M. C. Ruivo, *Europhys. Lett.* **60** (3), 356 (2002); P. Costa, M. C. Ruivo, C. A de Sousa and Yu. L. Kalinovsky, *Phys. Rev.* C **70**, 025204 (2004).
6. R.Alkofer, P. A. Amundsen and H. Reinhardt, *Phys. Lett.* B **218**, 75 (1989); T. Kunihiro, *Phys. Lett.* B **219**, 363 (1989).

7. J. Schaffner-Bielich, *Phys. Rev. Lett.* **84**, 3261 (2000).
8. K. Fukushima, K. Ohnishi and K. Ohta, *Phys. Rev.* C **63**, 045203 (2001).
9. B. Allés, M. D'Elia, M. P. Lombardo and M. Pepe, *Nucl. Phys.* (Proc. Suppl.) **94**, 441 (2001).
10. P. Costa, M. C. Ruivo, C. A de Sousa and Yu. L. Kalinovsky, *Phys. Rev.* D **70**, 116013 (2004).
11. P. Costa, M. C. Ruivo and Yu. L. Kalinovsky, *Phys. Lett.* B **560**, 171 (2003).

Cooling Evolution of Hybrid Stars

H. Grigorian

Institut für Physik, Universität Rostock, D-18051 Rostock, Germany
and Department of Physics, Yerevan State University, 375025 Yerevan, Armenia
E-mail: hovik.grigorian@uni-rostock.de

Abstract. The cooling of compact isolated objects for different values of the gravitational mass has been simulated for two alternative assumptions. One is that the interior of the star is purely hadronic[1] and second that the star can have a rather large quark core [2]. It has been shown that within a nonlocal chiral quark model the critical density for a phase transition to color superconducting quark matter under neutron star conditions can be low enough for these phases to occur in compact star configurations with masses below 1.3 $M_\odot$. For a realistic choice of parameters the equation of state (EoS) allows for 2SC quark matter with a large quark gap $\sim$ 100 MeV for u and d quarks of two colors that coexists with normal quark matter within a mixed phase in the hybrid star interior. We argue that, if in the hadronic phase the neutron pairing gap in $3P_2$ channel is larger than few keV and the phases with unpaired quarks are allowed, the corresponding hybrid stars would cool too fast.

Even in the case of the essentially suppressed $3P_2$ neutron gap if free quarks occur for $M < 1.3\ M_\odot$, as it follows from our EoS, one could not appropriately describe the neutron star cooling data existing by today.

It is suggested to discuss a "2SC+X" phase, as a possibility to have all quarks paired in two-flavor quark matter under neutron star constraints, where the X-gap is of the order of 10 keV - 1 MeV. Density independent gaps do not allow to fit the cooling data. Only the presence of an X-gap that decreases with increase of the density could allow to appropriately fit the data in a similar compact star mass interval to that following from a purely hadronic model.

Keywords: Dense baryon matter, Color superconductivity, Neutron stars, Heat transport
PACS: 98.80.-k, 98.80.Es, 98.56.-p

1. INTRODUCTION

The "standard" scenario of neutron star cooling is based on the main process responsible for the cooling, which is the modified Urca process (MU) $nn \to npe\bar{\nu}$ calculated using the free one pion exchange between nucleons, see [3]. However, this scenario explains only the group of slow cooling data. To explain a group of rapid cooling data "standard" scenario was supplemented by one of the so called "exotic" processes either with pion condensate, or with kaon condensate, or with hyperons, or involving the direct Urca (DU) reactions, see [4, 5] and refs therein. All these processes may occur only for the density higher than a critical density, $(2 \div 6)\ n_0$, depending on the model, where n_0 is the nuclear saturation density. An other alternative to "exotic" processes is the DU process on quarks related to the phase transition to quark matter.

Recently, the cooling of neutron stars has been reinvestigated within a purely hadronic model [1], i.e., when one suppresses the possibility of quark cores in neutron star interiors. We have demonstrated that the neutron star cooling data available by today can be well explained within the "nuclear medium cooling" scenario, cf. [6, 7], i.e., if one includes medium effects into consideration. In the "standard plus exotics" scenario for hadronic models the in-medium effects have not been incorporated, see [8, 9, 10]. Recently [10] called this approach the "minimal cooling" paradigm. Some papers included an extra possibility of internal heating that results in a slowing down of the cooling of old pulsars, see [11] and Refs. therein.

The necessity to include in-medium effects into the neutron star cooling is based on the whole experience of condensed matter physics, see [12, 13, 14]. The relevance of in-medium effects for the neutron star cooling problem has been shown by [7, 12, 15, 16, 17] and the efficiency of the developed "nuclear medium cooling" scenario for the description of the neutron star cooling was demonstrated within the cooling code by [6] and then by [1].

Each scenario puts some constraints on dense matter equation of state (EoS). In particular the density dependencies of the asymmetry energy and the pairing gaps are the regulators of the heat production and transport. The former dependence is an important issue for the analysis of heavy ion collisions especially within the new CBM (compressed baryon matter) program to be realized at the future accelerator facility FAIR at GSI Darmstadt.

The density dependence of the asymmetry energy also determines the proton fraction in neutron star matter and thus governs the onset of the very efficient direct Urca (DU) process. The DU process, once occurring, would lead to a very fast cooling of neutron stars. Within the "standard + DU" scenario the transition from slow cooling to the rapid cooling occurs namely due to the switching on the DU process. Thus the stars with $M < M_{\rm crit}^{\rm DU}$ cool down slowly whereas the stars with the mass only slightly above $M_{\rm crit}^{\rm DU}$ cool down very fast. Since it is doubtful that many neutron stars belonging to an intermediate cooling group have very similar masses, from our point of view such a scenario seems unrealistic, cf. [1, 18]. The modern EoS of the Urbana-Argonne group [19] allows for the DU process only for very high density $n > 5n_0$ (where n_0 is the saturation nuclear density) that relates to the neutron star masses $M \geq M_{\rm crit}^{\rm DU} \simeq 2\ M_\odot$. Thus, using mentioned Urbana-Argonne based EoS and the "standard +DU" scenario one should assume that the majority of experimentally measured cooling points relates to very massive neutron stars that seems us still more unrealistic.

The assumption about the mass distribution can be developed into a more quantitative test of cooling scenarios when these are combined with population synthesis models. The latter allow to obtain Log N – Log S distributions for nearby coolers which can be tested with data from the ROSAT catalogue [20]. Analysis [20] has supported ideas put forward in [1].

At high star masses the central baryon density exceeds rather large values $n > 5n_0$. At these densities exotic states of matter as, e.g., hyperonic matter or quark matter perhaps are permitted. Ref. [21] argued that the presence of the quark matter in massive compact star cores is a most reliable hypothesis.

The possibility of the existence of neutron stars with large quark matter cores is also not excluded [2, 22, 23, 24]. In the quark matter the DU process yielding the rapid cooling may arise on interacting but unpaired quarks [25]

In this review we want to sketch a scenario for the cooling of hybrid stars.

2. STRUCTURE OF HYBRID NEUTRON STARS

In describing the hadronic part of the hybrid star, as in [1], we exploit a modification of the Urbana-Argonne $V18 + \delta v + UIX^*$ model of the EoS given in [19], which is based on the most recent models for the nucleon-nucleon interaction with the inclusion of a parameterized three-body force and relativistic boost corrections. Actually we continue to adopt an analytic parameterization of this model by Heiselberg and Hjorth-Jensen [26], hereafter HHJ.

The HHJ EoS fits the symmetry energy to the original Argonne $V18+\delta v+UIX^*$ model in the mentioned density interval yielding the threshold density for the DU process $n_c^{DU} \simeq 5.19 \, n_0$ ($M_c^{DU} \simeq 1.839 \, M_\odot$).

We employ the EoS of a nonlocal chiral quark model developed in [27] for the case of neutron star constraints with a 2-flavor color superconductivity (2SC) phase. It has been shown in that work that the Gaussian formfactor ansatz leads to an early onset of the deconfinement transition and such a model is therefore suitable to discuss hybrid stars with large quark matter cores [28].

The quark-quark interaction in the color anti-triplet channel is attractive driving the pairing with a large zero-temperature pairing gap $\Delta \sim 100$ MeV for the quark chemical potential $\mu_q \sim (300 \div 500)$ MeV, cf. [29, 30], for a review see [13] and references therein. The attraction comes either from the one-gluon exchange, or from a non-perturbative 4-point interaction motivated by instantons [31], or from non-perturbative gluon propagators [32].

There may also exist a color-flavor locked (CFL) phase [33] for not too large values of the dynamical strange quark mass or large values of the baryon chemical potential [34]. In this phase the all quarks are paired. However, the 2SC phase occurs at lower baryon densities than the CFL phase, see [35, 36]. For applications to compact stars the omission of the strange quark flavor is justified by the fact that chemical potentials in central parts of the stars do barely reach the threshold value at which the mass gap for strange quarks breaks down and they appear in the system [37].

Following Refs [38] we omit the possibility of the hadron-quark mixed phase and found a tiny density jump on the phase boundary from $n_c^{hadr} \simeq 0.44 \, fm^{-3}$ to $n_c^{quark} \simeq 0.46 \, fm^{-3}$.

In Fig. 1 we present the mass-radius relation for hybrid stars with HHJ EoS vs. Gaussian nonlocal chiral quark separable model (SM) EoS. Configurations for SM model, given by the solid line, are stable, whereas without color super conductivity ("HHJ-SM without 2SC") no stable hybrid star configuration is possible. In the case "HHJ-SM with 2SC" the maximum neutron star mass proves to be 1.793 $M_\odot$.

Additionally, within the "HHJ-SM with 2SC" phase we will allow for the possibility of a weak pairing channel for all the quarks which were unpaired, with typical gaps $\Delta_X \sim 10$ keV $\div 1$ MeV, as in the case of the CSL pairing channel, see [41, 42]. Since we don't know yet the exact pairing pattern for this case, we call

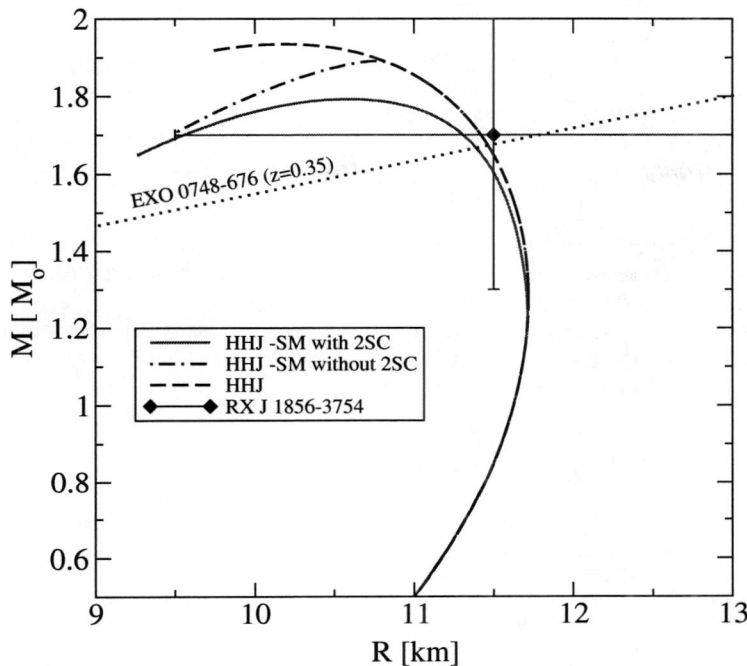

FIGURE 1. Mass - radius relations for compact star configurations with different EoS: purely hadronic star with HHJ EoS (dashed line), stable hybrid stars with HHJ - Gaussian nonlocal chiral quark separable model (SM) with 2SC phase (solid line) and with HHJ - SM, without 2SC phase (dash-dotted line). Data for two sources are also indicated, see [39, 40].

this hypothetical phase "2SC+X". In such a way all the quarks get paired, some strongly in the 2SC channel and some weakly in the X channel.

3. COOLING

We compute the neutron star thermal evolution adopting our fully general relativistic evolutionary code. This code was originally constructed for the description of hybrid stars by [24]. The main cooling regulators are the thermal conductivity, the heat capacity and the emissivity. In order to better compare our results with results of other groups we try to be as close as possible to their inputs for the quantities which we did not calculate ourselves. Then we add inevitable changes, improving EoS.

The density $n \sim 0.5 \div 0.7\, n_0$ is the boundary of the neutron star interior and the inner crust. The latter is constructed of a pasta phase discussed by [43], see also recent works of [44, 45].

Further on we need the relation between the crust and the surface temperature for neutron star. The sharp change of the temperature occurs in the envelope.

3.1. Cooling Evolution of Hadronic Stars

Here we will shortly summarize the results on hadronic cooling.

In framework of "minimal cooling" scenario the pair breaking and formation (PBF) processes may allow to cover an "intermediate cooling" group of data (even if one artificially suppressed medium effects)[6]. These processes are very efficient for large pairing gaps, for temperatures being not much less than the value of the gap.

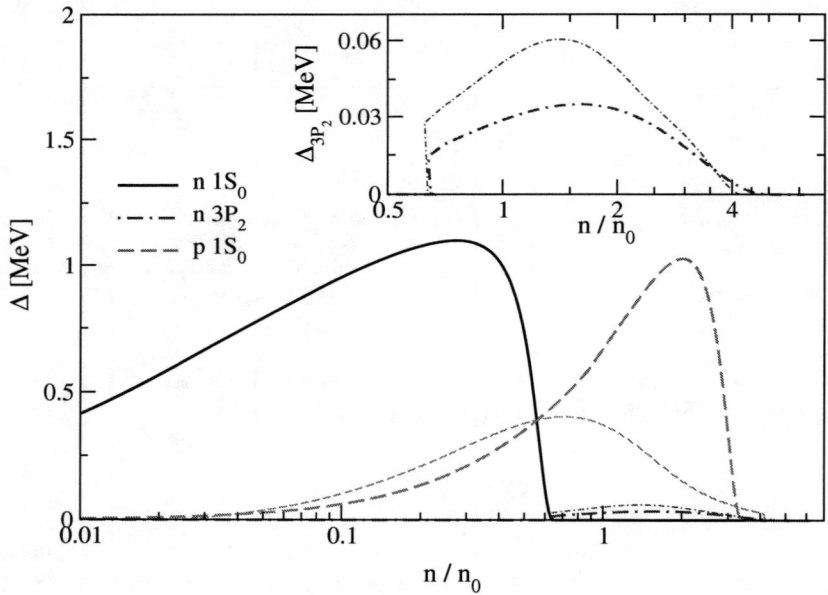

FIGURE 2. Neutron and proton pairing gaps according to model I (thick solid, dashed and dotted lines) and according to model II (thin lines), see text. The $1S_0$ neutron gap is the same in both models, taken from [46].

Gaps that we have adopted in the framework of the "nuclear medium cooling" scenario, see [1], are presented in Fig. 2. Thick dashed lines show proton gaps which were used in the work of [9] performed in the framework of the "standard plus exotics" scenario. We will call the choice of the "3nt" model from [9] the model

I. Thin lines show $1S_0$ proton and $3P_2$ neutron gaps from [48], for the model AV18 by [49] (we call it the model II). Recently [47] has argued for a strong suppression of the $3P_2$ neutron gaps, down to values ~ 10 keV, as the consequence of the medium-induced spin-orbit interaction.

These findings motivated [1] to suppress values of $3P_2$ gaps shown in Fig. 2 by an extra factor $f(3P_2, n) = 0.1$. Further possible suppression of the $3P_2$ gap is almost not reflected on the behavior of the cooling curves.

Contrary to expectations of [47] a more recent work of [50] argued that the $3P_2$ neutron pairing gap should be dramatically enhanced, as the consequence of the strong softening of the pion propagator. According to their estimate, the $3P_2$ neutron pairing gap is as large as $1 \div 10$ MeV in a broad region of densities, see Fig. 1 of their work. Thus results of calculations of [47] and [50], which both had the same aim to include medium effects in the evaluation of the $3P_2$ neutron gaps, are in a deep discrepancy with each other.

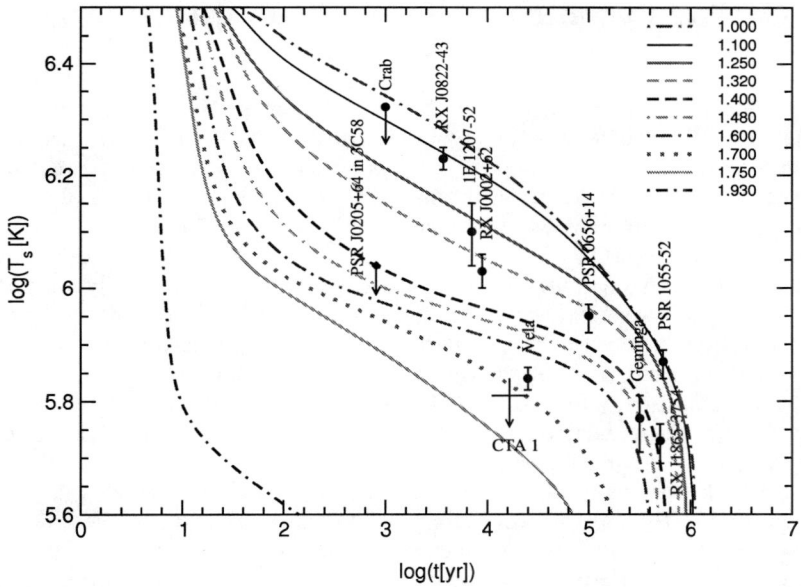

FIGURE 3. Adapted from Fig. 21 of [1]. Gaps are from Fig. 2 for model II. The original $3P_2$ neutron pairing gap is additionally suppressed by a factor $f(3P_2, n) = 0.1$. The $T_s - T_{in}$ relation is given by "our fit" curve of Fig. 4 in [1]. For more details see [1].

- Including superfluid gaps we see, in agreement with recent microscopic findings of [47], that $3P_2$ neutron gap should be as small as 10 keV or less. So the "nuclear medium cooling" scenario of [1] supports results of [47] and fails to appropriately fit the neutron star cooling data at the assumption of a strong enhancement of the $3P_2$ neutron gaps as suggested by [51].

- Medium effects associated with the pion softening are called for by the data. As the result of the pion softening the pion condensation may occur for $n \geq n_c^{PU}$ ($n \geq 3n_0$ in our model). Its appearance at such rather high densities does not contradict to the cooling data (see Fig. 3), but also the data are well described using the pion softening but without assumption on the pion condensation. This also means that the DU threshold density can't be too low that puts restrictions on the density dependence of the symmetry energy. Both statements might be important in the discussion of the heavy ion collision experiments.
- We demonstrated a regular mass dependence: for the neutron star masses $M > 1$ $M_\odot$ less massive neutron stars cool down slower, more massive neutron stars cool faster.

3.2. Cooling Evolution of Hybrid Stars with 2SC Quark Matter Core

For the calculation of the cooling of the quark core in the hybrid star we use the model [24]. We incorporate the most efficient processes: the quark direct Urca (QDU) processes on unpaired quarks, the quark modified Urca (QMU), the quark bremsstrahlung (QB), the electron bremsstrahlung (EB), and the massive gluon-photon decay (see [22]). Following [52] we include the emissivity of the quark pair formation and breaking (QPFB) processes. The specific heat incorporates the quark contribution, the electron contribution and the massless and massive gluon-photon contributions. The heat conductivity contains quark, electron and gluon terms.

The calculations are based on the hadronic cooling scenario presented in Fig. 3 and we add the contribution of the quark core. For the Gaussian form-factor the quark core occurs already for $M > 1.214$ $M_\odot$ according to the model [27], see Fig. 1. Most of the relevant neutron star configurations (see Fig. 3) are then affected by the presence of the quark core.

First we check the possibility of the 2SC+ normal quark phases Fig. 4.

The variation of the gaps for the strong pairing of quarks within the 2SC phase and the gluon-photon mass in the interval $\Delta, m_{g-\gamma} \sim 20 \div 200$ MeV only slightly affects the results. The main cooling process is the QDU process on normal quarks. We see that the presence of normal quarks entails too fast cooling. The data could be explained only if all the masses lie in a very narrow interval ($1.21 < M/M_\odot < 1.22$ in our case). In case of the other two crust models the resulting picture is similar.

The existence of only a very narrow mass interval in which the data can be fitted seems us unrealistic as by itself as from the point of view of the observation of the neutron stars in binary systems with different masses, e.g., $M_{\text{B1913+16}} \simeq 1.4408 \pm 0.0003$ $M_\odot$ and $M_{\text{J0737-3039B}} \simeq 1.250 \pm 0.005$ $M_\odot$, cf. [53]. *Thus the data can't be satisfactorily explained.*

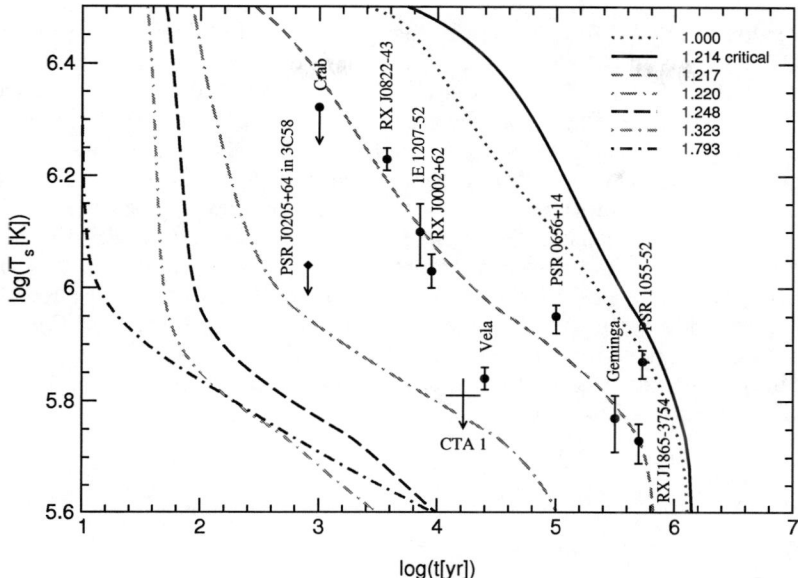

FIGURE 4. Model I. Cooling curves for hybrid star configurations with Gaussian quark matter core in the 2SC phase. The labels correspond to the gravitational masses of the configurations in units of the solar mass.

We first check the case Δ_X to be constant. For the $\Delta_X \simeq 1$ MeV cooing is too slow [2]. It is true for all three crust models. Thus the gaps for formerly unpaired quarks should be still smaller in order to obtain a satisfactory description of the cooling data.

For the $\Delta_X = 30$ keV the cooling data can be fitted but have a very fragile dependence on the gravitational mass of the configuration. Namely, we see that all data points, except the Vela, CTA 1 and Geminga, correspond to hybrid stars with masses in the narrow interval $M = 1.21 \div 1.22\ M_\odot$

Therefore we would like to explore whether a density-dependent X-gap could allow a description of the cooling data within a larger interval of compact star masses.

We employ the ansatz: X-gap as a decreasing function of the chemical potential

$$\Delta_X(\mu) = \Delta_c \, \exp[-\alpha(\mu - \mu_c)/\mu_c] \; , \qquad (1)$$

where the parameters are chosen such that at the critical quark chemical potential $\mu_c = 330$ MeV for the onset of the deconfinement phase transition the X-gap has its maximal value of $\Delta_c = 1.0$ MeV and at the highest attainable chemical potential $\mu_{max} = 507$ MeV, i.e. in the center of the maximum mass hybrid star configuration it falls to a value of the order of 10 keV. We choose the value $\alpha = 10$ for which $\Delta_X(\mu_{max}) = 4.6$ keV. In Fig. 5 we show the resulting cooling curves for the gap model II with gap anzatz 1, which we consider as the most realistic one.

We observe that the mass interval for compact stars which obey the cooling data constraint ranges now from $M = 1.32$ $M_\odot$ for slow coolers up to $M = 1.75$ $M_\odot$ for fast coolers such as Vela, cf. with that we have found with the purely hadronic model [1] with different parameter choices. Note that according to a recently suggested independent test of cooling models [20] by comparing results of a corresponding population synthesis model with the Log N - Log S distribution of nearby isolated X-ray sources the cooling model I did not pass the test. Thereby it would be interesting to see whether our quark model within the gap ansatz II could pass the Log N - Log S test.

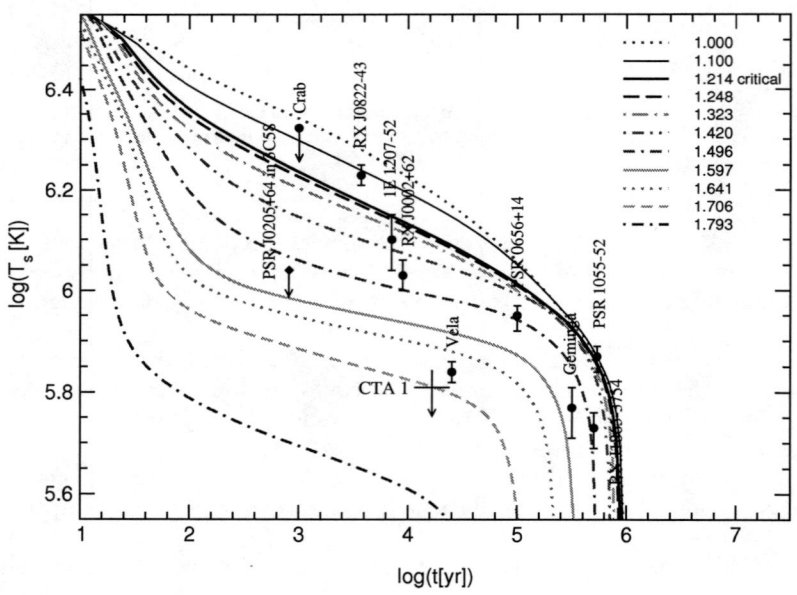

FIGURE 5. Cooling curves for hybrid star configurations with Gaussian quark matter core in the 2SC phase with a density dependent pairing gap according to Eq. (1) for model II.

4. CONCLUSION

- Within a nonlocal, chiral quark model the critical densities for a phase transition to color superconducting quark matter can be low enough for these phases to occur in compact star configurations with masses below 1.3 $M_\odot$.
- For the choice of the Gaussian form-factor the 2SC quark matter phase arises at $M \simeq 1.21\ M_\odot$.
- Without a residual pairing the 2SC quark matter phase could describe the cooling data only if compact stars had masses in a very narrow band around the critical mass for which the quark core can occur.
- Under assumption that formally unpaired quarks can be paired with small gaps $\Delta_X < 1$ MeV (2SC+X pairing), which values we varied in wide limits, only for density dependent gaps the cooling data can be appropriately fitted.

So the present day cooling data could be still explained by hybrid stars, however, when assuming a complex pairing pattern, where quarks are partly strongly paired within the 2SC channel, and partly weakly paired with gaps $\Delta_X < 1$ MeV, being rapidly decreasing with the increase of the density.

It remains to be investigated which microscopic pairing pattern could fulfill the constraints obtained in this work. Another indirect check of the model could be the Log N - Log S test.

ACKNOWLEDGMENTS

The research has been supported by the Virtual Institute of the Helmholtz Association under grant No. VH-VI-041 and by the DAAD partnership programme between the Universities of Yerevan and Rostock. In particular I acknowledge D. Blaschke for his active collaboration and support. The results reported in these Proceedings are obtained in collaboration with my colleagues D. Blaschke, D.N. Voskresensky and D.N. Aguilera. I thank the organizers of Spa HLPR2004 meeting.

REFERENCES

1. D. Blaschke, H. Grigorian, and D.N. Voskresensky, Cooling of Neutron Stars. Hadronic Model, *Astron. Astrophys.*, **424**, 979 (2004).
2. H. Grigorian, D. Blaschke, and D. Voskresensky, *arXiv:astro-ph/0411619* (2004).
3. B. Friman, and O. V. Maxwell, *Astrophys. J*, **232**, 541 (1979).
4. S. Tsuruta, *Phys. Rep.*, **56**, 237 (1979).
5. S. Shapiro, and S. A. Teukolsky,1983, *Black Holes, White Dwarfs, and Neutron Stars: The Physics of Compact Objects* (Wiley, New York), chapter 11
6. Ch. Schaab, D. Voskresensky, A.D. Sedrakian, F. Weber, and M. K. Weigel, *Astron. Astrophys.*, **321**, 591 (1997)
7. D. N. Voskresensky, in: "Physics of Neutron Star Interiors", *Lecture Notes in Physics*, (Eds.) D. Blaschke, N.K. Glendenning, A. Sedrakian, Springer, Heidelberg (2001), p. 467-502.
8. S. Tsuruta, M. A. Teter, T. Takatsuka, T. Tatsumi, and R. Tamagaki, *Astrophys. J*, 571, L143 (2002)

9. D. G. Yakovlev, O. Y. Gnedin, A. D. Kaminker, K. P. Levenfish, and A. Y. Potekhin, *Adv. Space Res.*, **33**, 523 (2004)
10. Page, D., Lattimer, J.M., Prakash, M., Steiner, A.W. [arXive: astro-ph/0403657]
11. S. Tsuruta, 2004, in: *Proceedings of IAU Symposium "Young Neutron Stars, and their Environments"*, F. Camilo, B.M. Gaensler (Eds.), 218
12. A. B. Migdal, E. E. Saperstein, M. A. Troitsky, and D. N. Voskresensky, *Phys. Rep.*, **192**, 179 (1990)
13. R. Rapp, J. Wambach, *Nucl. Phys.*, **A**, **573**, 626 (1994)
14. Yu. B. Ivanov, J. Knoll, H. van Hees, and D. N. Voskresensky, *Phys. Atom. Nucl.*, **64**, 652 (2001)
15. D. N. Voskresensky, and A. V. Senatorov, *JETP Lett.*, **40**, 1212 (1984)
16. D. N. Voskresensky, and A. V. Senatorov, *JETP*, **63**, 885 (1986)
17. D.N. Voskresensky, and A.V. Senatorov, 1987, *Sov. J. Nucl. Phys.*, **45**, 411; A. V. Senatorov, and D. N. Voskresensky, *Phys. Lett.*, **B 184**, 119 (1987)
18. E. E. Kolomeitsev, and D. Voskresensky, *arXiv:nucl-th/0410063* (2004)
19. A. Akmal, V.R. Pandharipande, and D.G. Ravenhall, *Phys. Rev.*, **C 58**, 1804 (1998)
20. S. Popov, H. Grigorian, R. Turolla, and D. Blaschke, *arXiv:astro-ph/0411618* (2004)
21. M. Baldo, G. F. Burgio, and H. J. Schulze, *arXiv:astro-ph/0312446* (2003)
22. D. Blaschke, T. Klähn, and D.N. Voskresensky, *Astrophys. J.* **533**, 406 (2000).
23. D. Page, M. Prakash, J.M. Lattimer, and A. Steiner, *Phys. Rev. Lett.* **85** 2048 (2000).
24. D. Blaschke, H. Grigorian, and D.N. Voskresensky, *Astron. Astrophys.*, **368**, 561 (2001).
25. N. Iwamoto, *Phys. Rev. Lett.* **44** 1637 (1980)
26. H. Heiselberg, and M. Hjorth-Jensen,*Astrophys. J*, **525**, L45 (1999).
27. D. Blaschke, S. Fredriksson, H. Grigorian, and A. Öztas, *Nucl. Phys.*, **A 736**, 203 (2004).
28. H. Grigorian, D. Blaschke, and D.N. Aguilera, *Phys. Rev.*, **C 69**, 065802 (2004).
29. M. Alford, K. Rajagopal, and F. Wilczek, *Phys. Lett.*, **B 422**, 247 (1998).
30. R. Rapp, T. Schäfer, E.V. Shuryak, and M. Velkovsky, *Phys. Rev. Lett.*, **81**, 53 (1998).
31. D. Diakonov, H. Forkel, and M. Lutz, *Phys. Lett.*, **B 373**, 147 (1996); G.W. Carter, and D. Diakonov, *Phys. Rev.* **D 60**, 016004 (1999); R. Rapp, E. Shuryak, and I. Zahed, *Phys. Rev.*, **D 63**, 034008 (2001).
32. D. Blaschke, and C.D. Roberts, *Nucl. Phys.*, **A 642**, 197 (1998); J.C.R. Bloch, C.D. Roberts, and S.M. Schmidt, *Phys. Rev.*, **C 60**, 65208 (1999).
33. M. Alford, K. Rajagopal, and F. Wilczek, *Nucl. Phys.*, **B 357**, 443 (1999); T. Schäfer, and F. Wilczek, *Phys. Rev. Lett.*, **82**, 3956 (1999).
34. M. Alford, J. Berges,, and K. Rajagopal, *Nucl. Phys.*, **B 558**, 219 (1999).
35. A.W. Steiner, S. Reddy, and M. Prakash, *Phys. Rev.* , **D 66**, 094007 (2002).
36. F. Neumann, M. Buballa, and M. Oertel, *Nucl. Phys.*, **A714**, 481 (2003).
37. C. Gocke, D. Blaschke, A. Khalatyan, and H. Grigorian, *arXiv:hep-ph/0104183*. (2002)
38. D.N. Voskresensky, M. Yasuhira, and T. Tatsumi, *Phys. Lett.* **B 541**, 93 (2002); *Nucl. Phys.*, **A 723**, 291 (2002).
39. M. Prakash, J.M. Lattimer, A.W. Steiner, and D. Page, *Nucl. Phys.*, **A 715**, 835 (2003).
40. J. Cottam, F. Paerels, and M. Mendez, *Nature*, **420**, 51 (2002).
41. T. Schäfer, *Phys. Rev.*, **D62**, 094007 (2000).
42. A. Schmitt, Q. Wang, and D. H. Rischke, *Phys. Rev.* , **D66**, 114010 (2002).
43. D. G. Ravenhall, C. J. Pethick, and J. R. Wilson, *Phys. Rev. Lett.*, **50**, 2066,(1983)
44. T. Maruyama,, et al., *arXiv:nucl-th/0402002*, (2004)
45. T. Tatsumi, T. Maruyama, D. N. Voskresensky, T. Tanigawa, S. Chiba. *arXiv: nucl-th/0502040* (2005)
46. Ainsworth, T., Wambach, J., and Pines, D., *Phys. Lett.* **B 222** 173 (1989)
47. A. Schwenk, B. Friman, *Phys. Rev. Lett.***92**, 082501 (2004)
48. T. Takatsuka, and R. Tamagaki *Prog. Theor. Phys.* **112** pp. 37-72 (2004)
49. R. B. Wiringa, V. G. Stoks, and R. Schiavilla, *Phys. Rev.* **C 51**, 38 (1995)
50. V. A. Khodel, J. W. Clark, M. Takano, and M. V. Zverev, Phys. Rev. Lett. **93** 151101 (2004)
51. H. Grigorian, and D. N. Voskresensky, *arXiv:astro-ph/0501678* (2005)
52. P. Jaikumar, and M. Prakash, *Phys. Lett.*, **B 516**, 345 (2001).
53. I.H. Stairs, *Science* **304**, 547 (2004).

Volume Dependence of the Pion Mass from Renormalization Group Flows

B. Klein[**,*,†], J. Braun[**] and H.J. Pirner[**,‡]

[*] *GSI, Planckstrasse 1, 64291 Darmstadt, Germany*
[†] *E-mail: b.klein@gsi.de*
[**] *Institute for Theoretical Physics, University of Heidelberg, Philosophenweg 19, 69120 Heidelberg, Germany*
[‡] *Max-Planck-Institut für Kernphysik, Saupfercheckweg 1, 69117 Heidelberg, Germany*

Abstract. We investigate finite volume effects on the pion mass and the pion decay constant with renormalization group (RG) methods in the framework of a phenomenological model for QCD. An understanding of such effects is important in order to interpret results from lattice QCD and extrapolate reliably from finite lattice volumes to infinite volume.

We consider the quark-meson-model in a finite Euclidean 3 + 1 dimensional volume. In order to break chiral symmetry in the finite volume, we introduce a small current quark mass. In the corresponding effective potential for the meson fields, the chiral $O(4)$-symmetry is broken explicitly, and the sigma and pion fields are treated individually. Using the proper-time renormalization group, we derive renormalization group flow equations in the finite volume and solve these equations in the approximation of a constant expectation value.

We calculate the volume dependence of pion mass and pion decay constant and compare our results with recent results from chiral perturbation theory in finite volume.

Keywords: Renormalization group, chiral perturbation theory, finite volume effects, lattice gauge theory
PACS: 12.38.Lg, 12.39.Fe

1. INTRODUCTION

Two important motivations inspire us to study QCD in a finite volume: a generic interest in finite volume effects, and the separation of scales afforded by the additional length scale. Among nonperturbative methods used in the study of QCD, lattice simulations have a very important place. However, they are necessarily performed in relatively small volumes, and a good understanding of finite volume effects is required for an extrapolation to infinite volume. Typically, lattice sizes are of the order of $L \lesssim 2-3$ fm, while pion masses are of the order $m_\pi \gtrsim 500-700$ MeV [1, 2]. Thus, we require tools to describe the volume dependence of observables in these regimes. In this work, our objective is to describe the volume dependence of two of the low energy observables, the pion mass and the pion decay constant, and to provide a model for finite volume effects across a wide range of length scales.

The low energy behavior of QCD is determined by spontaneous chiral symmetry breaking [3]: As a consequence, massless Goldstone bosons associated with the broken symmetries emerge, and the low energy limit can be described by an effective theory of these light, weakly interacting degrees of freedom. Such descriptions in terms of light degrees of freedom only become even better in finite Euclidean volumes: contributions of heavier particles are suppressed by e^{-ML}, where M is the typical separation of the hadronic mass scale from the Goldstone masses. Therefore, a second motivation to study finite volume QCD is that it allows isolation of the low-energy behavior and a description in terms of the Goldstone modes.

Different approaches based on this idea have proven very fruitful: Chiral perturbation theory makes predictions for the volume and temperature dependence of pion mass, pion decay constant and chiral condensate [4, 5, 6], finite volume partition functions [7] and random matrix theory [8, 9] predict eigenvalue spectra, and the dependence of the chiral condensate on volume and quark mass [10].

The most important results in the present context are those from chiral perturbation theory (chPT), which relies on an expansion in terms of the three-momentum $|\vec{p}|$ and the pion mass m_π, which are small compared to the chiral symmetry breaking scale $4\pi f_\pi$. Consequently, a finite volume places constraints on the expansion, since the smallest momentum is determined by the volume, $p_{\min} \sim \frac{2\pi}{L}$, and requires different expansion schemes, depending on the relative size of L and $1/m_\pi$.

A very useful result obtained by Lüscher [11] relates the leading corrections to the pion mass in finite Euclidean volume to the $\pi\pi$-scattering amplitude in infinite volume. The relative shift $R[m_\pi(L)]$ of the pion mass $m_\pi(L)$ in finite volume compared to its value $m_\pi(\infty)$ in infinite volume is according to this result given by

$$\begin{aligned} R[m_\pi(L)] &= \frac{m_\pi(L) - m_\pi(\infty)}{m_\pi(\infty)} \\ &= -\frac{3}{16\pi^2} \frac{1}{m_\pi} \frac{1}{m_\pi L} \int_{-\infty}^{\infty} dy\, F(iy) e^{-\sqrt{m_\pi^2 + y^2} L} + \mathcal{O}(e^{-\bar{m}L}). \end{aligned} \quad (1)$$

$F(s)$ is the forward $\pi\pi$-scattering amplitude as a function of the energy variable s, continued to complex values, and the sub-leading corrections drop at least as $\mathcal{O}(e^{-\bar{m}L})$ where $\bar{m} \geq \sqrt{3/2}\, m_\pi$. New results have been obtained by using a calculation of the $\pi\pi$-scattering amplitude in chPT to three loops (nnlo) as input for Lüscher's formula [12, 13]. The shift above the leading one-loop result from this approach can then be used to supplement the full one-loop calculation [5] for the mass shift in chPT.

However, this explanation relies purely on the "squeezing" of a "pion cloud" [1], whereas - albeit smaller - shifts of meson and hadron masses also appear in quenched lattice calculations, show a power law behavior for small L rather than the behavior expected from chPT [14, 15], and are generally underestimated [2]. This suggests that in addition to the pion effects, quark effects at higher momentum scales also contribute directly to the finite volume mass shifts.

In the present investigation, we thus use a phenomenological low-energy model which is suited to a description of dynamical chiral symmetry breaking and which incorporates heavy constituent quarks. As a drawback, this model is not a gauge theory and the constituent quarks are not confined, but decouple from the dynamics at low momenta only because of their large masses. Since spontaneous symmetry breaking does not occur in a finite volume, the inclusion of a small quark mass, which explicitly breaks the chiral symmetry, is essential.

We employ an RG method in order to cover the relevant range of length scales and pion masses. Thus, we do not have to rely on either the box size or the pion mass as an expansion parameter: the RG flow equations remain valid as long as the model used as input does. The great advantage is precisely that the RG flow equations describe the connection between different momentum scales, and in the present case also the dependence on the additional scale $1/L$ introduced by the finite volume.

2. RENORMALIZATION GROUP FLOW EQUATIONS

In the present work, we use the two-flavor quark-meson model, an $O(4)$-invariant linear sigma model where the meson fields $\phi = (\sigma, \vec{\pi})$ are coupled to two constituent quarks through an $SU(2) \otimes SU(2)$-invariant interaction term. It is an effective model for dynamical chiral symmetry breaking below a scale $\Lambda \simeq 1.5$ GeV, where a description of QCD in terms of hadronic degrees of freedom is valid. Although the linear sigma-model by itself is not compatible with the low-energy $\pi\pi$-scattering data, due to the presence of quarks the low energy constants of chiral perturbation theory are reproduced [16].

We consider the model in a four-dimensional Euclidean volume with compact Euclidean space and time directions. The bare effective action at the scale Λ is

$$\Gamma_\Lambda[\phi] = \int d^4x \left\{ \bar{q}\gamma \cdot \partial q + g\bar{q}(m_c + \sigma + i\vec{\tau} \cdot \vec{\pi}\gamma_5)q + \frac{1}{2}(\partial_\mu \phi)^2 + U_\Lambda(\phi) \right\}, \quad (2)$$

where the quark mass term gm_c explicitly breaks the chiral symmetry. At the scale Λ, the meson potential can be characterized by the values of two couplings:

$$U_\Lambda(\phi) = \frac{1}{2}m_{UV}^2 \phi^2 + \frac{1}{4}\lambda_{UV}(\phi^2)^2. \quad (3)$$

In Gaussian approximation, the one-loop effective action for the scalar fields is

$$\Gamma[\phi] = \Gamma_\Lambda[\phi] - \text{Tr} \log\left(\Gamma_F^{(2)}[\phi]\right) + \frac{1}{2} \text{Tr} \log\left(\Gamma_B^{(2)}[\phi]\right), \quad (4)$$

where $\Gamma_B^{(2)}[\phi]$ and $\Gamma_F^{(2)}[\phi]$ are the inverse two-point functions for the bosonic and the fermionic fields, evaluated at the expectation value of the mesonic field, ϕ. We consider an approximation where the field ϕ is constant over the entire volume and the effective action is reduced to an effective potential, $\Gamma[\phi] = \int d^4x \, U(\phi)$. To regularize the functional traces, we use a Schwinger proper-time representation of the logarithms. The dependence on a cutoff scale k is introduced through an infrared cutoff function $f_a(\tau k^2)$ of the form

$$k\frac{\partial}{\partial k}f_a(\tau k^2) = -\frac{2}{a!}(\tau k^2)^{a+1}\exp(-\tau k^2) \quad (5)$$

which satisfies all required regularization conditions [17, 18]. For reasons of convergence of the momentum integrals, we use $a = 2$. Such a cutoff function makes it possible to systematically integrate out only those quantum fluctuations around the expectation value which have momenta above the scale k.

We now wish to obtain a renormalization group flow equation for the effective action, which describes how the couplings in the action change with a change of the renormalization scale k. We arrive at such an equation by taking the derivative of the regularized expression for the effective action with respect to this scale k, and then replacing the bare two-point function from the original expression by the renormalized two-point functions, which contain the scale-dependent couplings:

$$k\frac{\partial}{\partial k}\Gamma_k[\phi] = \frac{1}{2}\text{Tr}\int_0^\infty \frac{d\tau}{\tau}\left[k\frac{\partial}{\partial k}f_a(\tau k^2)\right]\exp[-\tau\Gamma_{B,k}^{(2)}[\phi]]$$
$$-\text{Tr}\int_0^\infty \frac{d\tau}{\tau}\left[k\frac{\partial}{\partial k}f_a(\tau k^2)\right]\exp[-\tau\Gamma_{F,k}^{(2)}[\phi]]. \quad (6)$$

Note that boundary conditions for the fields are implicit in the traces. In Euclidean time direction, the sum over bosonic and fermionic Matsubara frequencies can be done analytically [19] before the zero temperature limit is taken. In infinite volume, the momentum integrations for the spatial directions can be performed as well, and the flow equation becomes

$$k\frac{\partial}{\partial k}U_k(\sigma,\vec{\pi}^2, L\to\infty) = \frac{k^6}{32\pi^2}\left(-\frac{4N_cN_f}{k^2+M_q^2}+\sum_{i=1}^4\frac{1}{k^2+M_i^2}\right). \quad (7)$$

In finite volume, the momentum integrations become sums over momentum modes. Choosing anti-periodic boundary conditions for the fermionic fields also in the spatial directions, we define

$$p_F^2 = \sum_{i=1}^{d-1}p_i^2 = \frac{4\pi^2}{L^2}\sum_{i=1}^{d-1}\left(n_i+\frac{1}{2}\right)^2, \qquad p_B^2 = \sum_{i=1}^{d-1}p_i^2 = \frac{4\pi^2}{L^2}\sum_{i=1}^{d-1}n_i^2. \quad (8)$$

The RG flow equation for the meson potential in finite volume then reads

$$k\frac{\partial}{\partial k}U_k(\sigma,\vec{\pi}^2,L) = \frac{3}{16}\frac{k^6}{L^3}\sum_{\vec{n}}\left(-\frac{4N_cN_f}{(k^2+p_F^2+M_q^2)^{5/2}}+\sum_{i=1}^4\frac{1}{(k^2+p_B^2+M_i^2)^{5/2}}\right). \quad (9)$$

The meson masses $M_i^2(\sigma,\vec{\pi}^2)$ are the eigenvalues of $[U_k''(\sigma,\vec{\pi}^2)]^{ij}$, the second derivative matrix of the meson potential. In the presence of a finite current quark mass gm_c, the quark mass $M_q^2(\sigma,\vec{\pi}^2) = g^2[(\sigma_0+m_c)^2 + 2m_c(\sigma-\sigma_0)+(\sigma^2+\vec{\pi}^2-\sigma_0^2)]$ contains explicitly symmetry breaking terms. Since symmetry breaking terms appear only here and are all of this form, we make for the meson potential the ansatz

$$U_k(\sigma,\vec{\pi}^2) = \sum_{i=0}^4\sum_{j=0}^{[\frac{1}{2}(4-i)]}a_{ij}(k)(\sigma-\sigma_0(k))^i(\sigma^2+\vec{\pi}^2-\sigma_0(k)^2)^j. \quad (10)$$

It depends only on $\vec{\pi}^2$ since the symmetry of the pion subspace remains unbroken.

TABLE 1. Values for the parameters at the UV-scale Λ which were used in the numerical evaluation. The parameters are obtained by fitting the results of the RG evolution to a particular pion mass and the corresponding pion decay constant in infinite volume. The physical current quark mass corresponds to gm_c.

Λ [MeV]	m_{UV} [MeV]	λ_{UV}	gm_c [MeV]	f_π [MeV]	m_π [MeV]
1500	779.0	60	2.10	90.38	100.8
1500	747.7	60	9.85	96.91	200.1
1500	698.0	60	25.70	105.30	300.2

3. RESULTS OF THE NUMERICAL CALCULATIONS

We solve the RG flow equations numerically for infinite and finite volume. In order to do this, we insert the ansatz (10) into the flow equations (7),(9) and obtain a set of coupled, ordinary differential equations for the couplings $a_{ij}(k)$ and the minimum $\sigma_0(k)$. We choose initial values for the couplings at the scale $k = \Lambda$ and evolve the couplings until at $k \to 0$ all quantum fluctuations have been integrated out. Pion mass and pion decay constant are extracted from the resulting effective potential. For the case of finite volume, the sums over the momentum modes need to be truncated at a finite number of modes. For the results presented here, where we use $n_{\max} = 40$, the effects of this truncation are negligible [20].

In TAB. 1, we give an overview over the three parameter sets we have used to obtain results for pion masses of 100, 200 and 300 MeV. These parameters are obtained by fitting so that the results of the RG evolution match a particular value of the pion mass $m_\pi(\infty)$ and the corresponding value of the pion decay constant $f_\pi(\infty)$ from chPT in infinite volume. With these fixed parameter values, we solve the finite volume RG equations, which gives a prediction for the volume dependence of $m_\pi(L)$ and $f_\pi(L)$. The coupling g does not evolve in our approximation and was set to $g = 3.26$, which leads to a reasonable constituent quark mass for physical values of f_π and m_π. We present our results for the volume dependence of the pion mass in FIGS. 1-3 and numerical values for select volumes in TAB. 2. We will focus on these results for the pion mass, since they are more easily accessible on the lattice than the pion decay constant. For the results on $f_\pi(L)$, we refer to [20].

FIGS. 1-3 show the relative change $R[m_\pi(L)] = \frac{m_\pi(L) - m_\pi(\infty)}{m_\pi(\infty)}$ of the pion mass in finite volume $m_\pi(L)$ compared to the value $m_\pi(\infty)$ in infinite volume as a function of the volume size L. We plot the results for the pion masses $m_\pi(\infty) = 100, 200, 300$ MeV on a logarithmic scale. For comparison, we plot in addition the one-loop chPT result of [5], and the one-loop chPT result with corrections to three-loop order obtained with the Lüscher formula [12].

The relative change of the pion mass decreases with increasing volume size. The RG results are consistently above the results from chPT. Higher loop order corrections in chPT increase the mass shift predicted by chPT, and the difference

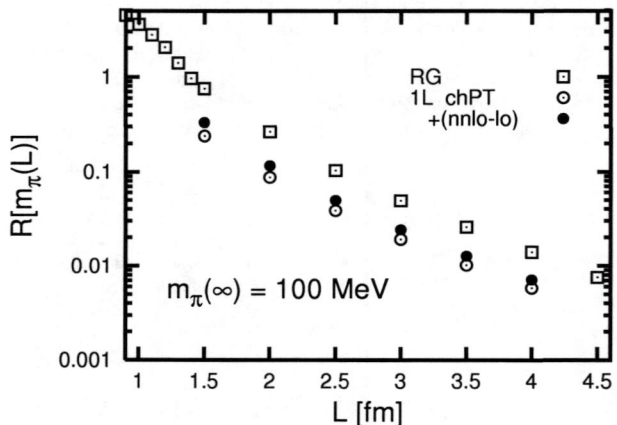

FIGURE 1. Volume dependence of the pion mass for a value $m_\pi(\infty) = 100$ MeV in infinite volume. We plot the relative shift of the pion mass from its infinite volume limit $R[m_\pi(L)] = (m_\pi(L) - m_\pi(\infty))/m_\pi(\infty)$ as a function of the size of the volume L. For comparison to the RG result, we also plot the results from chPT calculations taken from [12] for the full one-loop result (1L chPT) and with *nnlo*-corrections (1L chPT + (*nnlo-lo*)).

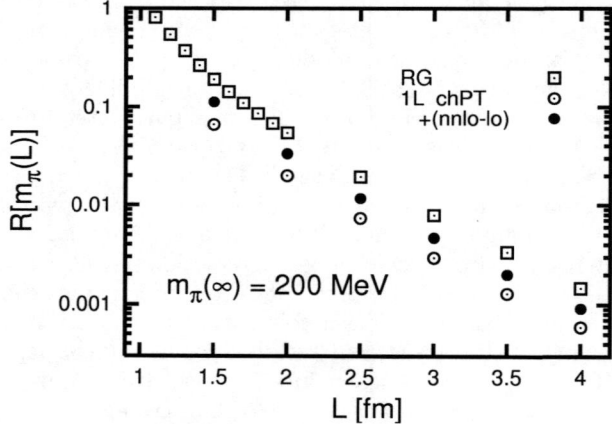

FIGURE 2. Volume dependence of the pion mass for a value $m_\pi(\infty) = 200$ MeV in infinite volume. For a detailed explanation, see caption of FIG. 1. Note the different scales on the axes for different values of $m_\pi(\infty)$.

to the RG result becomes smaller. We note that the relative size of the higher loop order corrections to the one-loop chPT result become larger with larger pion mass. This is not unexpected, since Lüscher's formula becomes an increasingly better approximation with increasing pion mass for a given volume size. For larger volumes, the differences between the RG result and the one-loop chPT

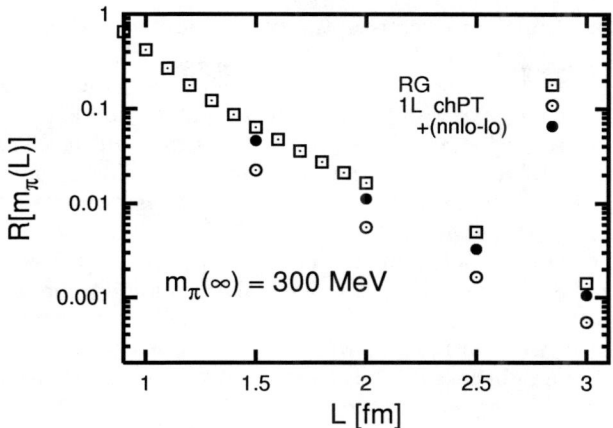

FIGURE 3. Volume dependence of the pion mass for a value $m_\pi(\infty) = 300$ MeV in infinite volume. For a detailed explanation, see caption of FIG. 1. Note the different scales on the axes for different values of $m_\pi(\infty)$.

TABLE 2. Values for $R[m_\pi(L)]$, the relative shift of the pion mass in finite volume compared to the value in infinite volume (cf. eq. (1)). We show results for volume sizes $L = 2.0, 2.5, 3.0$ fm, and for pion masses $m_\pi(\infty) = 100, 200, 300$ MeV. The RG results are compared to the exact one-loop chPT results of [5] (1L chPT), and to the exact one-loop calculation with corrections in three-loop order obtained with chPT and Lüscher's formula [12] (1L chPT + (*nnlo-lo*)). In the last column, the difference ΔR between the RG result and the three-loop corrected chPT result is given.

L	$m_\pi(\infty)$	$m_\pi L$	$R[m_\pi(L)]$			ΔR
[fm]	[MeV]		RG	1L chPT	+(*nnlo-lo*)	
2.0	100	1.01293	26.6×10^{-2}	8.74×10^{-2}	11.6×10^{-2}	15.0×10^{-2}
	200	2.02586	5.38×10^{-2}	2.00×10^{-2}	3.31×10^{-2}	2.07×10^{-2}
	300	3.03879	1.70×10^{-2}	0.56×10^{-2}	1.12×10^{-2}	0.58×10^{-2}
2.5	100	1.26616	10.37×10^{-2}	3.85×10^{-2}	4.97×10^{-2}	5.40×10^{-2}
	200	2.53233	1.95×10^{-2}	0.73×10^{-2}	1.17×10^{-2}	0.78×10^{-2}
	300	3.79849	5.31×10^{-3}	1.65×10^{-3}	3.27×10^{-3}	2.04×10^{-3}
3.0	100	1.5194	4.94×10^{-2}	1.91×10^{-2}	2.41×10^{-2}	2.53×10^{-3}
	200	3.03879	7.85×10^{-3}	2.95×10^{-3}	4.65×10^{-3}	3.20×10^{-3}
	300	4.55819	1.76×10^{-3}	0.54×10^{-3}	1.05×10^{-3}	0.71×10^{-3}

calculation with the corrections obtained with Lüscher's formula drop exponentially as $\exp(-cm_\pi L)$ with $c > 0$, compatible with the error estimate from Lüscher's result. As expected, for large volume the slopes of the chPT results and the RG results in the logarithmic plot are the same, since the mass shift is then due entirely to pion effects and thus should drop as $\exp(-m_\pi L)$. In contrast to chPT, the RG calculation can also be extended to very small volumes, where the chiral expansion is considered unreliable, and describe the transition into the regime where chiral symmetry is effectively restored.

In our model, two effects are responsible for the finite volume mass shift. Chiral symmetry is broken dynamically and light Goldstone modes appear when the quark fields form a condensate. The presence of the additional scale $1/L$ affects this condensation. In a finite volume, the lowest possible momentum mode of the fermions is $\sqrt{3}\pi/L$. For small volumes, this acts as an infrared cutoff, which effectively "freezes" the quark fields so that fewer modes can contribute to the condensate. The mesonic fields, on the other hand, are much less affected by the finite volume, since the scale $2\pi/L$ imposes only a minimum for the smallest non-zero momentum mode. Fluctuations due to these light mesonic modes lead to a further decrease of the condensate, which in turn leads to the observed increase in the masses of the light mesons. These pion effects dominate for larger volumes, which explains the convergence of the RG and chPT results for large L.

4. CONCLUSIONS

We have presented a study of finite volume effects on the pion mass and on the pion decay constant. We have applied renormalization group methods to a phenomenological model of low-energy QCD with mesons and constituent quarks as degrees of freedom. We solved the resulting RG flow equations in finite volume. We stress the importance of taking explicit symmetry breaking into account and of considering pions and the sigma as individual degrees of freedom.

We compared our results for the volume dependence of the pion mass to recent results obtained in chiral perturbation theory. While they are compatible for small pion masses and large volumes, we find from the RG generally a larger mass shift than the one predicted by chiral perturbation theory. We explain this for small volumes by the effect of the additional IR cutoff L on the quark condensation, while for larger volumes the fluctuations due to light pions are the dominating effect. This view is supported by the convergence between the RG and chPT results for large volumes. Thus, the RG approach in a model including quarks provides a mechanism to qualitatively understand the volume dependence. The RG results remain valid even to very small volumes, where the chiral expansion becomes unreliable.

An important point that still requires careful investigation is the influence of the choice of the boundary conditions for the fermionic fields in the spatial directions. As on the lattice, where anti-periodic boundary conditions for quarks lead to a larger finite volume shift than periodic boundary conditions [14], we observe a dependence of our results on this choice. This will be the topic of future work.

ACKNOWLEDGMENTS

We would like to thank the organizers for a very pleasurable and inspiring meeting, for the hospitality in Belgium and for a thoroughly enjoyable experience. J.B. would like to thank the GSI for financial support.

REFERENCES

1. A. Ali Khan *et al.* [QCDSF-UKQCD Collaboration], Nucl. Phys. B **689** (2004) 175.
2. S. Aoki *et al.* [JLQCD Collaboration], Phys. Rev. D **68** (2003) 054502.
3. J. Gasser and H. Leutwyler, Annals Phys. **158** (1984) 142.
4. J. Gasser and H. Leutwyler, Phys. Lett. B **184** (1987) 83.
5. J. Gasser and H. Leutwyler, Phys. Lett. B **188** (1987) 477.
6. J. Gasser and H. Leutwyler, Nucl. Phys. B **307** (1988) 763.
7. H. Leutwyler and A. Smilga, Phys. Rev. D **46** (1992) 5607.
8. E. V. Shuryak and J. J. M. Verbaarschot, Nucl. Phys. A **560**, 306 (1993).
9. J. J. M. Verbaarschot, Phys. Rev. Lett. **72**, 2531 (1994).
10. J. J. M. Verbaarschot, Phys. Lett. B **368**, 137 (1996).
11. M. Lüscher, Commun. Math. Phys. **104** (1986) 177.
12. G. Colangelo and S. Dürr, Eur. Phys. J. C **33** (2004) 543.
13. G. Colangelo and C. Haefeli, Phys. Lett. B **590** (2004) 258.
14. S. Aoki *et al.*, Phys. Rev. D **50** (1994) 486.
15. M. Fukugita, H. Mino, M. Okawa, G. Parisi and A. Ukawa, Phys. Lett. B **294** (1992) 380.
16. D. U. Jungnickel and C. Wetterich, Eur. Phys. J. C **2**, 557 (1998).
17. B. J. Schaefer and H. J. Pirner, Nucl. Phys. A **660** (1999) 439.
18. G. Papp, B. J. Schaefer, H. J. Pirner and J. Wambach, Phys. Rev. D **61**, 096002 (2000).
19. J. Braun, K. Schwenzer and H. J. Pirner, Phys. Rev. D **70** (2004) 085016.
20. J. Braun, B. Klein and H. J. Pirner, Phys. Rev. D **71** (2005) 014032.

Statistical physics on the light-front

J. Raufeisen

Institut für Theoretische Physik der Universität, Philosophenweg 19, 69120 Heidelberg, Germany
E-mail: J.Raufeisen@tphys.uni-heidelberg.de

Abstract. The formulation of statistical physics using light-front quantization, instead of conventional equal-time boundary conditions, has important advantages for describing relativistic statistical systems, such as heavy ion collisions. We develop light-front field theory at finite temperature and density with special attention to Quantum Chromodynamics. We construct the most general form of the statistical operator allowed by the Poincaré algebra and introduce the chemical potential in a covariant way. In light-front quantization, the Green's functions of a quark in a medium can be defined in terms of just 2-component spinors and does not lead to doublers in the transverse directions. A seminal property of light-front Green's functions is that they are related to parton densities in coordinate space. Namely, the diagonal and off-diagonal parton distributions measured in hard scattering experiments can be interpreted as light-front density matrices.

Keywords: Light-Front Quantization, Thermal Field Theory, Generalized Parton Distributions
PACS: 11.10.Wx, 12.38.Lg, 12.38.Mh

1. INTRODUCTION

It has been known for a long time that Dirac's front form of relativistic dynamics [1] has remarkable advantages in high energy and nuclear physics. Until now, however, most applications of the front form refer to the case of zero temperature. It is clearly important to exploit the advantages of light-front quantization also for thermal field theory.

There are three forms of Hamiltonian dynamics, as illustrated in Fig. 1. In the familiar instant form, initial conditions are defined on a hypersurface with $r^0 = 0$ and propagated into the future by the Hamiltonian $\hat{P}^0$. However, it is equally well possible to set initial conditions on a hypersurface defined by $r^+ = r^0 + r^3 = 0$, the so-called light front. The Poincaré generator $\hat{P}^-$ then propagates the system in light-front time r^+. The spatial coordinates are denoted by $\underline{r} = (r^-, \vec{r}_\perp)$, with $r^- = r^0 - r^3$. Note that the metric tensor is not diagonal in light-front coordinates. In particular, one has $r^\pm = 2r_\mp$.

Most appealing is the simplicity of the vacuum (the ground state of the free theory is also the ground state of the full theory) and the existence of boost-invariant light-cone wavefunctions (see Ref. [2] for a review.) This makes light-front quantization a natural candidate for the description of systems for which boost invariance is an issue, such as the fireball created in a heavy ion collision or the small-x features of a nuclear wavefunction.

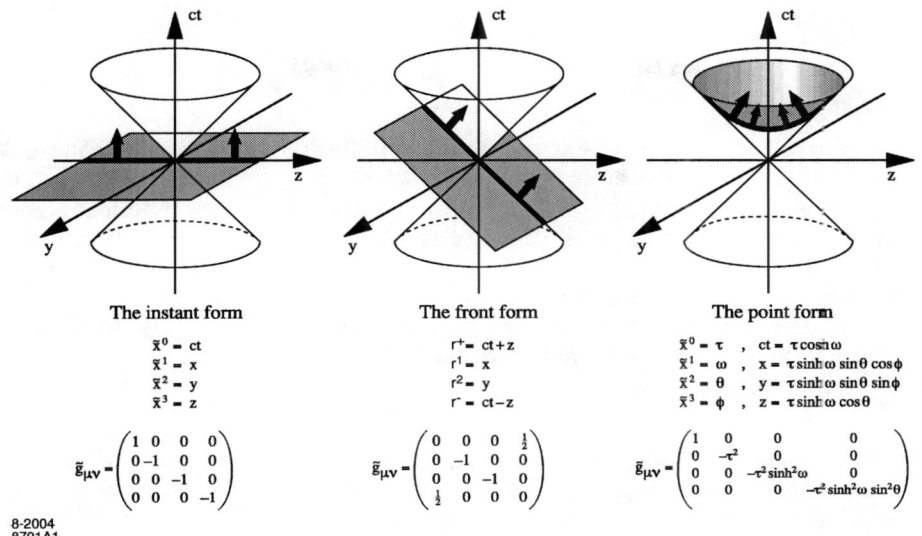

FIGURE 1. The three forms of Hamiltonian dynamics [1, 2]. In the front form (middle), initial conditions are defined on a light-like surface, and the system is the propagated in light-cone time r^+ by the light-cone Hamiltonian $\widehat{P}^-$. (Reprinted figure with permission from Ref. [3]. Copyright 2003 by the American Physical Society.)

In this talk, we report our recent findings, published in Ref. [3]. We applied frontform dynamics to statistical physics and investigated the prospects and challenges of this approach for quantum chromodynamic systems. Valuable work in this direction has already been done by several authors [4, 5, 6].

2. CONSTRUCTION OF THE STATISTICAL OPERATOR

The form of the statistical operator $\widehat{w}$ at finite temperature and density can be obtained from very general considerations. Our result for $\widehat{w}$, is compatible with the findings of [4, 5], i.e. $\widehat{w}$ is always the exponential of the equal time energy $\widehat{P}^0$ in the local rest frame of the system. Our derivation follows Ref. [7].

The light-front Liouville theorem [3] ($i\partial^-\widehat{w}=[\widehat{P}^-,\widehat{w}]$) requires that in equilibrium, $\widehat{w}$ is a function of only those Poincaré generators which commute with the light-front Hamiltonian $\widehat{P}^-$. In addition, since systems far apart from each other must be uncorrelated, the density operator of the combined system has to factorize into the density operators of the subsystems. Consequently, in equilibrium $\ln(\widehat{w})$ must be a linear combination of the additive constants of motion, namely the four components of the momentum $\widehat{P}^\mu$ and the 3-component of the angular momentum vector $\vec{J}$. Hence,

$$\ln(\widehat{w}) = \alpha - \beta \left(u_\nu \widehat{P}^\nu - \omega \widehat{J}_3 - \sum_l \mu_l \widehat{Q}_l \right). \tag{1}$$

Here, $\beta = 1/T$ is the inverse temperature and u_ν is the four velocity of the system, cf. Ref. [5]. In addition, ω is the angular velocity at which the body rotates. Additional conserved charges $\widehat{Q}_l$ are included along with their chemical potentials μ_l. In quantum mechanics, of course, one can simultaneously specify only charges which commute with each other, e.g. one cannot specify all four P_ν for systems with nonzero angular momentum.

The chemical potential may be included in a covariant way by modifying the translation generators as if μu^ν were gauge fields,

$$\widehat{\mathcal{P}}^\nu = \widehat{P}^\nu - \mu \widehat{Q} u^\nu. \tag{2}$$

These operators propagate the system along trajectories of constant charge and take into account that the medium moves as a whole as the "test particle" represented by the Green's function propagates from r_2 to r_1. The definition of the Heisenberg operators has to be changed accordingly.

We choose $\alpha = 0$ as normalization in Eq. (1), so that the partition function is given by $\mathcal{Z} = \text{Tr}\widehat{w}$. The grand-canonical ensemble can now be written in terms of light-cone wavefunctions $\phi_{n/h}(X)$ as (let $\vec{P}_\perp = \vec{0}_\perp$),

$$\begin{aligned} \widehat{w} &= \sum_h \sum_{n,n'} \sum_{X,X'} \exp\left\{-\beta \left[u^+ \frac{M_h^2}{2P^+} + \frac{u^-}{2} \sum_i p_i^+ - \mu Q \right] \right\} \\ &\times \phi_{n/h}(X) \phi_{n'/h}^*(X') |nX\rangle\langle n'X'|. \end{aligned} \tag{3}$$

The states $|nX\rangle$ are eigenstates of the free light-front Hamiltonian. The light-cone wavefunctions $\phi_{n/h}(X)$ are the coefficients in the expansion of an eigenstate h of the full Hamiltonian in eigenstates of the free Hamiltonian. Since the wavefunctions and the masses M_h of the (full) eigenstates can (in principle) be obtained from discretized light-cone quantization (DLCQ) [8], this expression shows that one can also calculate all thermodynamic properties of a field theory from DLCQ.

Furthermore, since $\mathcal{Z}$ is a Lorentz scalar, all thermodynamic potentials and the entropy transform as scalars, e.g. the Lorentz invariant generalization of the grand-canonical potential (or of the free energy in the case of $\mu = 0$) is $\Omega = -T \ln \mathcal{Z}(V, T, \mu)$, and the entropy is given by

$$S = -\left(\frac{\partial \Omega}{\partial T} \right)_{\mu,V} = -\frac{1}{\mathcal{Z}} \text{Tr}(\widehat{w} \ln \widehat{w}). \tag{4}$$

The role of the total energy of the system is now played by the expectation value of $u_\nu \widehat{P}^\nu$, $U = \langle u_\nu \widehat{P}^\nu \rangle$. As usual, $\Omega = U - TS - \mu Q$. All known relations between thermodynamic potentials remain valid.

The quantities β, u_ν, ω and μ_l, have the meaning of Lagrange multipliers that hold the mean values of the constants of motion fixed, while entropy is maximized.

In an ideal gas for example, the maximum entropy is attained for occupation numbers given by Fermi-Dirac and Bose-Einstein statistics [4, 5],

$$n(u_\nu p^\nu) = \frac{g}{e^{\beta(u_\nu p^\nu - \mu)} \pm 1} - \frac{g}{e^{\beta(u_\nu p^\nu + \mu)} \pm 1}, \quad (5)$$

assuming that particles carry charge +1 and antiparticles charge -1. (Note that $u_\nu p^\nu \geq 0$.) Here, p^ν is the 4-momentum of a single particle and the degeneracy factor for different spin states is denoted by g. The Lagrange multipliers define the equilibrium conditions for two systems. In complete equilibrium with each other, both systems must have the same values of temperature, u_ν, ω and μ_l, i.e. no internal motion of macroscopic parts of the system is possible in equilibrium (at least in the absence of vortex lines [7].)

The simplicity of the light-front vacuum, usually considered an advantage, seems to bear problems as far as phase transitions are concerned. However, the statistical weight of a configuration is maximized for minimal equal-time energy rather than for minimal light-front energy. Therefore, the ground state, i.e. the state the system is in at $T = 0$, is in general different from the light-front vacuum. For that reason, the authors of Ref. [5] obtain the standard pattern of spontaneous symmetry breaking in ϕ^4-theory with negative mass squared. No problem arises from $1/k^+$-poles. In addition, in Ref. [4] the chiral phase transition in the Nambu–Jona-Lasinio Model on the light-front is reproduced. We conclude that this approach is poised for the study of phase transitions in more complicated field theories, such as QCD.

3. GREEN'S FUNCTION OF A FERMION IN A MEDIUM

Until now one could get the impression that thermodynamics and statistical physics on the light front are identical to the usual instant form approach, except for a trivial change of variables. That this is not the case becomes most clear, when one studies fermions on the light-front.

It is a special property of front form dynamics that the Dirac equations are a set of two coupled spinor equations [2], only one of which contains a time derivative, $\partial^- = \partial/\partial r_- = 2\partial/\partial r^+$. This property of the light-front Dirac equation becomes most transparent in the following representation of the γ-matrices [9]. In 2×2 block matrix notation,

$$\gamma^+ = \gamma^0 + \gamma^3 = \begin{pmatrix} 0 & 0 \\ 2\imath & 0 \end{pmatrix}, \quad \gamma^- = \gamma^0 - \gamma^3 = \begin{pmatrix} 0 & -2\imath \\ 0 & 0 \end{pmatrix}, \quad \vec{\gamma}_\perp = \begin{pmatrix} -\imath\vec{\sigma}_\perp & 0 \\ 0 & \imath\vec{\sigma}_\perp \end{pmatrix}, \quad (6)$$

where $\vec{\sigma}_\perp = (\sigma_1, \sigma_2)$ are two of the three Pauli spin matrices. The operators projecting out the free (Λ_+) and the constrained (Λ_-) components of a 4-spinor take the simple form

$$\Lambda_+ = \frac{1}{4}\gamma^-\gamma^+ = \begin{pmatrix} 1 & 0 \\ 0 & 0 \end{pmatrix}, \quad \Lambda_- = \frac{1}{4}\gamma^+\gamma^- = \begin{pmatrix} 0 & 0 \\ 0 & 1 \end{pmatrix}. \quad (7)$$

Multiplying the Dirac equation $(i\not{\partial} - m)\tilde{\psi}(r) = 0$ with one of the two projection operators, one obtains

$$i\partial^- \psi(r) = \left(-i\vec{\sigma}_\perp \cdot \vec{\partial}_\perp - im\right) \eta(r) \qquad (8)$$

$$i\partial^+ \eta(r) = \left(-i\vec{\sigma}_\perp \cdot \vec{\partial}_\perp + im\right) \psi(r). \qquad (9)$$

Only one of these equations contains a time derivative, the other one is a constraint. As a consequence, the entire theory can be formulated in terms of 2-component spinors, very much like a non-relativistic theory.

The usual 4-component spinor $\tilde{\psi}$ is related to ψ and η by

$$\tilde{\psi}(r) = \begin{pmatrix} \psi(r) \\ \eta(r) \end{pmatrix}. \qquad (10)$$

The free particle solutions for positive $(\tilde{\psi}(r) = u(k)e^{-ikx})$ and negative energies $(\tilde{\psi}(r) = v(k)e^{+ikx})$ read,

$$u(k,\lambda) = \sqrt{k^+} \begin{pmatrix} \chi_\lambda \\ \frac{\vec{\sigma}_\perp \cdot \vec{k}_\perp + im}{k^+} \chi_\lambda \end{pmatrix} \quad , \quad v(k,\lambda) = \sqrt{k^+} \begin{pmatrix} \chi_{-\lambda} \\ \frac{\vec{\sigma}_\perp \cdot \vec{k}_\perp - im}{k^+} \chi_{-\lambda} \end{pmatrix}, \qquad (11)$$

where χ_λ is an eigenstates of σ_3 with eigenvalue λ.

The fermion field operators for the dynamical spinor components in the Schrödinger picture are expanded as

$$\hat{\Psi}(\underline{r}) = \sum_\lambda \int \frac{d^3k}{(2\pi)^3 2\sqrt{k^+}} \Theta(k^+) \left\{ \hat{b}(\underline{k},\lambda) \chi_\lambda e^{-i\underline{k}\cdot\underline{r}} + \hat{d}^\dagger(\underline{k},\lambda) \chi_{-\lambda} e^{+i\underline{k}\cdot\underline{r}} \right\}, \qquad (12)$$

$$\hat{\Psi}^\dagger(\underline{r}) = \sum_\lambda \int \frac{d^3k}{(2\pi)^3 2\sqrt{k^+}} \Theta(k^+) \left\{ \hat{b}^\dagger(\underline{k},\lambda) \chi_\lambda^\dagger e^{+i\underline{k}\cdot\underline{r}} + \hat{d}(\underline{k},\lambda) \chi_{-\lambda}^\dagger e^{-i\underline{k}\cdot\underline{r}} \right\}, \qquad (13)$$

with $\underline{r} = (r^-, \vec{r}_\perp)$, $\underline{k} = (k^+, \vec{k}_\perp)$ and $\underline{k} \cdot \underline{r} = k^+ r^-/2 - \vec{k}_\perp \cdot \vec{r}_\perp$. The creation and annihilation operators obey the anticommutation relations

$$\left\{\hat{b}(\underline{k},\lambda), \hat{b}^\dagger(\underline{k}',\lambda')\right\} = \left\{\hat{d}(\underline{k},\lambda), \hat{d}^\dagger(\underline{k}',\lambda')\right\} = (2\pi)^3 2k^+ \delta^{(3)}(\underline{k}-\underline{k}') \delta_{\lambda,\lambda'}, \qquad (14)$$

so that the anticommutator of the dynamical spinor components reads

$$\left\{\hat{\Psi}_\alpha(\underline{r}), \hat{\Psi}_\beta^\dagger(\underline{r}')\right\} = \delta_{\alpha,\beta} \delta^{(3)}(\underline{r}-\underline{r}'). \qquad (15)$$

The formal similarity between light-front field theory and non-relativistic many-body physics may occasionally be misleading. For example, the operator of the conserved charge is

$$\hat{Q} = \frac{1}{2} \int d^3r \overline{\tilde{\Psi}}(\underline{r}) \gamma^+ \tilde{\Psi}(\underline{r}) = \int d^3r \hat{\Psi}^\dagger(\underline{r}) \hat{\Psi}(\underline{r}) \qquad (16)$$

$$= \sum_\lambda \int \frac{d^3k}{(2\pi)^3 2k^+} \Theta(k^+) \left[\hat{b}^\dagger(\underline{k},\lambda)\hat{b}(\underline{k},\lambda) - \hat{d}^\dagger(\underline{k},\lambda)\hat{d}(\underline{k},\lambda)\right], \qquad (17)$$

which counts the number of fermions minus anti-fermions. Here $\hat{\bar{\Psi}}(\underline{r})$ is the operator of the usual 4-component spinor field The operator $\hat{\Psi}^\dagger \hat{\Psi}$ is not positive definite as is the case in non-relativistic systems. From the anticommutation relations

$$\{\hat{Q}, \hat{\Psi}(\underline{r})\} = \hat{\Psi}(\underline{r})(2\hat{Q}-1) = (2\hat{Q}+1)\hat{\Psi}(\underline{r}) \qquad (18)$$

$$\{\hat{Q}, \hat{\Psi}^\dagger(\underline{r})\} = \hat{\Psi}^\dagger(\underline{r})(2\hat{Q}+1) = (2\hat{Q}-1)\hat{\Psi}^\dagger(\underline{r}), \qquad (19)$$

one concludes that $\hat{\Psi}^\dagger(\underline{r})$ creates a fermion at position $\underline{r}$ in coordinate space, while $\hat{\Psi}(\underline{r})$ destroys one at the same point.

Furthermore, we display the expression for the light-front Hamiltonian of free fermions,

$$\hat{P}^- = \int d^3r\, \hat{\Psi}^\dagger(\underline{r}) \frac{-\vec{\partial}_\perp^2 + m^2}{i\partial^+} \hat{\Psi}(\underline{r}) \qquad (20)$$

$$= \sum_\lambda \int \frac{d^3k}{(2\pi)^3 2k^+} \Theta(k^+) \frac{\vec{k}_\perp^2 + m^2}{k^+} \left[\hat{b}^\dagger(\underline{k}, \lambda)\hat{b}(\underline{k}, \lambda) + \hat{d}^\dagger(\underline{k}, \lambda)\hat{d}(\underline{k}, \lambda)\right]. \qquad (21)$$

This operator is non-local because of the $(i\partial^+)^{-1}$. Boundary conditions have to be chosen such that positive energy particles propagate only into the forward light-cone, while negative energy solutions propagate only into the backward light-cone.

The time-ordered Green's functions of a fermion in a medium is defined in terms of the Heisenberg operators of the dynamical field components [5, 10],

$$iG_{\alpha,\beta}(r_1, r_2) = \langle \hat{\psi}_\alpha(r_1)\hat{\psi}_\beta^\dagger(r_2)\rangle \Theta(r_1^+ - r_2^+) - \langle \hat{\psi}_\beta^\dagger(r_2)\hat{\psi}_\alpha(r_1)\rangle \Theta(r_2^+ - r_1^+), \qquad (22)$$

where the average $\langle ... \rangle$ is to be taken with the appropriate ensemble, and $\alpha, \beta \in \{1, 2\}$. This definition of the Green's function includes the case of zero temperature. Therefore, the conventional light-front quantization at temperature $T = 0$ can be formulated in terms of $G_{\alpha,\beta}$ as well. The Green's function is the fundamental object of this approach. In addition, the retarded (R) and advanced (A) Green's functions are defined by the anticommutators

$$iG_{\alpha,\beta}^{R,A}(r_1, r_2) = \pm \langle \{\hat{\psi}_\alpha(r_1), \hat{\psi}_\beta^\dagger(r_2)\}\rangle \Theta(\pm(r_1^+ - r_2^+)), \qquad (23)$$

where the upper sign refers to $G_{\alpha,\beta}^R$ and the lower sign to $G_{\alpha,\beta}^A$. We remark, that in a gauge theory, it is also necessary to include a (path ordered) gauge link along the light-cone by redefining the fermion fields, see Ref. [3].

The Green's functions of a fermion in an ideal gas of temperature T were presented first in Ref. [5]. In momentum space, adjusted to our notation, they read

$$\widetilde{G}_{\alpha,\beta}^{(0)R,A}(k) = \delta_{\alpha,\beta} \frac{k^+}{k^2 - m^2 \pm i0\,\mathrm{sgn}(uk)}, \qquad (24)$$

$$\widetilde{G}_{\alpha,\beta}^{(0)}(k) = \delta_{\alpha,\beta} \left(\mathrm{P}\frac{k^+}{k^2 - m^2} - i\,\mathrm{sgn}(uk)\pi\tanh\left(\frac{uk}{2T}\right) k^+ \delta(k^2 - m^2)\right), \qquad (25)$$

where P refers to principal value prescription. The pole prescriptions $\pm i 0\mathrm{sgn}(uk)$ for the retarded and advanced Green's functions are another manifestation of the special meaning of the equal-time energy. These prescriptions ensure that $G^{(0)R}_{\alpha,\beta}(r)$ vanishes outside the forward lightcone, while $G^{(0)A}_{\alpha,\beta}(r)$ is non-vanishing only inside the backward lightcone. Most importantly, knowledge of the correct pole prescription eliminates ambiguities in the definition of the non-local operator $1/k^+$, which appears in the free light-cone Hamiltonian. However, the correct prescription for the $1/k^+$-pole depends on the type of Green's function and on the value of the other momentum components.

Another remarkable property of the light-front Green's functions is, that if the theory is discretized on a lattice in coordinate space, the factor k^+ in the numerator leads to only one pair of fermion doublers. For a transverse lattice lattice approach to finite temperature $SU(\infty)$, see Ref. [11]. Moreover, it is known that no fermion doubling problem occurs in DLCQ and one can perform DLCQ calculations for a fixed value of the total charge without any sign problem.

4. THE LIGHT-FRONT DENSITY MATRIX AND GENERALIZED PARTON DISTRIBUTIONS

In the limit $r^+ \to 0^\pm$, the time-ordered Green's function $G_{\alpha,\beta}$ is closely related to the one-particle density matrices for fermions and antifermions, $q_{\alpha,\beta}(\underline{r}_1,\underline{r}_2)$ and $\bar{q}_{\alpha,\beta}(\underline{r}_1,\underline{r}_2)$. In the 2-component theory, the separation of fermion and antifermion distributions requires the evaluation of a Fourier-integral of the Green's function. For quarks, one has

$$q_{\alpha,\beta}(k^+,\underline{R},\vec{r}_\perp) = -\frac{i}{4\pi}\int dr^- e^{+ik^+r^-/2} G_{\alpha,\beta}(r_2^+ \to 0^-, \underline{r}_1, r_2^+, \underline{r}_2). \quad (26)$$

For antiquarks, the limit $r^+ \to 0$ is taken from the other side to obtain the correct order of creation and annihilation operators [3]. Since $G_{\alpha,\beta}(r_1,r_2)$ often depends only on the difference $r = r_1 - r_2$, we introduce the variables $R = (r_1+r_2)/2$ and $r = r_1 - r_2$. The density matrix $q_{\alpha,\beta}$ is related to the so-called Wigner function by a Fourier transform over $r_\perp$. We remark that all properties of the quantum mechanical density matrix, such as hermiticity and positivity of the diagonal matrix elements, also apply to the light-front density matrix in $A^+ = 0$ gauge, since this gauge has only states with positive norm and no unphysical degrees of freedom. However, the light-front density matrix has matrix elements that are off-diagonal in Fock-space. The object defined in Eq. (26) is similar to the Wigner function introduced in Ref. [12].

The fermion density matrix contains all information about single-quark properties. It depends on 6 variables and is a 2×2 matrix in spinor space, which can be written as a linear combination of Pauli spin matrices. The coefficients are the density matrices for unpolarized, longitudinal, and transverse spin distributions. The diagonal matrix elements in coordinate space of $q_{\alpha,\beta}(k^+,\underline{R},\vec{r}_\perp)$, i.e. the ones with $\vec{r}_\perp = \vec{0}_\perp$, are closely related to the usual PDFs. For instance, the unpolarized

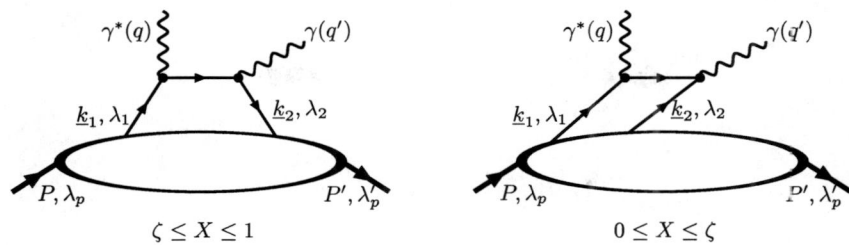

FIGURE 2. The off-diagonal matrix elements of the light-front density matrix are related to GPDs, which can be accessed in DVCS. The time-ordering shown on the left probes density matrix elements that are off-diagonal in momentum space, but diagonal in Fock-space. The other time-ordering (right) samples matrix elements that are off-diagonal in Fock-space. (Reprinted figure with permission from Ref. [3]. Copyright 2003 by the American Physical Society.)

collinear quark density is given by

$$q(k^+) = \frac{1}{2}\int d^3R \, \delta_{\alpha,\beta} q_{\beta,\alpha}(k^+, \underline{R}, \vec{r}_\perp = \vec{0}_\perp). \tag{27}$$

This parton density is normalized such that $\int_0^\infty dk^+ q(k^+) = q$, the total number of quarks in the system.

The off-diagonal matrix elements of $q_{\alpha,\beta}$ have the same operator structure as generalized parton distributions (GPDs) [13], which can be accessed e.g. in deeply virtual Compton scattering (DVCS), see Fig. 2. The generalized form factors H and E are defined by

$$\int_{-\infty}^{\infty} \frac{dr^-}{4\pi} e^{\imath k^+ r^-/2} \langle P', \lambda_p' | \psi^\dagger(0)\psi(r) | P, \lambda_p \rangle \Big|_{r^+=0, \vec{r}_\perp=\vec{0}_\perp}$$
$$= \frac{1}{2\overline{P}^+}\overline{U}(P',\lambda_p')\left[H(X,\zeta,t)\gamma^+ + E(X,\zeta,t)\frac{\imath}{2m_N}\sigma^{+\kappa}(-\Delta_\kappa)\right]U(P,\lambda_p) \tag{28}$$
$$= \frac{\sqrt{1-\zeta}}{1-\frac{\zeta}{2}}H(X,\zeta,t)\delta_{\lambda_p,\lambda_p'} - \left(\frac{\zeta^2}{4(1-\frac{\zeta}{2})}\delta_{\lambda_p,\lambda_p'} + \frac{\lambda_p \Delta^1 + \imath\Delta^2}{2m_N}\delta_{\lambda_p,-\lambda_p'}\right)\frac{E(X,\zeta,t)}{\sqrt{1-\zeta}}, \tag{29}$$

where $\overline{P}^\kappa = (P^\kappa + P'^\kappa)/2$. The meaning of the four-momenta P and P' is illustrated in Fig. 2. The momentum transfer to the target is $\Delta^\kappa = P^\kappa - P'^\kappa$ and $t = \Delta^2$. We use the kinematic variables introduced by Radyushkin [13], i.e. $X = k_1^+/P^+$ and $\zeta = 1 - P'^+/P^+$, and employ a frame with $q^+ = 0$, where q is the four-momentum of the virtual photon.

For the density matrix, one obtains

$$q_{\alpha,\beta}(k^+, \underline{R}, \vec{r}_\perp)$$
$$= \Bigg\langle \sum_{\lambda_1,\lambda_2} \int \frac{d^3k_1}{2(2\pi)^3\sqrt{k_1^+}}\Theta(k_1^+) \int \frac{d^3k_2}{2(2\pi)^3\sqrt{k_2^+}}\Theta(k_2^+)$$

$$\begin{aligned}&\left\{\hat{b}_2^\dagger\hat{b}_1\chi_\beta^\dagger(\lambda_2)\chi_\alpha(\lambda_1)\mathrm{e}^{+i(\underline{k}_2-\underline{k}_1)\cdot\underline{R}+\frac{i}{2}(\vec{k}_{2,\perp}+\vec{k}_{1,\perp})\cdot\vec{r}_\perp}\delta(k^+-\frac{k_1^++k_2^+}{2})\right.\\&+\ \hat{b}_2^\dagger\hat{d}_1^\dagger\chi_\beta^\dagger(\lambda_2)\chi_\alpha(-\lambda_1)\mathrm{e}^{+i(\underline{k}_2+\underline{k}_1)\cdot\underline{R}+\frac{i}{2}(\vec{k}_{2,\perp}-\vec{k}_{1,\perp})\cdot\vec{r}_\perp}\delta(k^+-\frac{k_2^+-k_1^+}{2})\\&+\ \hat{d}_2\hat{b}_1\chi_\beta^\dagger(-\lambda_2)\chi_\alpha(\lambda_1)\mathrm{e}^{-i(\underline{k}_2+\underline{k}_1)\cdot\underline{R}-\frac{i}{2}(\vec{k}_{2,\perp}-\vec{k}_{1,\perp})\cdot\vec{r}_\perp}\delta(k^+-\frac{k_1^+-k_2^+}{2})\left.\right\}\right\rangle.\end{aligned} \quad (30)$$

Here, we use the shorthand $\hat{b}_n = \hat{b}(\underline{k}_n, \lambda_n)$ with $n \in \{1, 2\}$ and similarly for $\hat{b}^\dagger$, $\hat{d}$ and $\hat{d}^\dagger$. As illustrated in Fig. 2, in DVCS different terms of the above expression are sampled, depending on the skewedness parameter ζ. In particular, for $\zeta = 0$, GPDs can be identified as impact parameter dependent parton densities [14]. In the case of no helicity flip we find,

$$q(k^+, \vec{b}) = \int dR \frac{1}{2} \delta_{\beta,\alpha} q_{\alpha,\beta}(k^+, R^-, \vec{b}, \vec{r}_\perp = \vec{0}_\perp) \quad (31)$$

$$= \frac{1}{4\pi k^+} \left\langle \sum_\lambda \hat{b}^\dagger(k^+, \vec{b}, \lambda) \hat{b}(k^+, \vec{b}, \lambda) \right\rangle, \quad (32)$$

$$q(k^+, \vec{b}) dk^+ = \int \frac{d^2 \Delta_\perp}{(2\pi)^2} \mathrm{e}^{i\vec{b}\cdot\vec{\Delta}_\perp} H(X, \zeta = 0, t), \quad (33)$$

We use the notation of Radyushkin here [13]. The creation operator of a quark at impact parameter $\vec{b}$ is given by

$$\hat{b}(k^+, \vec{b}, \lambda) = \int \frac{d^2 k_\perp}{(2\pi)^2} \mathrm{e}^{-i\vec{k}_\perp \cdot \vec{b}} \hat{b}(k^+, \vec{k}_\perp, \lambda). \quad (34)$$

The destruction operator $\hat{b}^\dagger(k^+, \vec{b}, \lambda)$ is defined analogously. We stress that for $\zeta \neq 0$, GPDs are in general not probability distributions but density matrices, which do not need to be positive. The light-front density matrix is a natural extension of the parton model to quantum mechanics: classical parton densities are replaced by a density matrix.

5. SUMMARY

In this talk, we have reported our recent findings of Ref. [3], where we investigated the prospects and challenges of light-front quantization in statistical physics and thermodynamics. Though most of the results of this paper apply to any field theory, we are especially interested in QCD.

We have constructed the most general form of the statistical operator $\hat{w}$, Eq. (1), using only the light-front Liouville theorem and the Poincaré algebra. Our results generalize earlier findings [4, 5, 6] to also include rotations and finite density. In particular, we treat the chemical potential in a covariant way. Remarkably, $\hat{w}$ is not the exponential of the light-front Hamiltonian $\hat{P}^-$, but of the operator

$\widehat{\mathcal{H}} = u_\nu \widehat{P}^\nu - \mu \widehat{Q}$, which propagates the system in eigentime $\tau = u_\nu r^\nu$, *i.e.* along its world line.

In light-front quantization, only one of the four translation generators, namely $\widehat{P}^-$, depends on the interaction. It is therefore possible to make use of the light-front Fock-state representation at finite temperature and density. The partition function of a theory for any T and especially any μ can therefore be obtained from DLCQ. Unlike in lattice gauge theory, there are no problems arising from dynamical fermions in DLCQ, but formidable numerical challenges still exist in $3+1$ dimensional field theories.

We regard the Green's functions as the fundamental quantities the theory is built upon, both at $T = 0$ or $T \neq 0$. The approach presented in this paper may therefore be regarded as a novel way of doing light-front quantization. The fermion Green's function in the front-form has properties very different from the equal-time Green's function. The light-front Green's function is a 2×2 matrix in spinor space and does not transform as a Lorentz scalar. It is closely related to GPDs, which can be interpreted as light-front density matrices.

ACKNOWLEDGMENTS

I am grateful to Stan Brodsky, with whom many of the results presented here have been achieved, and I thank the organizers of HLPR04 for arranging this stimulating meeting. This talk is based on research performed at SLAC with kind support from the Feodor Lynen program of the Alexander von Humboldt Foundation.

REFERENCES

1. P.A.M. Dirac, Rev. Mod. Phys. **21**, 392 (1949).
2. S. J. Brodsky, H. C. Pauli and S. S. Pinsky, Phys. Rept. **301**, 299 (1998).
3. J. Raufeisen and S. J. Brodsky, Phys. Rev. D **70**, 085017 (2004) [arXiv:hep-th/0408108].
4. M. Beyer, S. Mattiello, T. Frederico and H. J. Weber, Phys. Lett. B **521**, 33 (2001); arXiv:hep-ph/0310222; J. Phys. G **31**, 21 (2005).
5. V. S. Alves, A. Das and S. Perez, Phys. Rev. D **66**, 125008 (2002).
6. S. Lenz, *Statistische Mechanik auf dem Lichtkegel*, Diplom Thesis, Erlangen-Nuremberg U., 1990 (unpublished, in German); S. Elser and A. C. Kalloniatis, Phys. Lett. B **375**, 285 (1996); S. J. Brodsky, Nucl. Phys. Proc. Suppl. **108**, 327 (2002); H. A. Weldon, Phys. Rev. D **67**, 085027 (2003); *ibid.* 128701 (2003); A. Das and X. x. Zhou, Phys. Rev. D **68**, 065017 (2003); A. N. Kvinikhidze and B. Blankleider, Phys. Rev. D **69**, 125005 (2004).
7. E. M. Lifshitz and L. P. Pitaevskii, *Landau-Lifshitz Course of Theoretical Physics Vol. 5: Statistical Physics Part 1,* Pergamon Press, Oxford, UK, 1980.
8. H. C. Pauli and S. J. Brodsky, Phys. Rev. D **32**, 2001 (1985); *ibid.* 1993 (1985).
9. W. M. Zhang and A. Harindranath, Phys. Rev. D **48**, 4881 (1993).
10. P. P. Srivastava and S. J. Brodsky, Phys. Rev. D **64**, 045006 (2001).
11. S. Dalley and B. van de Sande, hep-ph/0409114.
12. A. V. Belitsky, X. d. Ji and F. Yuan, Phys. Rev. D **69**, 074014 (2004).
13. D. Müller, D. Robaschik, B. Geyer, F. M. Dittes and J. Horejsi, Fortsch. Phys. **42**, 101 (1994); X. D. Ji, Phys. Rev. D **55**, 7114 (1997); Phys. Rev. Lett. **78**, 610 (1997); A. V. Radyushkin, Phys. Rev. D **56**, 5524 (1997).
14. M. Burkardt, Phys. Rev. D **62**, 071503 (2000) [Erratum-ibid. D **66**, 119903 (2002)].

Light front NJL model at finite temperature[1]

S. Strauß*, M. Beyer* and S. Mattiello*

*Universität Rostock, Institut für Physik, D-18051 Rostock, Germany
E-mail: stefan.strauss@uni-rostock.de, michael.beyer@uni-rostock.de,
stefano.mattiello@uni-rostock.de

Abstract. We investigate the properties of qq and $q\bar q$ states in hot and dense quark matter in the framework of light-front finite temperature field theory. Presently we consider the Nambu Jona-Lasinio (NJL) model and derive the gap equation at finite temperature and density. We study pionic and scalar diquark dynamics in quark matter and compute the two-body masses and the Mott dissociation using a t-matrix approach. For the scalar diquark we determine the critical temperature of color superconductivity.

Keywords: light cone quantization, finite-temperature field theory, quark-gluon plasma, few-body physics, relativistic quantum field theory
PACS: 11.10.Wx 12.38.Mh 25.75.Nq

1. INTRODUCTION

A description of the dynamics of quark matter at finite temperature T and finite density (i.e. chemical potential μ) or even of quantum chromodynamics (QCD) under extreme conditions appears crucial in view of the newly planned relativistic heavy ion colliders such as LHC at CERN and SIS at the GSI and the possible interpretation of recent RHIC data [1, 2, 3, 4, 5, 6]. Furthermore an understanding of the plasma phase and the phase transitions are relevant for astrophysical scenarios such as the early universe and the interior of neutron stars. A promising novel approach to describe quark matter and hopefully QCD at finite temperature and density is provided by use of light-front quantization [7, 8]. To this end light-front quantization has been generalized in recent years to include effects of finite temperature and density, see [9] and refs. therein. As an example we apply light-front finite temperature field theory to the Nambu Jona-Lasinio (NJL) model that reproduces some low energy phenomenology of QCD, in particular its chiral properties, in a rather transparent way, and yet includes spontaneous symmetry breaking as a nontrivial feature to be handled by light-front quantization. We restrict ourselves to the two flavor case therefore the NJL Lagrangian reads

$$\mathcal{L}_{\mathrm{NJL}} = \bar\psi(i\gamma_\mu \partial^\mu - m_0)\psi + G\left((\bar\psi\psi)^2 + (\bar\psi i\gamma_5 \boldsymbol{\tau}\psi)^2\right) \tag{1}$$

---

[1] presented by S. Strauß.

with $\boldsymbol{\tau}=(\tau_1,\tau_2,\tau_3)$ the Pauli matrices. We show that it is possible to compute several phase boundaries known from instant form calculations [10] also in the framework of light-front quantization. These boundaries include the chiral restoration, the pion Mott dissociation, and the transition to color superconductivity, see e.g. [11] for a review of the phase diagram of quark matter.

2. CHIRAL SYMMETRY BREAKING

The Lagrangian (1) is approximately invariant under chiral symmetry transformation due to small current masses $m_0 \approx 5$ MeV of up and down quarks. In a meanfield approximation the quark condensate $\langle\bar{\psi}\psi\rangle$ can be understood as dynamical generated contribution to the mass, viz.

$$m = m_0 - 2G\langle\bar{\psi}\psi\rangle, \qquad (2)$$

where m denotes the constituent quark mass. One distinguishes the chiral symmetric and chiral broken phase by the condensate value $\langle\bar{\psi}\psi\rangle$. The light-front in-medium propagator relevant here has been given in Ref. [12], see also [9] this meeting,

$$\begin{aligned}\mathcal{G}(k) &= \frac{\gamma_\mu k^\mu_{\rm on}+m}{k^+}\left\{\theta(k^+)\left(\frac{1-f^+(k^+,\boldsymbol{k}_\perp)}{k^- - k^-_{\rm on}+i\varepsilon} + \frac{f^+(k^+,\boldsymbol{k}_\perp)}{k^- - k^-_{\rm on}-i\varepsilon}\right)\right.\\ &\left.+\theta(-k^+)\left(\frac{f^-(-k^+,\boldsymbol{k}_\perp)}{k^- - k^-_{\rm on}+i\varepsilon} + \frac{1-f^-(-k^+,\boldsymbol{k}_\perp)}{k^- - k^-_{\rm on}-i\varepsilon}\right)\right\}\end{aligned} \qquad (3)$$

expressed in terms of light-front momenta $\boldsymbol{k}_\perp = (k_1,k_2)$ and $k^\pm = k^0 \pm k^3$. The components of the on-shell momentum four vector read $k_{\rm on} = (k^-_{\rm on}, k^+, \boldsymbol{k}_\perp)$ with $k^-_{\rm on} = \left(\boldsymbol{k}_\perp^2 + m^2\right)/k^+$. The Fermi functions for a quark (f^+) or anti-quark (f^-) with the medium at rest are given by [13, 14]

$$f^\pm(k^+,\boldsymbol{k}_\perp) = \left[\exp\left\{\frac{1}{T}\left(\frac{1}{2}k^-_{\rm on}+\frac{1}{2}k^+ \mp \mu\right)\right\}+1\right]^{-1}. \qquad (4)$$

Evaluating (2) using (3) leads to the in-medium gap equation

$$m(T,\mu) = m_0 + 24G\int_0^\infty \frac{dk^+}{k^+(2\pi)^3}\int d^2\boldsymbol{k}_\perp m(1-f^+(k^+,\boldsymbol{k}_\perp)-f^-(k^+,\boldsymbol{k}_\perp)). \qquad (5)$$

We regularize divergent integrals as (5) utilizing the invariant Lepage-Brodsky (LB) cut-off scheme. The LB-condition

$$M^2_{20}(x,\boldsymbol{k}_\perp) = \frac{\boldsymbol{k}_\perp^2+m^2}{x(1-x)} \leq \Lambda^2_{\rm LB} \qquad (6)$$

implies the following integral boundaries

$$x_1 \leq x \leq x_2 \qquad (7)$$
$$0 \leq \boldsymbol{k}_\perp^2 \leq \Lambda^2_{\rm LB} x(1-x) - m^2 \qquad (8)$$

TABLE 1. The parameters used for the calculation. The LB cut-off Λ_{LB} is given for $T = \mu = 0$ MeV.

G (10^{-6} MeV)	m_0 (MeV)	Λ_{LB} (MeV)
5.51	5.67	1428

with $x_{1,2} = \frac{1}{2}\left(1 \mp \sqrt{1 - 4m^2/\Lambda_{LB}^2}\right)$.

There are three free paramters in the approach namely the quark current mass m, the coupling constant G, and the cut-off Λ_{LB}. The parameters are adjusted so that one reproduces the pion mass $m_\pi = 140$ MeV, the pion decay constant $f_\pi = 93$ MeV, and constituent quark mass $m = 336$ MeV in absence of the thermodynamical medium. The gap equation (5) and its correspondent instant form one are connected by Sawicki transformation. The resulting condensate value is in both cases $\langle \bar{q}q \rangle^{1/3} = -247$ MeV. Following Ref. [15] the gap equation with LB regularization in light-front form and the gap equation with 3-momentum (3M) cut-off ($\boldsymbol{k}^2 \leq \Lambda_{3M}^2$) in instant form are equivalent if one chooses

$$\Lambda_{LB}^2(T,\mu) = 4\left(\Lambda_{3M}^2 + m^2(T,\mu)\right). \quad (9)$$

Our parameters are listed in table 1. They correspond to the case I given in [10]. We define the chiral phase transition by $m_c(T,\mu) = m(0,0)/2$ where $m_c(T,\mu)$ denotes the critical mass. In Fig. 2 $m_c(T,\mu)$ is plotted as solid line and one reads off the critical temperature $T_c = 190$ MeV ($\mu = 0$) which is compatible with lattice calculations.

3. TWO-BODY CORRELATIONS

We investigate two-body bound states in a t-matrix approach. The BS equation for the two-body t-matrix $T(k)$ is given by

$$T(k) = K + \int \frac{d^4q}{(2\pi)^4} K S_F(q+k/2) S_F(q-k/2) T(k), \quad (10)$$

where K is an appropriate irreducible interaction kernel and $S_F(q)$ the dressed quark propagator. Bound states of mass M emerge as poles $k^2 = M^2$ of the t-matrix.

3.1. Pion

The interaction kernel in the pseudo-scalar channel can be read off (1)

$$K_{\alpha\beta,\gamma\delta} = -2iG(\gamma_5\tau_i)_{\alpha\beta}(\gamma_5\tau_i)_{\gamma\delta}. \quad (11)$$

Because of the separable structure of the kernel (11) the respective t-matrix equation simplifies drasticly. One introduces a reduced t-Matrix $t_\pi(k)$

$$T(k)_{\alpha\beta\gamma\delta} = (\gamma_5\tau_j)_{\alpha\beta} t_\pi(k) (\gamma_5\tau_j)_{\gamma\delta} \qquad (12)$$

in order to obtained the solution of (10)

$$t_\pi(k) = \frac{-2iG}{1 + 2G\Pi_\pi(k^2)}. \qquad (13)$$

Here $\Pi_\pi(k^2)$ denotes the loop diagram for the pion. Using (3) the loop $\Pi_\pi(k^2)$ takes the explicit form

$$\Pi_\pi(k^2) = -6 \int_{LB} \frac{dx d^2 q_\perp}{x(1-x)(2\pi)^3} \frac{M_{20}^2(x, q_\perp)(1 - f^+(M_{20}) - f^-(M_{20}))}{M_{20}^2(x, q_\perp) - k^2}, \qquad (14)$$

where the Fermi functions

$$f^\pm(M_{20}) = \left[\exp\left\{\beta\left(\frac{M_{20}}{2} \mp \mu\right)\right\} + 1\right]^{-1} \qquad (15)$$

depend on the mass $M_{20}^2(x, q_\perp) = (q_\perp^2 + m^2)/x(1-x)$ of the virtual two-body system. Via the pole condition for bound states $1 + 2G\Pi_\pi(m_\pi^2) = 0$ one determines the T and μ dependent pion mass $m_\pi(T,\mu)$. The results for different values of the chemical potential are shown in Fig. 1.A. Our calculations are limited by $m_\pi^2(T,\mu) < 4m^2 \leq M_{20}^2$ because otherwise the denominator on the r.h.s. of (14) becomes zero for some T, μ. The Mott dissociation line for the pion is given by the intersection pionts of $m_\pi(T,\mu)$ and the continuum $2m(T,\mu)$. In Fig. 2 the dashed line represents the pion dissocitation. One notices that the pion dissociation and the chiral phase transition are quit narrow together which reflects that the pion is the (pseudo) Goldstone boson of chiral symmetry.

3.2. Diquark

The quark-quark interaction is constructed by Fierz transformation of (1), cf. Ref. [16]. We focus on the scalar, isospin singulet, color-antitriplet channel. Therefore we choose the interaction kernel

$$K_{\alpha\beta,\gamma\delta} = 2iG_s(\gamma_5 C \tau_2 \lambda^a)_{\alpha\beta}(C^{-1}\gamma_5\tau_2\lambda^a)_{\gamma\delta} \qquad (16)$$

with $a = 2, 5, 7$ and $C = i\gamma^2\gamma^0$. The coupling constant for the quark-antiquark interaction is constrained by the mesonic spectrum. This might not be useful for the quark-quark coupling G_s because of higher order contributions that are not included in the meanfield approximation. Another possibility is to fit the effective quark-quark coupling to the baryonic spectrum. We require the diquark mass m_d for vanishing T and μ to be 550 MeV that leads to $G_s = 0.727G$.

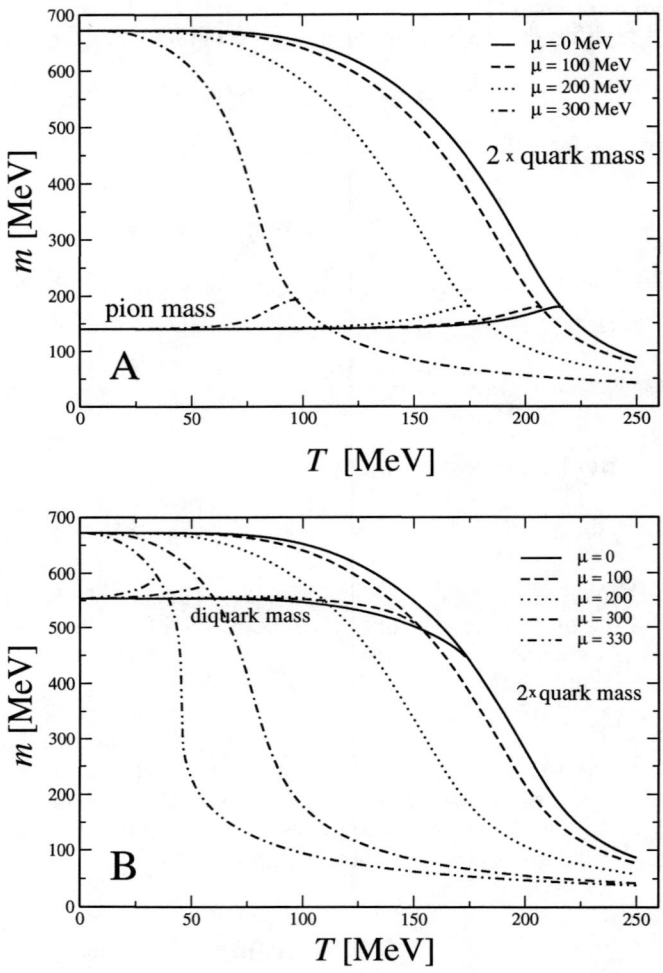

FIGURE 1. (A) The pion mass as a function of T for different μ. (B) The diquark mass as a function of T for different μ. The continuum is given by $2m$. The lines of the two-body masses end at the Mott dissociation points.

Analogous to the treatment of the pion we define the reduced t-matrix $\tau_s(k)$

$$T(k) = (\gamma_5 C \tau_2 \lambda^a)_{\alpha\beta} \tau_s(k) (C^{-1} \gamma_5 \tau_2 \lambda^a)_{\gamma\delta}. \tag{17}$$

The solution is given in terms of the diquark loop $\Pi_s(k^2)$

$$\tau_s(k) = \frac{2iG_s}{1 + 2G_s \Pi_s(k^2)} \tag{18}$$

where one computes the loop integral using (3) to

$$\Pi_s(k^2) = -4 \int_{LB} \frac{dx\, d^2\mathbf{q}_\perp}{x(1-x)(2\pi)^3} \frac{M_{20}^2(x,\mathbf{q}_\perp)\left(1-2f^+(M_{20})\right)}{M_{20}^2(x,\mathbf{q}_\perp)-k^2}. \tag{19}$$

Again we search for poles of $\tau_s(k)$ for various T and μ. Results for the medium dependence of the diquark mass m_d are presented in Fig. 1.B.

4. COLOR SUPERCONDUCTIVITY

The interaction in the above discussed quark-quark channel is attractive. Because of this attractive interaction between fermions one expects, by Cooper's theorem, the formation of a color superconducting (CSC) phase for sufficient high density and low temperatures, cf. [17] and references therein for a overview about CSC.

In the 2-flavor case one characterizes the CSC phase according to the condensate value

$$\Phi = \langle \psi^T C \gamma_5 \tau_2 \lambda_2 \psi \rangle. \tag{20}$$

We have chosen without loss of generality $\alpha = 2$ which means that the CSC condensate has the color antiblue and the Cooper pairs consist of red and green quarks. The $SU_c(3)$ color symmetry is broken down to $SU(2)$. The standard way is to derive a gap equation for the order parameter Φ but we emphasise a simpler treatment. In the previous section we obtained the t-matrix for the diquarks so that we can use the Thouless criterion [18]

$$m_d(T,\mu) = 2\mu \tag{21}$$

which picks the phase border to color superconductivity. One notices that (21) describes the condensation of bosonic diquarks but it also remains valid in domains of the phase diagram where the diquarks do not exist as bound states anymore. Inserting relation (21) the pole condition of (18) becomes

$$\frac{1}{2G_s} = 4 \int_{LB} \frac{dx}{x(1-x)} \int_{LB} \frac{d^2\mathbf{q}_\perp}{(2\pi)^3} \frac{M_{20}^2(x,\mathbf{q}_\perp)\left(1-2f^+(M_{20})\right)}{M_{20}^2(x,\mathbf{q}_\perp)-4\mu^2}. \tag{22}$$

Equation (22) enables one to compute the phase transition from deconfined to color superconducting quark matter. The result is given in Fig. 2. Also in this case we find general agreement with the instant form calculations.

5. CONCLUSION

The NJL model provides a rather transparent approach to low energy phenomena of QCD. We have used light-front quantization at finite temperature and density to calculate some features of the model, well known from instant form calculations. We find, as it should, that the results do not depend on the quantization form. So far,

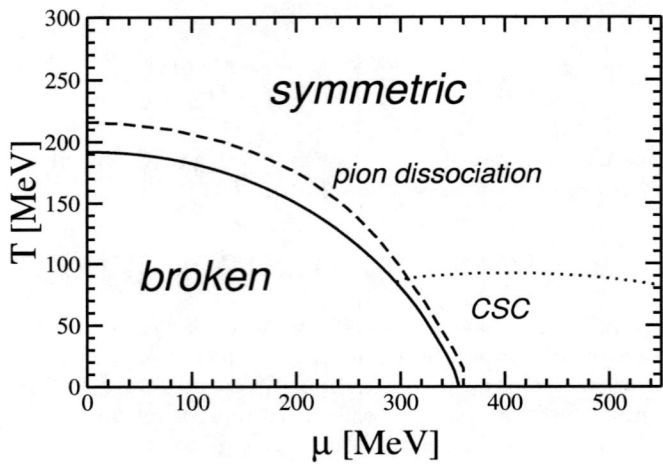

FIGURE 2. The phase diagram in the NJL model. The solid line shows the chiral phase transition. The transition between the deconfined and the CSC phase is given by the dotted line. And the dashed line is the Mott dissociation line of the pion.

only simple observables have been tested and by use of the Sawicki transformation the different forms can be related to each other. In conclusion, the light front form leads to the same phenomena as the instant form. This is not trivial, since spontaneous symmetry breaking is described in a conceptionally different way on the light front than in the instant form and also there is no frame of reference where the four velocity of the heat bath is equivalent to the light-front time direction (which is the case in the instant form). More on the latter point has been given by [9, 19] during this meeting. A next logical step along this line towards a treatment of full QCD would be to go beyond the NJL model and tackle (simple) gauge theories. Another possible step would be to calculate the influence of three-quark correlations provided in this approach [9] to the critical temperature of the transition to CSC quark matter, which has not been considered so far.

ACKNOWLEDGMENTS

We thank the organizers of HLPR04 in Spa for this inspiring meeting. S.S. thanks for local support. This work is supported by the Deutsche Forschungsgesellschaft.

REFERENCES

1. M. Gyulassy and L. McLerran, arXiv:nucl-th/0405013.
2. M. Gyulassy, arXiv:nucl-th/0403032.
3. H. Stoecker, arXiv:nucl-th/0406018.
4. X. N. Wang, arXiv:nucl-th/0405017.

5. E. V. Shuryak, arXiv:hep-ph/0405066.
6. U. W. Heinz, AIP Conf. Proc. **739** (2005) 163 [arXiv:nucl-th/0407067].
7. P.A.M. Dirac, Rev. Mod. Phys. **21**, 392 (1949)
8. S. J. Brodsky, H. C. Pauli and S. S. Pinsky, Phys. Rept. **301** (1998) 299 [arXiv:hep-ph/9705477].
9. M. Beyer, S. Strauß, S. Mattiello, T. Frederico, and H.J. Weber these proceedings.
10. S. P. Klevansky, Rev. Mod. Phys. **64** (1992) 649.
11. D. H. Rischke, Prog. Part. Nucl. Phys. **52** (2004) 197 [arXiv:nucl-th/0305030].
12. M. Beyer, S. Mattiello, T. Frederico and H. J. Weber, J. Phys. G **31** (2005) 21.
13. M. Beyer, S. Mattiello, T. Frederico and H. J. Weber, Phys. Lett. B **521** (2001) 33;
14. J. Raufeisen and S. J. Brodsky, Phys. Rev. D **70** (2004) 085017 [arXiv:hep-th/0408108].
15. W. Bentz, T. Hama, T. Matsuki and K. Yazaki, Nucl. Phys. A **651** (1999) 143.
16. N. Ishii, W. Bentz and K. Yazaki, Nucl. Phys. A **587** (1995) 617;
17. M. Buballa, arXiv:hep-ph/0402234.
18. D.J. Thouless, Ann. Phys. **10**, 553 (1960).
19. J. Raufeisen, these proceedings.

Random matrix models for chiral and diquark condensation

B. Vanderheyden *,† and A. D. Jackson**

University of Liège, Departement of Electrical Engineering and Computer Science, Bât. B28, Sart Tilman, B-4000 Liège 1, Belgium
†*E-mail: B.Vanderheyden@ulg.ac.be*
**The Niels Bohr Institute, Blegdamsvej 17, DK-2100 Copenhagen Ø, Denmark*

Abstract. We consider random matrix models for the thermodynamic competition between chiral symmetry breaking and diquark condensation in QCD at finite temperature and finite baryon density. The models produce mean field phase diagrams whose topology depends solely on the global symmetries of the theory. We discuss the block structure of the interactions that is imposed by chiral, spin, and color degrees of freedom and comment on the treatment of density and temperature effects. Extension of the coupling parameters to a larger class of theories allows us to investigate the robustness of the phase topology with respect to variations in the dynamics of the interactions. We briefly study the phase structure as a function of coupling parameters and the number of colors.

Keywords: Chiral symmetry, color superconductivity, random matrix theory
PACS: 11.30. Fs, 11.30. Qc, 11.30. Rd, 12.38. Aw

1. INTRODUCTION

Some thirty years ago, it was observed that dense and cold quark matter might exhibit Cooper pairing as a result of an attractive quark-quark interaction in the color antitriplet channel [1]. More recent models, based on non-perturbative effective interactions or on diagrammatic calculations of single-gluon exchange interactions, indicate that a color superconducting phase might develop pairing gaps as large as $\Delta \sim 100$ MeV for quark chemical potentials on the order of 300 MeV [2]. This interesting possibility has direct consequences on the physics of dense stars and is certainly important for the determination of the phase diagram of nuclear matter under extreme (ultrarelativistic) conditions [3, 4].

Different order parameters have been proposed in the literature and studied as a function of, e.g., the quark masses, the number of flavors and colors, the quark chemical potentials, and temperature [3, 4]. In the limit of QCD with two flavors of light quarks (the 2SC limit), the order parameter has the form

$$\langle \psi_{f\alpha\sigma}(p)\psi_{g\beta\sigma'}(-p)\rangle = \Phi(p^2)\varepsilon_{fg}\varepsilon_{\alpha\beta 3}\varepsilon_{\sigma\sigma'}, \qquad (1)$$

where p is a four-momentum and where we have displayed the flavor (f,g), color (α, β), and spin (σ, σ') indices. The tensors ε ensure that the condensate is antisymmetric in flavor, color, and spin. Color is broken from $SU(3)$ to $SU(2)$, but the flavor symmetry $SU(2)_L \times SU(2)_R$ and the baryon symmetry $U(1)_B$ remain intact. In the other limit of three degenerate flavors of light quarks, the favored

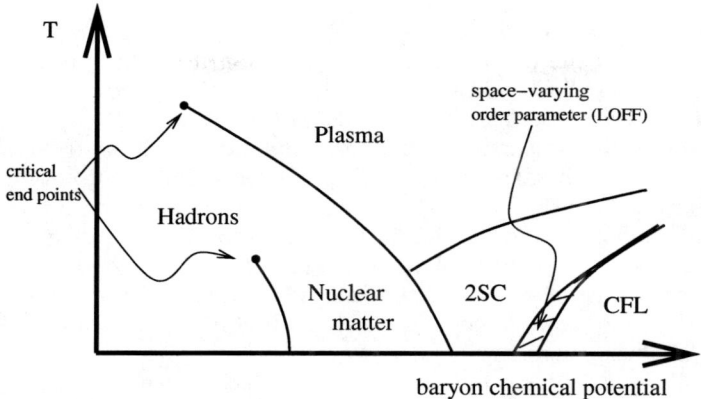

FIGURE 1. Sketch of the phase diagram for three flavor QCD with realistic quark masses. After Ref. [4].

order parameter exhibits a coupling between color and flavor rotations (called color-flavor locking or CFL) and has the approximate form [3]

$$\langle \psi_{f\alpha\sigma}(p)\psi_{g\beta\sigma'}(-p)\rangle = \Phi(p^2)\,\varepsilon_{fgA}\varepsilon_{\alpha\beta A}\varepsilon_{\sigma\sigma'}. \qquad (2)$$

Color, flavor, and baryon symmetries are now broken down to $SU(3)_{\text{color}+L+R} \times Z_2$. Since both $SU(3)_L$ and $SU(3)_R$ are locked to $SU(3)_{\text{color}}$, the CFL condensate also breaks chiral symmetry.

Following Ref. [4], the conjectured phase diagram for QCD with three flavors and realistic quark masses is given in Fig. 1. At asymptotically high densities, the scale is set by the quark chemical potential, μ. For $\mu \gg m_s$ (m_s is the strange quark mass), the more symmetric CFL phase is favored. As μ decreases, the increasing difference between u- (or d-) and s-quark Fermi momenta weakens $\langle us \rangle$ and $\langle ds \rangle$ condensates and eventually leads to a transition to the 2SC phase [5], possibly via a so-called LOFF phase characterized by a spatially varying gap [6].

Most of these results have not been confirmed by lattice simulations, which are difficult to perform and interpret for QCD with three colors at finite baryon densities. The difficulty resides in the fact that the fermion determinant in the partition function is complex for non-zero μ, and sampling weights are no longer positive definite. This difficult problem represents a significant barrier to understanding the phase structure of QCD in lattice simulations.

Random matrix theory offers models that are capable of distinguishing those physical properties that are determined by global symmetries from those that depend on the detailed dynamics of the interactions. The random matrix Hamiltonian mimics the true interactions by adopting a block structure that is solely determined by the global symmetries under consideration. The matrix elements are drawn on a random distribution, usually a Gaussian distribution, so that the model can be solved exactly. Such an approach leads to universal results that can be studied at two different levels. At the microscopic level, the statistical properties of the lowest eigenvalues of the Dirac operator are determined by the spontaneous breaking of chiral symmetry, as indicated for example by the universality of the spectral density near zero virtuality and related sum rules [7, 8, 9, 10]. At the macroscopic level, many of the properties of the phase diagram (such as its topology and the presence of given critical lines or points) are independent of the detailed form of the interactions and are thus symmetry protected. The random matrix approach treats low-lying fermion excitations and, in its usual form, neglects their momentum and kinetic energy. The resulting phase diagram is mean-field. Near critical regions, it produces results similar to those obtained in a Landau-Ginzburg approach [11, 12]. In general, random matrix models provide a useful tool for studying systems with non-trivial phase diagrams.

This talk focuses on random matrix models that implement coexisting chiral and color symmetries in QCD with two light degenerate flavors (2SC). These models are extensions of the chiral random matrix models, which we introduce in Sec. 2. We present the construction of the color and spin block structure of the interactions in Sec. 3. The phase diagram is discussed in Sec. 4. This talk is a summary of previous work on the question, see Refs. [13, 14, 15].

2. CHIRAL RANDOM MATRIX MODELS

We first consider chiral symmetry alone and turn to chiral random matrix models [16, 17]. For QCD in the sector of zero topological charge with N_f flavors and zero chemical potential, the partition function is given as

$$Z = \int DW \prod_{i=1}^{N_f} D\psi_i^* D\psi_i \, \exp\left[i \sum_{i=1}^{N_f} \psi_i^* \mathcal{D} \psi_i\right] \exp\left(-\frac{n\beta\Sigma^2}{2} \text{Tr}[WW^\dagger]\right), \quad (3)$$

where W is an $n \times n$ matrix which models the interactions. Its elements are drawn on a Gaussian distribution of mean zero and inverse variance Σ. Here, n is a measure of the number of low-lying degrees of freedom and is to be taken to infinity at the end of the calculations (i.e., in the thermodynamic limit). In the chiral limit, $m = 0$, the Dirac operator, $\mathcal{D}$, has a block structure imposed by the chiral symmetry of QCD, $\{D, \gamma_5\} = 0$. In the basis of the eigenstates of γ_5, this leads to the block structure

$$\mathcal{D} = \begin{pmatrix} 0 & iW \\ iW^\dagger & 0 \end{pmatrix}. \quad (4)$$

For random matrix models of QCD with $SU(3)$ and fermions in the fundamental representation, W is complex. This choice corresponds to the chiral unitary ensemble, which is characterized by a Dyson index $\beta = 2$. The QCD Dirac operator in $SU(2)$ with fermions in the fundamental representation satisfies an additional anti-unitary symmetry,

$$[C(\sigma_2)_{\text{color}} K, i\mathcal{D}] = 0, \quad (5)$$

with $(C(\sigma_2)_{\text{color}}K)^2 = 1$. ($C$ is the charge conjugation operator, σ_2 is the antisymmetric color matrix, and K is the complex conjugation operator.) This implies that it is possible to find a particular basis of states in which $i\mathcal{D}$ is real. Accordingly, W is chosen real in chiral random matrix models for $SU(2)$, and this leads to the chiral orthogonal ensemble with an index $\beta = 1$ [17]. For fermions in the adjoint representation of the gauge group and any number of colors, the Dirac operator obeys the anti-unitary symmetry $C^{-1}K$ with $(C^{-1}K)^2 = -1$. This leads to the symplectic chiral ensemble with $\beta = 4$ and quaternion real matrix elements [10, 17]. We will not consider this ensemble further in this talk.

The essential difference among the ensembles lies in the number of independent random variables that are allowed per matrix element. (I.e., A complex number has two degrees of freedom; a real number has only one.) This difference is important in determining the statistical properties of the Dirac operator and, in turn, affects the phase diagram.

3. RANDOM MATRIX MODELS WITH AN EXPANDED BLOCK STRUCTURE

In order to study the competition between chiral, $\langle \bar{q}q \rangle$, and diquark, $\langle q^T q \rangle$, condensates, we introduce an explicit dependence in the spin and color quantum numbers appearing in the 2SC order parameter of Eq. (1). We are thus lead to a Dirac operator with the chiral block structure of Eq. (4), where W now has the following expanded color and spin sub-block structure:

$$W = \sum_{\mu=0}^{3} \sum_{a=1}^{N_c^2-1} \lambda_a \otimes \sigma_\mu \otimes A_{\mu a}. \tag{6}$$

Here, the deterministic matrices represent spin and color degrees of freedom: $\sigma_\mu = (1, i\vec{\sigma})$ with $\vec{\sigma}$ the Pauli matrices, whereas λ_a are color matrices (Gell-Mann matrices for $N_c = 3$). The random matrices, $A_{\mu a}$, are $N \times N$ and represent gluon fields. They are chosen real. Their elements are drawn on a Gaussian distribution with an inverse variance that is independent of μ and a in order to respect the Lorentz and $SU(N_c)$ invariance in the vacuum.

Including a quark chemical potential, μ, a quark mass, m, and a temperature dependence, the partition function is now written as

$$Z = \int D\psi_1^\dagger D\psi_1 D\psi_2^* D\psi_2^T \{\prod_{\mu a} DA_{\mu a}\} \exp\left(-2N\Sigma^2 \sum_{\mu a} \text{Tr}[A_{\mu a} A_{\mu a}^T]\right)$$
$$\times \exp\left[i \begin{pmatrix} \psi_1^\dagger \\ \psi_2^T \end{pmatrix}^T \begin{pmatrix} i\mathcal{D} + im + C_+ & 0 \\ 0 & -i\mathcal{D}^T - im - C_-^T \end{pmatrix} \begin{pmatrix} \psi_1 \\ \psi_2^* \end{pmatrix}\right], \tag{7}$$

where the dependence on the external parameters is given by $C_\pm = (\mp \sigma_3 \pi T + i\mu)\gamma_4$ with

$$\gamma_4 = \begin{pmatrix} 0 & 1 \\ 1 & 0 \end{pmatrix}. \tag{8}$$

Note that we have used a Gorkov representation and have transposed the flavor-2 fields so that the diquark condensate appears as an off-diagonal component of the inverse Dirac operator.

Several comments are in order regarding the μ and T dependences. The μ dependence mimics the $\mu\psi^\dagger \gamma_4 \psi$ term in the Euclidean QCD Lagrangian [11]. The T dependence is based on two assumptions. First, temperature is introduced via a sum over the fermion Matsubara frequencies, $\omega_n = in\pi T$ with n odd. As random matrix theory only treats low-lying degrees of freedom, much of the critical physics can be captured if one restricts the frequency sum to its two lowest terms [18], hence the two-dimensional form $\sigma_3 \pi T = \text{diag}(\pi T, -\pi T)$. The second assumption comes from the physical observation that the diquark order parameter in Eq. (1) couples fields of opposite four-momenta. We explicitly impose this coupling in Eq. (7) by taking opposite Matsubara frequencies for each flavor. Different temperature dependences have been proposed in closely related random matrix models [19, 20]; we discuss their relationship to the present models in [21].

4. PHASE DIAGRAMS

The partition function can be evaluated following a classical method. An integration over $A_{\mu a}$ produces a four-fermion interaction which can be Fierz-transformed to obtain the chiral and diquark channel terms. The resulting four-fermion potential can be expressed as a fermion field bilinear via a Hubbard-Stratonovitch transformation. These successive steps yield [13]

$$Z \sim \int d\sigma d\Delta \, e^{-4N\Omega(\sigma,\Delta)}, \tag{9}$$

where the thermodynamical potential, $\Omega(\sigma, \Delta)$, is given as

$$\begin{aligned}
\Omega(\sigma, \Delta) &= A\Delta^2 + B\sigma^2 - \frac{N_c - 2}{2} \sum_\pm \log\left((\sigma + m \pm \mu)^2 + \pi^2 T^2\right) \\
&\quad - \sum_\pm \log\left((\sigma + m \pm \mu)^2 + \pi^2 T^2 + \Delta^2\right).
\end{aligned} \tag{10}$$

Here, Δ is the auxiliary field associated with the diquark channel (with condensates $\langle \psi_{2R}^T (i\sigma_2)_{\text{spin}} \lambda_2 \psi_{1R}\rangle = \langle \psi_{2L}^T(i\sigma_2)_{\text{spin}}\lambda_2 \psi_{1L}\rangle$), while σ is related to the chiral channel (with condensates $\langle \psi_1^\dagger \psi_1 \rangle = -\langle \psi_2^T \psi_2^*\rangle$). The partition function can be treated exactly in the thermodynamic limit $N \to \infty$ by a saddle point method. The equilibrium values of the field then satisfy the two gap equations

$$\frac{\partial \Omega}{\partial \sigma} = 0, \tag{11}$$

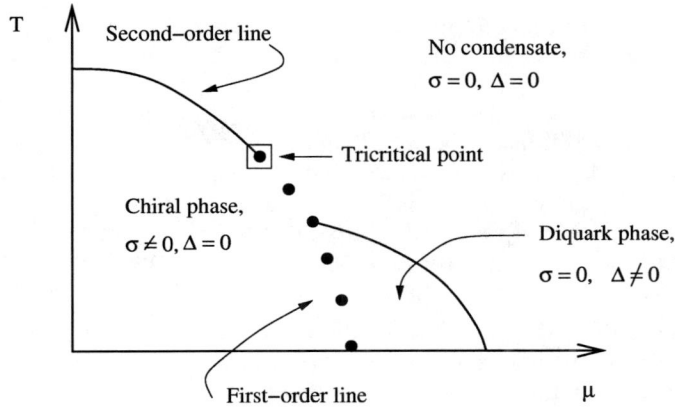

FIGURE 2. Phase diagram in the random matrix model for QCD with three colors and massless quarks.

$$\frac{\partial \Omega}{\partial \Delta} = 0, \quad (12)$$

which constitute a system of polynomial equations that can be solved analytically or numerically.

The form of the potential in Eq. (10) has a straightforward interpretation. The quadratic terms correspond to the energy cost for having constant auxiliary fields. The logarithmic terms are to be related to the single quasiparticle spectrum in a given set of fields σ and Δ. Keeping in mind that the fermion four-momenta are neglected, these energies are given as $\varepsilon = ((\sigma + m \mp \mu)^2 + \Delta^2)^{1/2}$ for the two pairing colors (with the $\mp$ sign standing for either quark or antiquark excitations). For the $N_c - 2$ unpaired colors, $\varepsilon = |\sigma + m \mp \mu|$.

We now turn to a chiral symmetric theory with $m = 0$. Note that, because it is possible to rescale all fields and external parameters by a constant, the topology of the phase diagram depends only on the ratio B/A. This is in fact a ratio of the Fierz constants in the chiral and diquark channel, and it measures the relative importance of the two symmetries. For example, small ratios are obtained with $A \gg B$; the energy cost of $A\Delta^2$ then prohibits diquark condensation. Similarly, large ratios B/A disfavor chiral symmetry breaking. For $N_c = 3$, the interactions in Eq. (6) lead to a ratio $B/A = 3/4$, which corresponds to the phase diagram of Fig. 2. One observes a low-density chiral broken symmetry phase separated from a higher density diquark phase by a first-order line. As the temperature is raised, the diquark phase makes a transition to the chiral symmetric phase across a second-order line. The transition from broken to restored chiral symmetry is second-order at low densities and first-order at intermediate densities. The two lines are separated by a tricritical point, in the vicinity of which the thermodynamical potential reduces to a ϕ^6 theory [11, 13].

It is interesting to ask how the phase structure evolves if one changes the channel couplings A and B from the values representative of QCD. To this end, we have considered Hermitean Dirac $i\mathcal{D}$ operators spanning an exhaustive set of combinations of helicity, spin, and color block structures. Remarkably, this set produces ratios B/A in the *bounded* range $[0, N_c/2]$. This result differs from what would have been obtained in a pure Landau-Ginsburg approach, where there is no *a priori* knowledge of existing constraints among the coefficients of the effective thermodynamical potential. Here, however, having started at a more microscopic level, we are capable of discovering possible bounds on the coupling ratios. Each ratio corresponds to a separate phase structure. The major conclusions resulting from the study of phase diagrams for ratios in the allowed range are as follows:

- there is only a finite number of different topologies. As B/A is varied continuously, the evolution from one topology to another is marked by the emergence or the vanishing of new critical points or lines;
- it takes moderate — but finite — alterations of the theory to depart from the topology of Fig. 2. In that sense, the phase structure of Fig. 2 is protected by symmetry.

5. DISCUSSION

5.1. Comparison with a microscopic theory

In order to appreciate the approximations involved in the formulation of the random matrix model, consider the gap equation in the limit $\sigma = 0$,

$$\frac{\partial \Omega}{\partial \Delta} = 0 \Rightarrow A\Delta = \frac{2\Delta}{\Delta^2 + \mu^2 + \pi^2 T^2}, \quad (13)$$

and compare with that obtained in a microscopic mean-field theory such as that of Ref. [22], based on an effective interaction modeled by that induced by instantons. Approximately, we find

$$\Delta \simeq G \int \frac{d^4 p}{(2\pi)^4} \left(\frac{2\Delta}{\Delta^2 + (p-\mu)^2 + p_4^2} + \frac{2\Delta}{\Delta^2 + (p+\mu)^2 + p_4^2} \right), \quad (14)$$

where we have dropped the form factors needed for the convergence of the integral. It is also implied that the integral over p_4 is a sum over Matsubara frequencies, $p_4 = n\pi T$ with n odd.

Because we have neglected the fermion four-momenta, the gap equation in the random matrix approach does not contain an integral. This has two consequences. First, the right term in Eq. (13) does not exhibit the logarithmic divergence observed in Eq. (14) for $p \simeq \mu$ and $\Delta = 0$. This divergence arises for values of p near the Fermi momentum in the limit $\Delta \to 0$ and implies that a non-zero Δ must develop at all μ. In contrast, the diquark phase in Fig. 2 does not exist for asymptotically high values of μ. The random matrix interactions saturate in the diquark channel for μ larger than $\sim \Sigma$. The second consequence of the absence of a momentum integral is that the gap, Δ, evolves monotonically as a function of μ in contrast to results from a microscopic approach, see Refs. [22, 23].

These discrepancies should not be considered as weaknesses of the random matrix approach. In the first case, Δ vanishes in a region where microscopic theories tend to produce small gaps which probably do not survive fluctuations beyond mean field. What random matrix cannot reproduce, however, is the $\Delta \sim e^{-1/g}$ behavior as a function of the QCD coupling constant, g. This behavior is due to the magnetic gluon interactions [24]. The random matrix approach is also unable to reproduce the unbounded increase of Δ at asymptotically high μ, which is related to the running of g [25]. Both these features are predicted from diagrammatic theory and are related to particular dynamic processes. The second observed difference, the variation of gaps at moderate values of μ, arises from the detailed dynamics of the interactions and depends on the choice of the regulating form factors. This behavior is not dictated by symmetry, and it is thus no surprise that it is not revealed by the random matrix approach.

5.2. QCD with two colors

In the limit of $N_c = 2$, the gauge group is pseudoreal, and quark and antiquark states transform similarly under global color rotations. They can be combined into spinors which obey an extended flavor symmetry $SU(2N_f)$ for which $\langle \bar{q}q \rangle$ mesons and $\langle qq \rangle$ baryons belong to the same multiplets. The random matrix model (with $N_f = 2$) reproduces this extended symmetry in the vacuum and its breaking pattern as a function of μ, T, and m. In the vacuum and in the chiral limit, the thermodynamic potential depends on the condensation fields through the combination $\sigma^2 + \Delta^2$. The extended $SU(4)$ symmetry is here apparent since a state with $(\sigma, \Delta) = (\Sigma, 0)$ is indistinguishable from its rotated version with $(\sigma, \Delta) = (0, \Sigma)$. For $m > 0$ and low temperature, the phase $(\Sigma, 0)$ undergoes a second order phase transition to a diquark phase at $\mu_c \simeq m_\pi/2$ where $m_\pi \sim (m\Sigma)^{1/2}$ is the pion mass [14]. Many results of the random matrix approach agree with chiral perturbation theory in this case [26].

5.3. Two more questions

The random matrix model is a theory of low-lying modes. We argued earlier that restricting the sum over Matsubara frequencies leads to a model which nicely captures the critical physics. The neglect of high-energy modes leads however to unphysical results, such as a negative baryon density and a variation of the chiral field as a function μ, both in a theory with $N_c = 2$ and in the region $\mu < m_\pi/2$. We have shown that the inclusion of appropriate high-energy terms (which should not describe the critical physics), either in the form of correction terms or as a sum over all Matsubara frequencies, fixes these anomalies while leaving the topology of the phase diagram intact [14].

Another question is whether the interactions in Eq. (6) preserves the statistical properties of the eigenvalues of the Dirac operator that are expected from chiral symmetry alone. These properties should follow the predictions of the chiral unitary ensemble for QCD with three colors and those of the chiral orthogonal ensemble for QCD with two colors. This question is related to the number of random degrees of freedom allowed for the matrix elements. We have shown that, even if the interaction matrices are complex for both $N_c = 3$ and $N_c = 2$, the deterministic spin and color dependences lead to the properties that are expected for the spectrum of the Dirac operator [15].

6. SUMMARY

We have considered random matrix models of QCD with two flavors that are capable of developing chiral and diquark condensation. We have studied the symmetry breaking patterns as a function of temperature and quark chemical potential. The phase diagrams can be established exactly from an analytical evaluation of the partition function. The resulting thermodynamical potential has a straightforward interpretation in terms of elementary excitations. Upon arbitrary variations of the coupling constants in the two condensation channels, the phase structure evolves continuously but can only adopt a fixed set of topologies.

This study shows that random matrix theory provides useful tools for studying systems with non-trivial phase diagrams and for distinguishing those properties that are protected by symmetry.

REFERENCES

1. B. Barrois, Nucl. Phys. **B129** (1977) 390; D. Bailin and A. Love, Phys. Rept. **107** (1984) 325.
2. M. Alford, K. Rajagopal, and F. Wilczek, Phys. Lett. B **422** (1998) 247; R. Rapp, T. Schäfer, E. V. Shuryak, and M. Velkovsky, Phys. Rev. Lett. **81** (1998) 53.
3. K. Rajagopal and F. Wilczek, in B. L. Ioffe Festschrift, *At the Frontier of Particle Physics/Handbook of QCD*, M. Shifman ed., (World Scientific 2001); M. Alford, Ann. Rev. Nucl. Part. Sci. **51** (2001) 131; H.-c. Ren, hep-ph/0404074.
4. T. Schäfer, hep-ph/0304281.
5. M. Alford, J. Berges, and K. Rajagopal, Nucl. Phys. **558** (1999) 219; T. Schäfer and F. Wilczek, Phys. Rev. D **60** (1999) 074014.
6. M. Alford, J. Bowers, and K. Rajagopal, Phys. Rev. D **63** (2001) 074016.
7. J. J. M. Verbaarschot and I. Zahed, Phys. Rev. Lett. **70** (1993) 3852.
8. J. J. M. Verbaarschot, Nucl. Phys. **B427** (1994) 534.
9. J. J. M. Verbaarschot, Phys. Lett. **B329** (1994) 351.
10. See for instance the review by J. J. M. Verbaarschot and T. Wettig, Ann. Rev. Nucl. Part. Sci. **50** (2000) 343.
11. M. A. Halasz, A. D. Jackson, R. E. Shrock, M. A. Stephanov, and J. J. M. Verbaarschot, Phys. Rev. D **58** (1998) 096007.
12. K. Iida and G. Baym, Phys. Rev. D **63** (2001) 074018; Phys. Rev. D **66** (2002) 059903(E); Phys. Rev. D **66** (2002) 014015; I. Giannakis and H.-C. Ren, Nucl.Phys. **B669** (2003) 462.
13. B. Vanderheyden and A. D. Jackson, Phys. Rev. D **61** (2000) 076004; Phys. Rev. D **62** (2000) 094010.
14. B. Vanderheyden and A. D. Jackson, Phys. Rev. D **64** (2001) 074016.
15. B. Vanderheyden and A. D. Jackson, Phys. Rev. D **67** (2003) 085016.

16. E. V. Shuryak and J. J. M. Verbaarschot, Nucl. Phys. **A560** (1993) 306.
17. J. J. M. Verbaarschot, Phys. Rev. Lett. **72** (1994) 2531.
18. A. D. Jackson and J. J. M. Verbaarschot, Phys. Rev. D **53** (1996) 7223.
19. B. Klein, D. Toublan, and J. J. M. Verbaarschot, Phys. Rev. D **68** (2003) 014009.
20. B. Klein, D. Toublan, and J. J. M. Verbaarschot, [arXiv:hep-ph/0405180].
21. B. Vanderheyden and A. D. Jackson (to be published).
22. J. Berges and K. Rajagopal, Nucl. Phys. **B538** (1999) 215.
23. See also the contribution by D. Blaschke, these proceedings.
24. D. T. Son, Phys. Rev. D **59** (1999) 094019.
25. T Schaëfer and F. Wilczek, Phys. Rev. D **60** (1999) 114033.
26. J. B. Kogut, M. A. Stephanov, D. Toublan, J. J. M. Verbaarschot, and A. Zhitnitsky, Nucl. Phys. **B582** (2000) 477.

FUNDAMENTAL QUESTIONS

The future of nuclear energy

J. Cugnon

*Physique théorique fondamentale, Département de Physique, Université de Liège,
allée du 6 Août 17, bât. B5, B-4000 Liège 1, Belgium
E-mail: J.Cugnon@ulg.ac.be*

Abstract. Various aspects of the World energy problem indicate that nuclear energy will still be needed in the future. Conditions for a continued valuable use are discussed. Special attention is focused on the nuclear waste problem.

Keywords: Nuclear energy, Nuclear waste, Transmutation
PACS: 28.50.Hw, 28.41.Kw, 28.41.Te, 28.90.+i

1. INTRODUCTION

Although nuclear energy is an example of hadronic interactions, it is unfrequent to discuss such a matter in a meeting like this one. Particle physicists use to get together and discuss pure science, with little regard to the applications. Setting up a discussion session on nuclear energy in this meeting reveals that the issue of the future of nuclear energy appears more and more as a society problem, that cannot be ignored by nuclear and particle physicists.

In this introductory talk, I will present an overview of the main features that will influence the use of nuclear energy in the future. I will successively discuss the World energy problem, the necessary conditions for a valuable use of nuclear energy, the safety aspects, in particular the nuclear waste problem, and the strategy for the future.

2. THE WORLD ENERGY PROBLEM

1. World energy consumption. In year 2000, the world energy consumption raises to ~10 Gtoe[1] and is steadily increasing, as shown in Fig.1. The energy consumption is unevenly distributed among the regions of the World (see Table 1). The breakdown of the energy sources is given in Table 2. By far, the main energy sources are fossil fuels. Energy resources are not uniformly distributed, as it is well known.

2. Prospectives. An important point is the (proved) energy resources. An indicative account is given in Table 3. The striking feature is that oil is running out, although the resources are probably underestimated. Oil shale may provide with another period of 40 years, but the extraction of this oil is still to be demonstrated. Nuclear energy based on present technology is in a better shape, but not that much.

---

[1] 1 toe = one ton of oil equivalent.

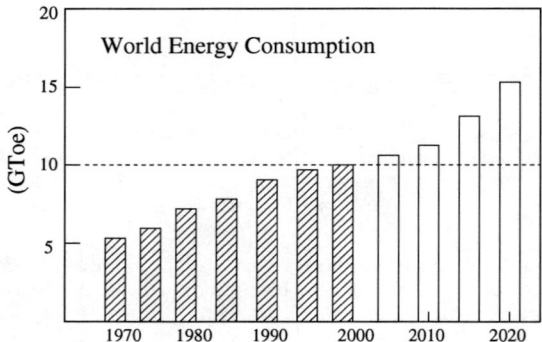

FIGURE 1. World energy consumption for the period 1970-2000 and projections for 2000-2020. Based on data from the International Atomic Energy Agency [1].

TABLE 1. Energy consumption in various regions of the World in year 2001.

	Energy consumption (Gtoe)	Population (Billions)	Consumption indice (toe/capita)
EU	1.50	0.380	3.95
Africa	0.52	0.812	0.64
Latin America	0.45	0.422	1.06
Asia (- China)	1.15	1.935	0.59
China	1.15	1.278	0.90
Former USSR	0.92	0.289	3.18
Middle East	0.38	0.169	2.31
USA+Canada	2.50	0.317	7.88
Rest	1.15	0.500	
World	10.20	6.102	1.67

If one includes the "estimated additional resources of type I", according to the Nuclear Energy Agency (NEA) classification [3], the horizon widens to 400 ans. If one turns to the fast reactor (FR) technology, which is not well developed up to now, another factor 10 would be gained.

According to the United Nations organisation, the World population will be of 7.5 ± 0.4 Billions of people in 2020 and is expected to increase further. The low variant scenarios predict a maximum around 9 Billions in the second half of this century. The energy demand is expected to increase from 10 Gtoe to 13-17 Gtoe in the year 2020 (see Table 1). The main increase will come from developing countries, in which 1.6 Billion people are in "energy poverty".

It is hard to say what kind of energy resources will be developed or will be prominently chosen in the future. Three kinds of factors, economical, technological and societal, are influential. First of all, the trends will mainly depend on the supply and demand mechanism, which, however, may be strongly affected by price policy

TABLE 2. Share (in percents) of total primary energy supply in year 2001.

Source	Oil	Coal	Gas	Renewable and waste	Nuclear	Hydro	Geothermal and wind
Percentage	35.8	23.0	20.9	10.8	6.8	2.2	0.5

TABLE 3. Proved recoverable resources, established in 1999 [2]. Note that they are given in metric tons, except for natural gas. The last column gives the number of years after which these resources will be exhausted if their respective present consumption rates are maintained.

Source	Amount	Number of years
Coal	796 Gt	220
Lignite	189 Gt	237
Oil	143 Gt	42
Oil shale	~100 Gt	
Natural gas	151 Gm^3	63
Uranium	3.28 Mt	100
Nuclear (fast reactors)		21000

(we will disregard this aspect in the rest of this paper). Technological developments may drastically change the respective importances of the various energy sources. Solar energy is often cited as an example, but experts do not foresee a rapid breakthrough. Fusion energy is another quoted example, but the first full-scale reactor is not expected before fifty years.

The pattern may also be strongly influenced by slow but profound changes which are presently reshaping our societies. Let me just mention three of them:

- *Environmental concerns.* The most obvious one concerns the reduction of greenhouse effect gases, in order to prevent a global climate change. This presumably implies a reduced use of fossil fuels.
- *Sustainable development.* An obvious application of this principle calls for a development of renewable (and soft) energy sources. The latter account for 15% of the present energy consumption (see Table 2) and they are not expected to rapidly contribute for substantially more. Let me shortly comment on each of them separately. It is estimated that the hydroelectric capacity can be increased by a factor 5, but investments are slow and present other problems. Wood already contributes to about 7% and can hardly be doubled, hitting evident environmental problems. Biomass has an enormous potential (~6 Gtoe/year), but tapping it would divert its use from agricultural purposes. Solar energy is very diffuse. Solar photovoltaic technology is developing well, but the installed power amounts to only 600 MW and is not expected to amplify rapidly. Wind energy is expanding very rapidly and is promising: it has passed from 2 GW to 18 GW in the last 30 years. Geothermal energy is

not important and is developing very slowly. Tidal energy and other sources are only in the experimental stage.

Of course, the principle of sustainable development has other facets. For instance, it advocates looking for increased efficiency of technological devices. In western countries, a strong effort has been made in this direction during the last years, both in the industry and in domestic use. But at the same time the energy demand has kept growing. More profoundly, the principle of sustainable development challenges our model of development and calls for other less energy-demanding ones.

- *Changes in society management.* Western countries have undergone a strong mutation in the last 25 years, almost unknown of the populations. They have shifted from a centralized undebatable management (governments were considered to work from the wealth of nations and to do the right technological choices without referring to the populations) to a local participative style of management. This mutation will have, for sure, implications for future options concerning energy, although it is hard to figure out which ones [4].

Concluding this Section, one can state that there is a strong chance that the energy demand will still be increasing for many years, especially due to the rise of developing countries. The weight that will be given to the various energy resources is hard to predict. Due to the foreseen pressure for a reduced use of fossil fuels and the presumably slow development of so-called alternative sources, it is expected that we will have to rely to all energy sources, including nuclear energy.

3. CONDITIONS FOR A VALUABLE USE OF NUCLEAR ENERGY

Applying the notion of "sustainability" to energy production by nuclear means, one can formulate these conditions as follows:

1. No rapid exhaustion of resources. It would not be wise to mutiply the numbers of reactors of the present technology by a factor of, say 10, in view of the proved U resources. Furthermore, a light water reactor (LWR) working in "open cycle", i.e. without a recycling of useful matter in spent fuels, consumes less than 1% of the potential fission energy (^{238}U is practically untouched). This argues in favour of using fast neutron reactors (FR), where rapid neutrons can fission ^{238}U. This technology exists, but is not well developed and more complex. It probably cannot compete valuably as long as U is cheap and abundant.

2. No unacceptable risk. This point mainly covers the operational risks and the problem of proliferation. I will elaborate on these points in the next Section.

3. No generation of untractable problems for future generations. This implies taking care of the waste problem. The latter will be discussed below.

4. SAFETY PROBLEMS

There are three main safety problems.

1. Reactor safety. Because it escapes to our senses, radioactivity is frightening. However, in normal operation, a nuclear reactor is among the large energy-producing installations the one whose impact on workers, population and environment is the weakest [5]. What really scares the population is a major accident, especially after the Chernobyl catastrophy, even if it can be argued that this has been a Soviet as much as a nuclear accident [6]. To dissipate the fears of the public, one has to demonstrate the feasibility of a technology that would not release radioactivity even in the case of melt-down of a reactor core. This requirement is at the base of the future so-called Generation III reactors, of which the French-German EPR reactor (whose conceptual project is currently in the optimisation phase) is a example.

2. Proliferation. ^{239}Pu or other radioactive materials can be diverted from the fuel cycle for military or terrorist purposes. The solution to this problem rests on the control by the International Atomic Energy Agency (IAEA). However, the recent Iranian example indicates that this control is not without limit. Another acute problem is generated from the rise of centrifuge-based technology for separation of ^{235}U, which seems to be accessible to countries with medium technology industry. Surely, the solution to this problem passes by an increased role of the IAEA and possibly by new treaties of non-proliferation.

3. Wastes. The highly-radioactive products consitute the real problem, since the reprocessing of low and medium-radioactive wastes is now industrially mastered. In addition, the absence of any real application of accepted and durable method for storing the highly-radioactive wastes leaves the impression that there is no solution to this problem. This is not the case, as it is shown in the next Section.

5. WASTE PROCESSING. SOLUTIONS FOR THE FUTURE

5.1. A short reminder about waste classification

To simplify the presentation, one may distinguish between low-level (LLW) and high-level wastes (HLW). LLW contain radio-elements of short lifetime (less than 30 years) and mainly originate from laboratories, hospitals and industry. HLW are mainly coming from spent fuel and some structural materials of nuclear reactors. Spent fuel contains $U+Pu$, minor actinides and fission products. In addition, one has to include wastes from U extraction and from enrichment process. Sometimes, U and Pu are not considered as wastes, since they can be used as fuel for future FR's.

Almost everywhere, LLW are conditioned in special containers and stored in surface or near-surface depositories. HLW pose the most serious problem, because they release heat, have a large activity and contain many long-lived isotopes. I will

TABLE 4. Inventory of the main isotopes yearly produced by an typical LWR reactor, three years after unloading of the reactor.

	Isotopes	Period (years)	Loading (kg)	Unloading (kg)
Uranium	^{235}U	$7.08\ 10^8$	751	221
	^{236}U	$2.34\ 10^7$	-	88
	^{238}U	$4.47\ 10^9$	20734	20204
Plutonium	^{238}Pu	87.7		3.3
	^{239}Pu	24119		123.1
	^{240}Pu	6569		47.5
	^{241}Pu	14.4		25.4
	^{242}Pu	$3.7\ 10^5$		10.5
Minor actinides	^{237}Np	$2.14\ 10^6$		8.8
	^{241}Am	432.2		4.4
	^{243}Am	7380		2.2
	^{244}Cm	18.1		0.5
	^{245}Cm	8500		0.06
Fission products (medium-lived)	^{90}Sr	28		10.5
	^{137}Cs	30		24.3
Fission products (long-lived)	^{79}Se	70000		0.11
	^{93}Zr	$1.5\ 10^6$		15.5
	^{99}Tc	$2.1\ 10^5$		17.7
	^{107}Pd	$6.5\ 10^6$		4.4
	^{126}Sn	10^5		0.44
	^{129}I	$1.57\ 10^7$		3.9
	^{135}Cs	$2\ 10^6$		7.7
	^{151}Sm	93		0.33

concentrate on these wastes in the following.

5.2. The size of the problem

Nuclear wastes have a rather limited volume. A typical reactor (LWR, 900 MWe, burning rate of 33000MWd/t, 3.5 % enrichment) produces on the average about 20 tons of spent fuel per year. In a country like France, this means 20gr/year/capita compared to the 2.5t/year/capita of ordinary wastes. On the other hand, nuclear wastes are truly highly radioactive: three months after a shut-down, the activity of the core of a typical reactor is of the order of 1 GCi, corresponding to a release of heat at a rate of ~8 MW.

It is interesting to look at the elements contained in the wastes. The inventory is given in Table 4. The largest part corresponds to U and Pu isotopes. If the latter are considered as fuel (see below), the weight is considerably reduced, but the radiotoxicity is not reduced in the same proportions. Let us look at the problem at the European scale. About 145 LWR's producing 880TWh per year (about 35% fo EU's electricity) generate about 2500t/year of spent fuel. Up to now these wastes are stored. The present total stockpile amounts to 37000 t, among which 330 t of Pu, 52 t of minor actinides, 1500 t of fission products including 46 t of long-lived isotopes. Less data are available concerning military wastes (mainly Pu and tritium). The stockpile of military Pu is estimated to 260 t, worldwile [7].

HLW are highly radiotoxic. Each ton corresponds to an equivalent dose of about 10^8 Sv. This should be compared with the radiation workers limiting dose, which has recently been reduced from 50 mSv to 20 mSv. The problem becomes more accute when the time evolution of the radiotoxicity is considered. This is illustrated by Fig.2, inspired from Ref. [8]. It needs more than 10 thousands of years for the radiotoxicity of HLW to reach down the radiotoxicity of natural uranium ore, the level which is generally considered to be necessary for a harmless release in the environment. If U and Pu are removed, this time reduces to a thousand years. If in addition minor actinides are also extracted, the necessary time is now of a few hundred years. The dashed curve indicates that this time is further diminished if long-lived fission products are removed.

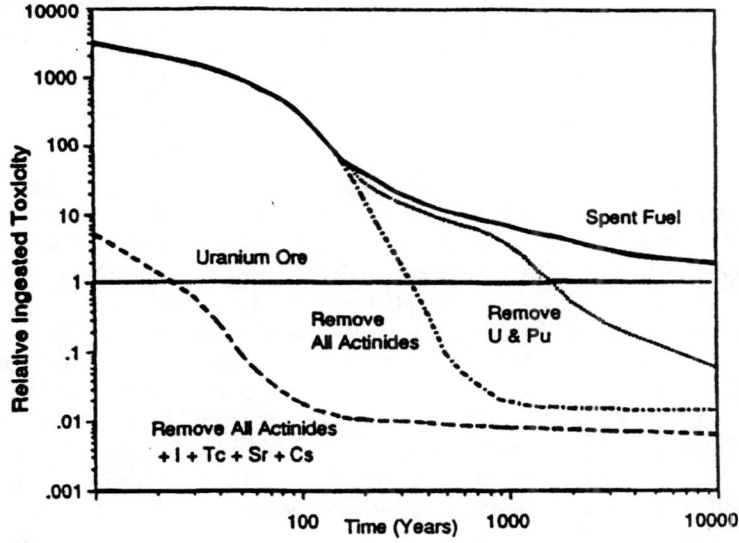

FIGURE 2. Evaluation of the (ingested) radiotoxicity of nuclear wastes coming from a typical reactor, relative to the uranium ore radiotoxicity. The different curves correspond to different scenarios of partitioning. See text for detail.

5.3. Waste processing

Obviously, HLW coming from spent fuel require a very drastic protection against direct radiation exposure of population, release of radioactive fluids, excessive heating and criticality. Furthermore a strict inventory is of uttermost importance.

For the moment, spent fuel is kept on site, in water pools, for some time. Afterwards, it is sent either to a storage facility or to a reprocessing plant. In the first case, HLW are consolidated and stored in special containers. In the second case, HLW are reprocessed. This means that U, Pu, Np and some fission products are separated by a solvent extraction (PUREX) process from other HLW. U, Pu and Np are further separated. U can be sent to an enrichment plant to be re-used. Pu is partly incorporated in MOX (mixed oxyde) fuel to be burned in LWR's. Such a reprocessing is performed in the La Hague plant, in France, and in Sellafield, in United Kingdom. Reprocessing is often considered as a serious threat for proliferation, since separated Pu can be stolen. On the other hand, it is also claimed that this Pu is not "military grade". It indeed contains a substantial amount of ^{240}Pu, which emits neutrons in spontaneous fission (Pu for nuclear weapons is generally made in special reactors). But civil Pu is probably suitable to build some "nasty" bombs. Reprocessing presents some other advantages. It reduces the volume of HLW. Furthermore some isotopes, such as ^{85}Kr, ^{90}Sr or ^{137}Cs, which have industrial or medical applications, can be reclaimed from the wastes. It is also the case for some rare elements, for example Rd, Ru and Pd.

5.4. Solutions for the future.

The long run management of HLW is still a question of debate. Some countries, e.g. USA and Sweden, which do not reprocess their wastes, have opted for disposal in deep underground repositories. In USA, the site of Yucca Mountain is being prepared. Such repositories are arranged with a multibarrier approach: the first level is the waste conditioning itself (a glassy material), the second one is the container which should be compatible with the surrounding material, the third one is a layer of clay which should prevent intrusion of water and the fourth one is the geological site itself. It has to be suitable to minimize water flow and effects of heat generation. Repositories may be arranged in a reversible way, so that wastes may be reclaimed and reprocessed if future techniques make this possibility favourable.

When reprocessing is adopted, recycling of waste is envisaged. We have already indicated that U is recycled and that Pu is burned in ordinary reactors. However, Pu burning in LWR's is limited for reasons of reactor stability. Actually only about 7% of Pu can be incorporated in MOX. Furthermore, in LWR's, Pu is partly transformed into Cm and Am, by low energy neutron capture. The burning (by fission) of Pu (and of Am and Cm) is more advantageous in dedicated or "ordinary" FR's. The capture to fission ratio is much smaller for fast neutrons. However, fission cross-sections are not tremendously high, so that burning takes more time.

Another possibility is the transmutation by so-called accelerator-driven systems (ADS). Transmutation is the transformation of long-lived isotopes to shorter-lived ones. For actinides this can be achieved by fission. An ADS is an assembly made of a subcritical reactor, a spallation source and a proton accelerator. The reactor works owing to the continuous supply of neutrons emitted by a spallation source (basically of piece of Pb-Bi) bombarded by high-energy protons. Since the reactor runs in a subcritical mode, it accomodates exotic fuels. Pu and minor actinides can thus be loaded in greater quantities than in the examples above. Several projects exist (see Ref. [9] for a review and a discussion of the merits of ADS's). The EU has recently launched the EUROTRANS project which has to evaluate the feasibility of the partitioning-transmutation cycle and to start the technical studies for a future demonstrator of ADS, that could be elaborated starting from the Belgian MYRRHA project [11].

Transmutation of fission products cannot be made by thermal neutrons because the capture cross-sections are too low. However many of these products show intense narrow resonances in the epithermal domain. They can then be transmuted by placing them in suitable locations in a reactor, corresponding to the appropriate neutron energies. The TARC experiment has shown that this method is promising [10].

Evidently combinations of these possibilities are foreseen. Even plans for the deployment of future FR's and ADS's in EU are drawn. ADS's are expected to start operating around 2050 and the amount of transuranic wastes is expected to stabilize at a level lower than the actual one in year 2070 [12].

One has to keep in mind that there are losses in the partitioning-transmutation procedures and that a small fraction of HLW (2-4%) will have to put in repositories, anyway.

6. CONCLUSION

The foreseeable World energy needs for the XXIst century, the environmental concerns and the long-waited and ethically justified access of poor countries to development make plausible a continued, if not enlarged, use of nuclear energy (unless our development model is radically revised), in spite of the fact that that some countries have opted, perhaps hastily, to a phasing out of their nuclear power. Of course, this choice should be accepted by the populations, which should be convinced of the advantages of nuclear energy and of the mastering of security and waste problems. This short overview indicates that there exist solutions to the last problems, even if they still need further investigations.

In my opinion, academic institutions have an important role to play concerning these issues. Basic research in nuclear physics should be pursued to support future technological developments. Studies on energy development scenarios should be refined. These institutions have also a role for public education. Especially, owing to their independence, they have to explain the advantages and disavantages of the energy options, in order to secure truly democratic choices. Finally, they have to form experts in nuclear sciences. Some countries, in view of a possible phasing out, are progressively neglecting these formations.

ACKNOWLEDGMENTS

I would like to thank Profs. David Blaschke and Jorg Hüfner, who asked me to present this lecture. I am also thankful to Dr. Hamid Aït Abderrahim for having provided me with some material and to Dr. Sylvie Leray for interesting discussions.

REFERENCES

1. *World Energy Outlook 2004* International Energy Agency publications, ISBN 92-64-10750-9 (paper) 92-64-10752-5 (CD-ROM), Paris, 2004.
2. World Energy Council, http://www.worldenergy.org/
3. Nuclear Energy Agency, http://www.nea.fr/
4. B. Barré, *Nucl. Phys. A*, **654**, 409c–416c (1999).
5. R. Wolfson, *Nuclear choices*, The MIT Press, Cambridge, USA, 1991, Chap.9.
6. B. Barré, L. F. Durret and B. Tinturier, *Le nucléaire durable*, Institut National des Sciences Appliquées, Paris, 2004.
7. D. Albright and K. Kramer, *Bulletin of the Atomic Scientists*, November/December 2004.
8. H. Takahashi and H. Rief, *Proceedings of the Specialists' Meeting on Acclerator-Based Transmutation*, PSI Zürich, March 1992.
9. W. Gudowski, *Nucl. Phys. A*, **654**, 436c–457c (1999).
10. S.Andriamonje et al., *Physics Letters B*, **458**, 167–180 (1999).
11. SCK-CEN Mol, Belgium, The MYRRHA project, http://www.sckcen.be.
12. H. Aït Abderrahim, *Proceedings of the International Conference on Nuclear Data for Science and Technology*, Santa Fe, USA, Sept. 2004, to be published.

How to Become a Nobel Laureate
J. Hüfner

University of Heidelberg, Germany
E-mail: joerg.huefner@tphys.uni-heidelberg.de

Abstract. We discuss the family background and the educational careers of Nobel laureates in science as well as the age, at which they perform the Nobel work. As an example, we describe the discovery of the nuclear shell model.

Keywords: Nobel laureates in science, Family background, Education, Nuclear shell model
PACS: 01.65.+g, 01.75.+m

1. INTRODUCTION

As may be allowed for an after-dinner-speech, I have chosen a rather catchy title, "How to Become a Nobel Laureate", suggesting that I shall provide recipes for becoming a Laureate. This is not quite so. Rather, I will discuss several common features in the biographies of Nobel laureates, and thereby demonstrate that a genius does not fall from heaven, but develops, if conditions are right. I have studied this topic lately for the following reason: For a lecture series at Heidelberg University on various aspects of intellectual elites, I had been asked to give a talk about "Elites in the Sciences", i.e., about the leading scientists in physics, chemistry, and the life sciences. Since I did not know how to properly define elites, I chose to talk about the Nobel laureates in the sciences. It is generally agreed that they are part of the group of top scientists, though it is equally agreed that there are also top scientists who did not receive the Nobel prize. My research was helped by the fact that extensive literature on Nobel laureates exists. In this talk, I will mainly rely on two books. Harriet Zuckerman's "Scientific Elite: Nobel Laureates in the United States" [1] is a sociological study of the 95 American laureates from the years 1901 to 1972. Biographies of female laureates are presented in "Nobel Prize Women in Science, Their Lifes, Struggles and Momentous Discoveries" by Sharon Bertsch McGrayne [2].

2. NOBEL'S TESTAMENT

The Swedish industrialist Alfred Nobel (1833 - 1896) had made a fortune with the production of dynamite. Although this new explosive improved the safety of civilian applications, it was equally useful in the production of weapons. Nobel was unhappy about the latter consequence of his invention and he decided not to keep any money that accrues from this application of his product, but to put it into a fund, whose annual income was to be divided among five prizes, three in the sciences (physics, chemistry and physiology or medicine), one in literature, and one to advance the cause of world peace. In the sciences he wanted that the "most important discoveries or inventions" of the preceding year should be honored, irrespective of the nationality of the recipient. After Nobel's death the Nobel foundation was set up and the rules for administering the prizes were laid down. For the sciences these are essentially the following:

(i) The laureates in physics and chemistry are chosen by the Royal Swedish Academy of Sciences and the prize in the life sciences by the Royal Caroline Institute. A selection committee of five periodically changing members in each science invites and evaluates proposals for candidates.
(ii) Candidates may be proposed by members of the selection committee, former laureates, professors of Scandinavian universities, and professors of relevant faculties of six yearly selected foreign universities.
(iii) The prizes are given by the Swedish king in an impressive ceremony each year on December 10, the day of Nobel's death.
(iv) While Nobel wanted the prize to be given to one person in each science, the committee has the ability to split the prize among at most three scientists.
(v) Nobel intended that the prize money should enable the recipient to continue the successful research. Therefore, he wanted the prize to be given to the most important work of the preceding year. This turned out not to be practicable, since it takes quite some time until the correctness and importance of a discovery or invention is proven beyond doubt. Therefore Nobel's will has not been followed in this aspect. This has the advantage that the decisions of the committee are seldom wrong.

In the meantime, the Nobel prize has become the prize in the sciences. Its high prestige arises mainly from the impressive list of past laureates and less from the amount of the money, which is surpassed by other prizes by now.

3. THE FAMILY BACKGROUND OF THE LAUREATES

In her book, Zuckerman investigates American Nobel laureates, and defines as an American laureate any scientist, who has conducted the prize winning research in the USA, independent of whether he or she is a native American or an immigrant. However, for the study of their social origin, Zuckerman restricts herself to the subgroup of those who were born and raised in the US. These are 72 laureates in the peroid 1901 to 1972. What is their family background? Zuckerman has investigated the profession of the fathers and the economic situation of the families. The results are listed in Table 1, and are compared with two other groups of American society: science doctorates, and employed males in general. According to this table, about 80% of the laureates come from families from the upper middle class, where the fathers are professionals (academically trained; teachers, medical doctors etc.) or managers and proprietors. This group represents a rather small fraction of the American society of this time, namely, only 11%. The results of

TABLE 1. Socioeconomic origins of American reared laureates (1901-1972) compared with those of American scientists receiving doctorates (1935-1940) and employed males (1910). From [1].

Father's occupation	Nobel laureates 1901-1972	Science doctorates 1935-1940	Employed males 1910
Professionals	53.5%	29.1%	3.5%
Managers and proprietors	28.2%	18.7%	7.7%
Farmers	2.8%	19.5%	34.7%
Sales, service and clerical workers	7.0%	13.1%	12.8%
Workers: skilled and unskilled	8.5%	18.0%	41.3%
Total	71	2695	29 847 000

Table 1 for the laureates confirm what is common knowledge: the intellectual level of the family into which a child is born, is extremely important for future intellectual achievements. Various conceivable causes come together, among which are (i) the child receives significant intellectual stimulations at an early age, and (ii) parents of the higher strata of the society see to it that their children receive a good education.

Also, the religious tradition of a family can have an important impact on the children. I quote only one result from Zuckerman [1]. While only 3% of the American population are Jewish, 27% of the American laureates come from Jewish families. A similar situation is known from Germany before 1933. Various explanations are proposed, among them: (i) learning is an important value in the Jewish tradition and (ii) Jewish families in the US are predominantly part of the upper middle class, where laureates most often originate.

The statistics about female Nobel laureates is depressing. Of the 500 prizes which have been awarded worldwide since 1901, only 12 have been given to a woman, i.e., only 2.5% of all prizes. This result is in line with the observation that the fraction of women in the sciences is significantly below 50%, depending somewhat on the country in question. The reasons for this are discussed in detail in the book by Bertsch McGrayne [2].

4. EDUCATION

The intellectual development of the future laureates is characterized by a continuous "accumulation of advantages": already priviliged by the family background, these children pass the entrance examinations to the better colleges and universities, are accepted as doctoral students by the top professors of a field, receive fellowships etc. The result that better students get better chances, is also called the "Matthew effect", according to Matt. 13:12, where it is said that "Those who have, will receive more ...". This rather surprising verse in the Bible seems to ex-

TABLE 2. Percentage of future Nobel laureates who obtained their degrees from Harvard, Columbia, Berkeley, John Hopkins, Princeton, Chicago, and Caltech as compared with American science students in general. Calculated on the basis of numbers given in [1].

	Nobel laureates 1901-1972	Science students
Bachelor	25.7%	4.3%
Doctorate	67.6%	25.7%

press an old wisdom. The French proverb "On ne prête qu'aux riches" states the same. In our times, the accumulation of advantages is a political aim intended to foster the development of elites in a society. As is well known, there is a strong hierarchy of prestige and quality in the educational institutions in the US. Students try to get into the best universities in order to receive a good education as well as to profit from the prestige of the institution from which he or she has received a degree. Future Nobel laureates are no exception to this rule. It turns out that 67% of all American raised laureates received their Ph.D. degrees from Harvard, Columbia, Berkeley, John Hopkins, Princeton, Chicago, and Caltech, while only 25% of all science students finish their Ph.D. at these seven institutions.

The adviser of a future laureate plays an even more decisive role than the institution. The adviser may be the Ph.D. adviser or some elderly scientist during the post-doctoral work. The investigations by Zuckerman [1] have revealed that in about 50% of all cases, future laureates had worked with advisers who had won the Nobel prize already, or would later receive it. Thus, future laureates choose the best teachers in their field. Otto Warburg, the German laureate in medicine (1931), once reminisced about his own experience in this regard: "The most important event in the life of a young scientist is in the personal contact with the great scientists of his time. Such an event happened in my life, when Emil Fischer, the second scientist to be awarded the Nobel prize in chemistry 1902, accepted me in 1903 as a co-worker in protein chemistry, which at that time was at the height of its development. During the following years, I met Fischer almost daily, ... I learned that a scientist must have the courage to attack the great unsolved problems of his time, and that solutions usually have to be forced by carrying out innumerable experiments without much critical hesitation" [1]. Zuckerman calls this decisive relation "master-apprentice", using rather old-fashioned words. We know of similar relationships between great musicians, who emphasize the importance of having worked with this or that master. On a much smaller scale, I also look back with gratitude to those teachers, from whom I have learned so much during my work as a Ph.D. student in Heidelberg and at the Weizmann Institute, and later as a postdoc at MIT.

TABLE 3. Age distribution of Nobel laureates at time of prize winning research.

Age at prize winning research	Single discovery	Line of research	All laureates
20 - 29	16%	6%	12%
30 - 34	25%	19%	22%
35 - 39	22%	18%	20%
40 - 44	21%	28%	24%
45 - 49	7%	20%	13%
50 - 54	7%	9%	8%
55+	-	2%	1%
Mean age	37.6	39.9	38.7
Total number	67	55	122

5. AGE OF LAUREATE AT TIME OF PRIZE WINNING RESEARCH

In the sciences, especially in physics, there exists the following popular idea about a genius: a brilliant wild young man comes, overthrows a dogma, and is awarded with the Nobel prize. I must admit that I also held this idea, and was surprised, when I saw the statistics in Zuckerman's book. It is presented in Table 3. Although there is a fraction of 12% of young scientists who do their prize winning research before the age of 30, the bulk of scientists is in the age bracket 30 to 44 years, when they are most productive and find the breakthrough. We explain this as follows. In most cases research needs a lot of work, careful studies and measurements, and also careful checking in order not to be fooled by some artefact. Also in the sciences one has the saying "A genius is made by 5% inspiration and 95% perspiration". It is the perspiration phase which explains the unexpected high average age of 39 years when the prize winning research is done.

6. THE PRIZE WINNING RESEARCH

I think there is no general rule as to how a scientist finds the topic of research, which eventually leads to the Nobel prize. Therefore, I tell you a particular story and choose the discovery of the nuclear shell model by Hans Jensen and Maria Goeppert-Mayer. I choose this story in part because I am quite familiar with the details, since Jensen worked in Heidelberg. The story of how the subject of the magic numbers, whose explanation brought the Nobel prize, had been "forced" on Jensen, is beautifully told by O. Haxel, later a professor of experimental physics at the University of Heidelberg. In the year 1949, when the story happens, Haxel was a research associate in Göttingen, where he investigated various properties of nuclei. He observed that nuclei, which had particular numbers of protons and/or neutrons, were particularly stable and abundant in nature. Since the origin of these numbers could not be explained, they were called magic. Haxel looked for some theoretician's help. He went to see Heisenberg and told him of the phenomenon. After a very short time, Heisenberg interrupted him and told him that he knew everything, since he had already spent hundreds of hours on the problem, without being able to solve it. And he had promised to himself, that he would not waste any more time on it. But he encouraged Haxel to continue. So Haxel went to see Jensen, who at that time was a professor of theoretical physics at the university in the nearby town of Hannover. When Haxel explained the problem of the magic numbers, Jensen told him squarely: "Sorry, I am not interested. I am engaged in science and not magic." Yet, Haxel did not give up, and succeeded in convincing Jensen that this was indeed physics, and important physics. After a few weeks, Jensen came to Göttingen and told Haxel: "I solved it". Jensen had discovered that a spin-orbit force in the average potential of the nucleus could explain the magic numbers. Why was it so relatively easy for him to find this anwer? I think that Jensen's previous extensive work on problems of the atomic structure (where the spin-orbit force is essential and well understood) had given him the idea to try this force also in the nuclear case. Jensen never forgot that it had been Haxel, who had "forced" him, and stated in his testament that the gold medal of his Nobel prize should be given to Haxel. This was the discovery of the shell model on this side on the Atlantic.

In Chicago, Maria Goeppert-Mayer was also working on the properties of nuclei, and the systematics of the magic numbers had also not escaped her attention. Yet, she did not find the clue. Then, after a casual discussion with Fermi, in which he asked "Did you ever consider a spin-orbit force?" it took only minutes, as she reports, until everything fell into place. Also she had discovered the nuclear shell model, as we know it now. At that time, in the year 1949, the communication systems were not well developed, and Jensen and Goeppert-Mayer did not know of each others' work. It was an American scientist (I think it was V. Weisskopf), who visited Jensen and learned from him about his discovery. Then back in the US he learned that Goeppert-Mayer had made the same discovery. Weisskopf was not only a great physicist, but also a great man. In order to avoid any struggles of priority, he proposed that Jensen and Goeppert-Mayer publish their findings in the same edition of the Physical Review Letters. This was not so easy. The Germans had no dollars to pay the page charge and Goeppert-Mayer was not allowed to publish her classified research. Friends gave the dollars and the relevant agency declassified the results after it had been revealed that these results had already been discovered by other people as well. Later Goeppert-Mayer and Jensen worked together on various occasions and eventually summarized their results in the book "Elementary Theory of Nuclear Shell Structure". They shared the Nobel prize in physics of the year 1963 with E. Wigner.

How to become a Nobel Laureate? This was the question we started from. Do we know it now? Let me try to give some necessary conditions. You are gifted in science (genetically?). You are born into the "right" family, where you receive the proper education, not only intellectually, but also in your character. For instance, you learn to stick with a problem and not give up easily. You receive the proper education in school and at the university. The advisers, whom you choose are extremely important. And then with some luck you stumble upon the important problem. And last but not least, we must remember the 95% perspiration needed to solve it.

ACKNOWLEDGMENTS

I thank W. Müller, University of Mannheim, for several discussions and help with the references, B. Witzler and M. Peachin for their careful reading of the paper.

REFERENCES

1. H. Zuckerman, *Scientific Elite: Nobel Laureates in the United States*, New Brunswick 1996, Transaction Publishers
2. S. Bertsch McGrayne, *Nobel Prize Women in Science, Their Lifes, Struggles and Momentous Discoveries*, New York 1993, Birch Lane Press

QCD – NLC

H.J. Pirner[1]

University of Heidelberg, Germany
E-mail: pir@tphys.uni-heidelberg.de

Abstract. We give a status report of our current theoretical work on QCD near the light-cone.

Keywords: QCD
PACS: 11.15-q

1. INTRODUCTION

A simple extension of the quantum mechanics of bound states to a relativistic field theory of massless quanta bound into hadrons is not possible. The light-cone Hamiltonian approach attacks this problem from a quite different point of view. Take a cube of length 2 fm filled with quarks and gluons and boost it in the 3− direction with a Lorentz factor of $\gamma = 1000$. This gedanken experiment is well suited to imagine a proton moving with fast speed in the laboratory. The box will contract on one side, valence quark momenta will be high, and valence states will have very high energies. Naively vacuum properties of QCD are not important because of the high energies. By some suitable kinematic choices of coordinates one can construct invariants. Commonly, the light-cone energy $P^- = \frac{E-P_z}{\sqrt{2}}$ and the light-cone momentum $P^+ = \frac{E+P_z}{\sqrt{2}}$ are chosen and $M^2 = 2P_+P_- - P_\perp^2$ is invariant. With these variables all light-cone energies are positive and increase as $P^- = \frac{P_\perp^2 + m^2}{\sqrt{2}P^+}$ for small light-cone momenta. Only fluctuations with small P^+ momenta may pose a problem. Their light-cone energies are very high. In light-cone physics the ultraviolet problem gets mixed up with the infrared problem. Formally, the problem reappears in the context of constraint equations for x^- independent fields [1]. These constraint equations arise in the light-cone Hamiltonian framework, since the Lagrangian contains the velocities in linear form $L = \partial_- \phi \partial_+ \phi$. The momenta related to these velocities obey constraint equations including $\partial_- \phi$. Therefore, integrals of the equations of motion over the spatial light-cone distance x^- become operator equations of reduced dimensionality (two transverse spatial dimensions and one time dimension). These equations are called zero-mode equations. For example, in equal time theory zero-mode equations determine the condensate of a scalar field. The x^- independent zero-mode field couples to the transverse fluctuations of all other fields, consequently these equations depend on the cutoff and are

---

[1] Supported by the EU-Program Computational Hadron Physics

involved in the whole renormalization procedure. This feature is often overlooked in naïve pictures assuming either superrenormalizable models or models with a simple cutoff. In Nambu-Jona-Lasinio models one has been able to solve [2, 3] these zero-mode equations e.g. in large N_c approximation giving a view of chiral symmetry breaking on the light-cone, which is quite special. These zero-mode equations have not been solved in QCD.

A common argument goes as follows: Zero modes decouple from the rest of the theory, because their energies lie beyond the cutoff. Naïvely, the light-cone momentum $P^+ = 0$ means that the light-cone energy $P^- = \infty$. If, however, the mass m of the zero mode is zero, the mode does not disappear into infinity for very small transverse momenta. How is the situation in QCD? Can we just ignore this problem, buy a big computer, use some suitable Fock truncation, put all transverse gluon modes into a Hamiltonian matrix and diagonalize it? Pauli and Brodsky [4] and many others have solved successfully 1+1 dimensional theories. QCD on the light-cone is a tremendously seductive field theory, since the Euler-Lagrange equation for time-like light-cone potential can be solved directly in a gauge, where the potential along the spatial light-cone direction vanishes. The resulting Hamiltonian contains the light-cone Coulomb energy plus the kinetic energies of the transverse gluons and nothing else. The light-cone Coulomb energy is already in a form which linearly confines sources separated along the spatial light-cone directions. This is a simple consequence of the massless gluon propagator in one spatial dimension.

The massless gluon interaction has to be implemented also correctly for colored line sources smeared over the spatial light-cone direction. Otherwise, we violate the equal treatment of all spatial directions. This necessity can be demonstrated rather easily in perturbation theory, where the rotational invariance of the gluon exchange is reconstituted via the exchange of one transverse gluon. I think, one can be easily misled by the experience that QCD will always favor a finite correlation mass for color sources moving along time-like directions. At finite energies one sees this phenomenon in the hadronic cross sections which are given by the geometrical sizes of the hadrons, the low light-cone momentum partons do not matter at finite (small) energies. There is a natural transverse scale of the moving proton. The energy dependence of the high Q^2-structure functions indicate, however, an abnormal increase of "size" in transverse direction. The proton first gets blacker, but then its transverse radius has to increase. Purely theoretical arguments point towards conformal invariance at high energies, a conjecture, which supports the view, that partons with small light-cone momenta sampling large spatial light-cone distances correlate over large transverse distances compared to normal hadronic scales.

We have analyzed QCD approaching the light-cone with a tilted near-light-cone coordinate reference system [5, 6, 7] containing a parameter $\eta \neq 0$ giving the distance away from the light-cone. The constraint equations appear in the near-light-cone Hamiltonian as terms proportional to $1/\eta^2$. We then multiply the light-cone energy with η, considering $\tilde{P}_+ = \eta T_+^+$ and divide the light-cone momentum by η, defining $\tilde{P}_- = \frac{1}{\eta} T_-^+$. The invariant masses remain unchanged up to terms higher order in η. By the trick with near-light-like coordinates we can derive a full quantum Hamiltonian for the zero modes which now depends on the QCD coupling

g, the extension $L_\|$ of the spatial light-cone distance compared to some lattice size a (or ultraviolet cutoff $\Lambda = 1/a$) and the parameter η which gives the nearness to the light-cone. Having fixed the QCD coupling g which determines the lattice size a, we would like to study in this Hamiltonian the physics at large longitudinal distances $L_\|/a \to \infty$ close to the light-cone $\eta \to 0$. Because of dimensional reduction the product

$$s = \frac{\eta L_\|}{a} \qquad (1)$$

appears as a coupling in the Hamiltonian. Its limit is not defined. The order of the limiting process is important as one knows from simple superrenormalizable models. One first has to let $L_\|/a \to \infty$ and then $\eta \to 0$ in order not to lose the nonperturbative properties of the vacuum. For QCD an analytical limiting process is impossible. Therefore, the only way out is to start for large s, corresponding to fixed η and large $L_\|$ and then approach smaller values of s.

This procedure ends, when we have found a fixed point $s^* = \frac{\eta L_\|}{a}$ where the mass gap of the zero mode theory vanishes. Approaching this fixed point from the correct side which corresponds to a large longitudinal extension of the lattice, we include the nonperturbative dynamics of the zero modes. The trivial, wrong other side where s is arbitrarily small would be disconnected from the large $L_\|$ limit. When the (2+1)-dimensional system has an infinite correlation length, both the infrared limit of large longitudinal distances and of nearness to the light-cone is realized. For a simplified zero mode theory in SU(2) we have demonstrated such a possibility on the lattice [5]. In principle the full (3+1)-dimensional theory can be solved for any η as long $g, L_\|/a$ are chosen in such a way that we have asymptotic scaling. But in order to synchronize the infrared behavior encoded in the zero mode system correctly with the ultraviolet behavior of small lattice size, the choice of η is no longer free for a given length of the longitudinal direction, one must choose η in agreement with the fixed point found in the zero mode calculation, i.e. in the (3+1)-dimensional calculation the number of slices $L_\|/a$ in spatial light-cone direction determines η

$$\eta = \frac{s^*}{L_\|/a}. \qquad (2)$$

It has to be demonstrated numerically that with decreasing QCD-coupling g the value s^* becomes smaller in such a way that we approach the light-cone $\eta \to 0$ having a reasonable number of slices $L_\|/a$ in spatial light-cone direction. The reduced calculation in SU(2) [5] was done without the inclusion of transverse gluons, so we still have to prove that this procedure works. Phenomenologically [7] we have conjectured that the increase of the high-energy electron-proton cross section is due to this critical point s^*. At infinite energies when this point is approached, the correlation length of near-light-like Wilson lines of the partons increases with a critical index from $Z(3)$ symmetry. The photon density remains power behaved beyond the short distance scale given by the resolution of the photon. According to our conjecture this critical opalescence phenomenon is the cause of the increase of the virtual photon cross section with high energies.

2. THE QCD HAMILTONIAN AND MOMENTUM

We use the near light-cone coordinates defined in ref. [6]

$$x^+ = \frac{1}{\sqrt{2}}((1+\frac{\eta^2}{2})x^0 + (1-\frac{\eta^2}{2})x^3)$$
$$x^- = \frac{1}{\sqrt{2}}(x^0 - x^3) \qquad (3)$$

with the metric

$$g_{\mu\nu} = \begin{pmatrix} 0 & 0 & 0 & 1 \\ 0 & -1 & 0 & 0 \\ 0 & 0 & -1 & 0 \\ 1 & 0 & 0 & -\eta^2 \end{pmatrix}, g^{\mu\nu} = \begin{pmatrix} \eta^2 & 0 & 0 & 1 \\ 0 & -1 & 0 & 0 \\ 0 & 0 & -1 & 0 \\ 1 & 0 & 0 & 0 \end{pmatrix} \qquad (4)$$

with $\mu,\nu = +,1,2,-$, $\det g = -1$, which defines the scalar product

$$\begin{aligned} x_\mu y^\mu &= x^- y^+ + x^+ y^- - \eta^2 x^- y^- - \vec{x}_\perp \vec{y}_\perp \\ &= x_- y_+ + x_+ y_- + \eta^2 x_+ y_+ - \vec{x}_\perp \vec{y}_\perp. \end{aligned} \qquad (5)$$

We refer to p_+ as "light-cone" energy and to p_- as "light-cone" momentum. We restrict ourselves to the color gauge group $SU(2)$ and to gluons. The Lagrangian density in the near light-cone coordinates reads:

$$\mathcal{L} = \frac{1}{2}F^a_{+-}F^a_{+-} + \sum_{i=1,2}(F^a_{+i}F^a_{-i} + \frac{\eta^2}{2}F^a_{+i}F^a_{+i})$$
$$- \frac{1}{2}F^a_{12}F^a_{12}. \qquad (6)$$

The energy momentum tensor has the general form:

$$T^{\mu\nu} = \sum_a (g^{\mu\alpha}g^{\rho\beta}F^a_{\alpha\beta}F^a_{\rho\nu} - \delta^\mu_n u g^{\rho\alpha}g^{\sigma\beta}F^a_{\alpha\beta}F^a_{\sigma\rho}) \qquad (7)$$

We introduce dimensionless gauge fields and coordinates

$$\begin{aligned} \tilde{A}^a_\pm &= ga_\parallel A^a_\pm, & \tilde{x}_\pm &= x_\pm/a_\parallel; \\ \tilde{A}^a_i &= ga_\perp A^a_i, & \tilde{x}_\perp &= x_\perp/a_\perp. \end{aligned} \qquad (8)$$

The dimensionless Lagrange density $\mathcal{L}$ has the form

$$\int \mathcal{L} d^4 x = \int \tilde{\mathcal{L}} d^4 \tilde{x} \qquad (9)$$

with

$$\tilde{\mathcal{L}} = \frac{1}{2}\frac{1}{g^2}\left(\frac{a_\perp}{a_\parallel}\right)^2 \tilde{F}^a_{+-}\tilde{F}^a_{+-}$$
$$+\frac{1}{g^2}\sum_i (\tilde{F}_{+i}\tilde{F}_{-i} + \frac{\eta^2}{2}\tilde{F}_{+i}\tilde{F}_{+i})$$
$$-\frac{1}{2}\frac{1}{g^2}\left(\frac{a_\parallel}{a_\perp}\right)^2 \tilde{F}^a_{12}\tilde{F}^a_{12}. \tag{10}$$

The field momenta conjugate to $\tilde{A}_i$ and $\tilde{A}_-$ are

$$\tilde{\Pi}^a_i = \frac{\partial \tilde{\mathcal{L}}}{\partial \tilde{F}_{+i}} = \frac{1}{g^2}(\tilde{F}^a_{-i} + \eta^2 \tilde{F}^a_{+i})$$
$$\tilde{\Pi}^a_- = \frac{\partial \mathcal{L}}{\partial \tilde{F}_{+-}} = \frac{1}{g^2}\left(\frac{a_\perp}{a_\parallel}\right)^2 \tilde{F}_{+-}. \tag{11}$$

¿From now on we drop the tilde symbol from all coordinates, fields and momenta to facilitate the writing and reading of all formulas. The commutation relations between fields and momenta are standard.

$$[\Pi^a_i(\tilde{x}_-, x_\perp, x_+), A^b_j(y_-, y_\perp, x_+)] =$$
$$i\delta^{ab}\delta_{ij}\delta(x_- - y_-)\delta(x_\perp - y_\perp)$$
$$[\Pi^a_-(x_-, x_\perp, x_+), A^b_-(y_-, y_\perp, y_+)] =$$
$$i\delta^{ab}\delta(x_- - y_-)\delta(x_\perp - y_\perp). \tag{12}$$

The dimensionless light-cone energy density T^+_+ and light-cone momentum density T^+_- are obtained from the energy momentum tensor $T^{\mu\nu}$ and the skewed metric

$$P_+ = \int T^+_+ dx^- dx_\perp \tag{13}$$
$$P_- = \int T^+_- dx^- dx_\perp \tag{14}$$

with

$$T^+_+ = \frac{1}{2}g^2\left(\frac{a_\parallel}{a_\perp}\right)^2 \Pi^2_-$$
$$+\frac{1}{2}\frac{1}{g^2}\left(\frac{a_\parallel}{a_\perp}\right)^2 F_{12}F_{12}$$
$$+\frac{1}{2}g^2\frac{1}{\eta^2}(\Pi_i - \frac{1}{g^2}F_{-i})^2$$
$$T^+_- = \frac{1}{2}(\Pi_i F_{-i} + F_{-i}\Pi_i). \tag{15}$$

The light cone energy and momentum have an obvious symmetry. They have electric-magnetic duality of the transverse fields which any solution to the problem must respect.

$$\Pi_i \rightarrow \frac{1}{g^2} F_{-i}$$
$$\frac{1}{g^2} F_{-i} \rightarrow \Pi_i. \tag{16}$$

Furthermore, since in the Lagrangian $\mathcal{L}$ there are no terms with time derivatives of A_+, the field A_+^a acts like a Lagrange multiplier for G^a, the Gauss law. For any wavefunction $|\Phi>$ of the system the following identity must hold.

$$\begin{aligned} G^a|\Phi> &= (\frac{1}{g^2}\left(\frac{a_\perp}{a_\parallel}\right)^2 D_-^a F_{+-}^a \\ &+ \frac{1}{g^2}\sum_i D_i\left(F_{-i}^a + \eta^2 F_{+i}^a\right))|\Phi> \\ &= (D_-^a \Pi_-^a + \sum_i D_i \Pi_i^a)|\Phi>= 0. \end{aligned} \tag{17}$$

This Gauss law is fulfilled as long as only closed loops exist in the wave function, or in the case of excited links the electric flux must be conserved at each site, i.e. there are also multiple connected flux loops possible.

If one chooses $a_\parallel \ll a_\perp$ and uses the same number of sites in x_- and $x_\perp$ directions, one ends up with a real system, which is contracted in the longitudinal directions. Verlinde and Verlinde [8], and Arefeva [9], have advocated such a set up to describe high energy scattering. A contracted lattice means the minimal momenta become high in longitudinal direction and this looks a promising starting point for high energy scattering.

One sees from the Lagrangian $\mathcal{L}$ in eq. (10) that the limit $a_\parallel/a_\perp \rightarrow 0$ enhances the terms with $F_{+-}F_{+-}$ and suppresses transverse $F_{12}F_{12}$.

Because of the enhanced couplings Verlinde and Verlinde conclude that the curvature in longitudinal directions is zero. One ends up with only one term which in the Hamiltonian is the term $\propto \frac{1}{\eta^2}$ and fixes the dual symmetry of the electric and magnetic fields.

Our current project [10] is to introduce lattice variables into this framework and solve the $1/\eta^2$ part of the Hamiltonian exactly.

3. A VALENCE-QUARK LIGHT-CONE HAMILTONIAN

In this section, I would like to present a derivation where the near-light-cone method and the field strength correlators work nicely together. This example demonstrates their practicality as a calculational and heuristic tool. Firstly, one can analytically do the calculation in the stochastic vacuum model and secondly, the result is so close to reality that one can see the model-independent result. In our application of the stochastic vacuum model to high-energy scattering we always use Wilson loops which are on the light-cone. The expectation values of a Wilson loop along the light-cone is unity, because the area of a light-like Wilson loop is zero. I was always disturbed by this fact, because I thought that a color dipole moving with the speed of light should feel confining forces. The wavefunction renormalization due to single loops cancels out in the S-matrix, but the puzzle remained to me. So recently, Nurpeissov and myself [11] have looked into this problem again using a tilted Wilson loop corresponding to a fast moving dipole in Euclidean and in Minkowski space, i.e. we applied the near light-cone trick.

In Euclidean space the Wegner-Wilson loop can be represented with the help of the Casimir operator in the fundamental representation $C_2(3) = t^2 = 4/3$

$$\langle W[C] \rangle_G = \exp\left[-\frac{C_2(3)}{2}\chi_{ss}\right]. \tag{18}$$

We calculate χ_{ss} as the double area integral of the correlation function over the surface of the loop. Let us consider the χ_{ss} function for large separations R_0 of the quark and antiquark, where the confinement term plays the main role. For the nonperturbative (NP) confining (c) component χ_{ss}^{NPc} we get the following expression for large distances $R_0 \alpha \gg 2a$

$$\chi_{ss}^{NPc} = \lim_{T \to \infty} \left[\frac{2\pi^3 a^2 G_2 \kappa T}{3(N_c^2 - 1)} \cdot R_0 \alpha\right]. \tag{19}$$

Here G_2 denotes the gluon condensate, κ the weight of the confining correlator compared to the nonconfining correlator, a gives the correlation length and T the extension of the loop in Euclidean time.

The geometry of the arrangement enters into the factor α. The angle θ gives the tilting of the loop in the X_3, X_4 plane. The angle ϕ defines the angle of the $q\bar{q}$ connection in the X_1, X_3 plane.

$$\alpha = \sqrt{1 - \cos^2\phi \sin^2\theta}. \tag{20}$$

One recognizes that the confining interaction leads to a VEV of the tilted Wilson loop which is consistent with the area law for large distances R_0

$$<W[C]> = e^{-\sigma R_0 \alpha T} \tag{21}$$

$$\sigma = \frac{\pi^3 G_2 a^2 \kappa}{18}, \tag{22}$$

where σ is the string tension and the area is obtained from

$$Area = TR_0 \int_{-1/2}^{1/2} du \int_0^1 dv \sqrt{\left(\frac{dX_\mu}{du}\right)^2 \left(\frac{dX_\mu}{dv}\right)^2 - \left(\frac{dX_\mu}{du}\frac{dX_\mu}{dv}\right)^2} \tag{23}$$

$$= TR_0 \alpha. \tag{24}$$

For the Wegner-Wilson loop in Minkowski space-time we define χ_{ss} in the following way

$$\langle W[C]\rangle_G = \exp\left[-i\frac{C_2(3)}{2}\chi_{ss}\right]. \tag{25}$$

Minkowskian geometry enters via the factor

$$\alpha_M = \sqrt{1+\cos^2\phi\sinh^2\psi}, \tag{26}$$

which is consistent with the analytical continuation of the Euclidean expression $\alpha = 1-\cos^2\phi\sin^2\theta$ into Minkowski space by transforming the angle $\theta \to i\psi$. This analytical continuation is similar to the analytical continuation used in high-energy scattering [12, 13, 14], where the angle between two Wilson loops transforms in the same way.

The confining contribution to χ_{ss} reads in Minkowski space:

$$\chi_{ss}^{NPc} = \lim_{T\to\infty}\left[\frac{2\pi^3 a^2 G_2 \kappa T}{3(N_c^2-1)}\cdot R_0\alpha_M\right]. \tag{27}$$

In order to interpret this result, one must define the four velocities of the particles described by the tilted loop

$$u_\mu = (\gamma, 0_\perp, \gamma\beta). \tag{28}$$

The exponent giving the expectation value of the Wilson loop acquires a new meaning now, since $-ig\int d\tau A^\mu u_\mu = -ig\int d\tau(\gamma A^0 - \gamma\beta A^3)$, which leads in the VEV to a value for $\beta \approx 1$

$$<W_r[C]> = e^{-i\gamma(P^0-P^3)T}. \tag{29}$$

The light-cone energy arising from the confining part of the correlation function has the form

$$P^- = \frac{1}{\sqrt{2}}\left(\sigma R_0\sqrt{\cos(\phi)^2 + \sin(\phi)^2/\gamma^2}\right). \tag{30}$$

One sees that the Wilson loop for boosts with large γ indicates that the light-cone energy does not depend on the transverse distance $R_0 \sin\phi$ between the quarks. We introduce the relative + momentum k^+ and transverse momentum $k_\perp$ for the quarks with mass μ. By adding the above "potential" term to the kinetic term of relative motion of the two particles we complete the Hamiltonian P^-

$$P^- = \frac{(\mu^2+k_\perp^2)P}{2(1/4P^2-k^{+2})} + \frac{1}{\sqrt{2}}\sigma\sqrt{x_3^2+x_\perp^2/\gamma^2}. \tag{31}$$

Next, we multiply P^- with the plus component of the momentum P^+ and use that $P^+/M = \sqrt{2}\gamma M$ to eliminate the boost variable from the Hamiltonian. Further, we introduce the fraction $\xi = k^+/P^+$ with $|\xi| < 1/2$ and its conjugate the scaled longitudinal space coordinate $\sqrt{2}\rho = P^+ x_3$ as dynamical variables. For our configuration the relative time of the quark and antiquark is zero. Then we get the light-cone Hamiltonian in a Lorentz invariant manner, because the variables $\xi, \rho, k_\perp$ and $x_\perp$ are invariant under boosts

$$M^2 = 2P^+P^- = \frac{(\mu^2 + k_\perp^2)}{1/4 - \xi^2} + 2\sigma\sqrt{\rho^2 + M^2 x_\perp^2}. \tag{32}$$

To solve the M^2 operator one has to replace the square root operator by introducing an auxiliary parameter s of dimension mass squared and minimize M^2 with respect to variations of s. Final self consistency must be reached with a guessed mass eigenvalue M_0

$$M^2 = \frac{(\mu^2 + k_\perp^2)}{1/4 - \xi^2} + \frac{1}{2}\left(4\sigma^2 \frac{\rho^2 + M_0^2 x_\perp^2}{s} + s\right). \tag{33}$$

In addition, one has to put the self-energy correction calculated by Simonov [15] which is $\Delta\mu^2 = -4\sigma * f(m_q)/\pi$ and get

$$M^2 = \frac{(\mu^2 - \frac{4\sigma f}{\pi} + k_\perp^2)}{1/4 - \xi^2} + \frac{1}{2}\left(4\sigma^2 \frac{\rho^2 + M_0^2 x_\perp^2}{s} + s\right). \tag{34}$$

For light quarks the function $f(m_q)$ is close to unity. We have used the above equation with a simple trial function:

$$\psi(\xi, x_\perp) = N\cos(\xi\pi) e^{-\frac{x_\perp^2}{2x_0^2}}. \tag{35}$$

We obtain two solutions [16] with positive masses due to the s-minimization. One solution is very low in mass and the other rather high. By tuning $f(mq) = 0.8615$ away from unity the lower solution is pion-like with a really low mass, whereas the other solution lies at a typical hadronic scale

$$M_{low} = 0.138 GeV \tag{36}$$

$$M_{high} = 1.1 GeV. \tag{37}$$

Since on the light-cone the mechanism of chiral symmetry breaking is of particular interest, we would like to understand this result better. In the approach given here confinement plays an important role in contrast to Nambu-Jona-Lasinio effective models, which give an adequate description of spontaneous chiral symmetry breaking but do not include confinement.

The confining interaction in the light Hamiltonian was derived in the specific model of the stochastic vacuum. But it also can be inferred from the simple Lorentz transformation properties of the phase in the Wilson loop and a lattice determination of the tilted Wilson expectation values. In this respect the final Hamiltonian is model independent.

The inclusion of confining forces in the initial and final state wave functions can put all scattering cross sections calculated with the stochastic vacuum model on a much safer base, when wave functions and cross sections are derived consistently. For low Q^2 photon wave functions the long-distance part of the wave function matters strongly and confinement is important cf. [17]. Especially the diffractive cross section has a large contribution from large dipole sizes and a correct behavior can only be expected when the problem of the large dipole wave function is treated adequately. Another extension of the above calculation is the coupling of the initial $q\bar{q}$ state to higher Fock states $q\bar{q}g$ with gluons which can be calculated with Wilson loops near the light-cone in Minkowski space.

4. DISCUSSION AND CONCLUSIONS

I have tried to give some impression how QCD appears near the light-cone. I think we have now a calculational framework to approach the light-cone in a systematic way. It does not look much easier than equal time lattice gauge theory. One may hope that some simplifications arise in the process of studying it. The work on a Wilson loop near the light-cone looked very complicated and intransparent at the beginning, but it reduced to some simple form. I like this example because it shows how the vacuum acts near the light-cone.

REFERENCES

1. V. A. Franke, Y. V. Novozhilov, S. A. Paston and E. V. Prokhvatilov, arXiv:hep-th/0404031, Department of Theoretical Physics, St. Petersburg University (2004).
2. K. Itakura and S. Maedan, Prog. Theor. Phys. **105**, 537 (2001).
3. F. Lenz, K. Ohta, M. Thies and K. Yazaki, arXiv:hep-th/0403186, Theoretische Physik, Universität Erlangen (2004).
4. K. Hornbostel, S. J. Brodsky and H. C. Pauli, Phys. Rev. D **41**, 3814 (1990).
5. E. M. Ilgenfritz, Y. P. Ivanov and H. J. Pirner, Phys. Rev. D **62**, 054006 (2000).
6. H. W. L. Naus, H. J. Pirner, T. J. Fields and J. P. Vary, Phys. Rev. D **56**, 8062 (1997).
7. H. J. Pirner and F. Yuan, Phys. Rev. D **66**, 034020 (2002).
8. H. Verlinde and E. Verlinde, arXiv:hep-th/9302104.
9. I. Y. Arefeva, Phys. Lett. B **328** (1994) 411 [arXiv:hep-th/9306014].
10. M. Ilgenfritz, H.J. Pirner and E. Prokhvatilov, work in progress.
11. H. J. Pirner and N. Nurpeissov, arXiv:hep-ph/0404179.
12. E. Meggiolaro, Z. Phys. C **76**, 523 (1997).
13. E. Meggiolaro, Nucl. Phys. B **625**, 312 (2002).
14. A. Hebecker, E. Meggiolaro and O. Nachtmann, Nucl. Phys. B **571**, 26 (2000).
15. Y. A. Simonov, Phys. Lett. B **515**, 137 (2001).
16. O. Schlaudt and H.J.Pirner, work in progress.
17. H. G. Dosch, T. Gousset and H. J. Pirner, Phys. Rev. D **57**, 1666 (1998).

Confinement of quarks in dual superconductor models

Georges Ripka

Service de Physique Théorique, Centre d'Etudes de Saclay
F-91191 Gif-sur-Yvette Cedex, France
E-mail: ripka@cea.fr

Abstract. We review some aspects and problems of quark confinement models which are based on a dual super-conductor description of the QCD vacuum.

Keywords: confinement, quark confinement, dual superconductor, color flux tube
PACS: 12.38.Lg, 12.39.Ki

1. ELECTRIC CHARGES EMBEDDED IN A DUAL SUPERCONDUCTOR

Consider first a magnetic dipole composed of a positive (north) magnetic pole g and a negative (south) magnetic pole $-g$, separated by a distance R. The magnetic field lines stem from the north positive magnetic pole and terminate on the south negative magnetic pole. If the dipole rests in a normal (non-superconducting) medium, then, as the distance R increases, the magnetic field lines extend further and further away from the line joining the magnetic poles and the energy of the system varies as $-g^2/R$. Now imagine that the system is embedded in a superconductor. The Meissner effect attempts to eliminate the magnetic field. However, because of Gauss's law, the flux of the magnetic field through a surface surrounding one of the poles must equal the magnetic charge inside the sphere. So the Meissner effect cannot completely expel the magnetic field from the superconductor. The best it can do, to minimize the energy, is to compress the magnetic field lines into a minimal space, thereby creating a thin flux tube joining the magnetic poles in a straight line. As the distance between the poles increases, the flux tube becomes longer but it maintains a minimal thickness. A magnetic field runs parallel to the flux tube and maintains a constant profile in the perpendicular direction. The mere geometry of the flux tube ensures that the energy increases linearly with R thereby creating a linearly confining potential between the magnetic charges. This is, in essence, one way in which confinement of quarks in QCD was modeled in the mid-seventies, in pioneering works of Nielsen and Olesen [1], Nambu [2], Creutz [3], 't Hooft [4] and Mandelstam [5]. For a recent review, see [6]. It is only in the past few years that lattice calculations have provided some evidence for such a scenario.

In QCD, the scenario needs to be adapted because quarks carry color-electric (not magnetic) charges. Color-charged quarks will be confined if the QCD ground state is a *dual* superconductor, possibly caused by a condensation of color-magnetic monopoles. By dual superconductor, we mean a superconductor, in which the roles of the electric and magnetic fields are exchanged: $\vec{E} \to \vec{H}$, $\vec{H} \to -\vec{E}$. The Meissner effect will then suppress the electric (rather than the magnetic) field so that, when two equal and opposite *electric* charges are embedded in a dual superconductor, a flux tube will form, thereby producing a linearly rising potential between the electric charges. The superconducting phase can be described by the Landau and Ginzburg model [7] with one important difference. The Landau-Ginzburg theory of superconductivity of metals is expressed in terms of the (photon) gauge field A^μ associated to the electromagnetic field tensor $F^{\mu\nu} = \partial^\mu A^\nu - \partial^\nu A^\mu$. If we want to use a Landau-Ginzburg lagrangian to describe a *dual* superconductor, we need to express the lagrangian in terms of a gauge field B^μ which is associated to the *dual* field tensor $\overline{F}^{\mu\nu} = \frac{1}{2}\varepsilon^{\mu\nu\alpha\beta} F_{\alpha\beta} = \partial^\mu B^\nu - \partial^\nu B^\mu$. But here we run into a problem. In the presence of both electric and magnetic charges, the Maxwell equations read:

$$\partial_\nu F^{\nu\mu} = j^\mu \qquad \partial_\nu \bar{F}^{\nu\mu} = j^\mu_{mag} \qquad (1)$$

But if we write $\overline{F}^{\mu\nu} = \partial^\mu B^\nu - \partial^\nu B^\mu$, then $\partial_\nu F^{\nu\mu} = -\frac{1}{2}\partial_\nu \varepsilon^{\mu\nu\alpha\beta}(\partial_\alpha B_\beta - \partial_\beta B_\alpha) = 0$ so that the electric current $j^\mu = \partial_\nu F^{\nu\mu}$ vanishes identically. In other words, the theory cannot accommodate the electric charges we wished to confine.

This problem was solved by Dirac in 1948 in his famous paper on magnetic monopoles [8]. He faced the opposite problem: when electrodynamics is expressed in terms of a vector potential A^μ associated to the field tensor $F^{\mu\nu}$, we have $\partial_\nu \bar{F}^{\nu\mu} = 0$ and the magnetic current vanishes. Usual electromagnetism cannot accommodate magnetic charges and, indeed, they have never been observed. However, Dirac refused to give up the beautiful symmetry of Maxwell equations with respect to electric and magnetic currents and he solved the problem with the so-called Dirac strings. We can adopt his solution, *mutatis mutandis*, by modifying the expression $\overline{F}^{\mu\nu} = \partial^\mu B^\nu - \partial^\nu B^\mu$ by the adjunction of a so-called *Dirac string term*. The dual field tensor $\bar{F}$ is written:

$$\overline{F}^{\mu\nu} = \partial^\mu B^\nu - \partial^\nu B^\mu + \bar{G}^{\mu\nu} \qquad (2)$$

where the Dirac string $G^{\mu\nu}$ is independent of the gauge field B^μ and is required to satisfy the equation $\partial_\nu G^{\nu\mu} = j^\mu$. This solves our problem. Indeed, the dual of equation (2) reads $F^{\mu\nu} = -\frac{1}{2}\varepsilon^{\mu\nu\alpha\beta}(\partial_\alpha B_\beta - \partial_\beta B_\alpha) + G^{\mu\nu}$ so that the first Maxwell equation reads $\partial_\nu F^{\nu\mu} = \partial_\nu G^{\nu\mu} = j^\mu$. This equation must be satisfied by $G^{\mu\nu}$, independently of B^μ. The gauge potential B^μ can be calculated in the presence of magnetic charges, by solving the second Maxwell equation, which now reads:

$$\partial_\nu (\partial^\nu B^\mu - \partial^\mu B^\nu) + \partial_\nu \bar{G}^{\nu\mu} = j^\mu_{mag} \qquad (3)$$

A solution of the equation $\partial_\nu G^{\nu\mu} = j^\mu$ which has the familiar form $G^{\mu\nu} = \partial^\mu A^\nu - \partial^\nu A^\mu$ exists, but it will not do, because if such a solution is inserted into the lagrangian, the magnetic charges decouple from the system. Dirac invented another solution, namely a string solution. Consider the case of two static equal and opposite electric charges situated at the positions $\vec{R}_1$ and $\vec{R}_2$. The electric current is then:

$$j^\mu = (\rho, \vec{j}), \quad \vec{j} = 0, \quad \rho(\vec{r}) = e\delta(\vec{r} - \vec{R}_1) - e\delta(\vec{r} - \vec{R}_2) \qquad (4)$$

The string term $G^{\mu\nu}$ is an antisymmetric tensor which may be expressed in terms of two vectors $\vec{E}_{st}$ and $\vec{H}_{st}$:

$$\vec{E}^i_{st} = -G^{0i} \quad \vec{H}^i_{st} = -\frac{1}{2}\varepsilon^{0ijk}G_{jk} \qquad (5)$$

The string solution of equation $\partial_\nu G^{\nu\mu} = j^\mu$ is then:

$$\vec{E}^i_{st}(\vec{r}) = e\int_L d\vec{R}\,\delta\left(\vec{r}-\vec{R}\right) \quad \vec{H}^i_{st}(\vec{r}) = 0 \qquad (6)$$

where the integral runs along a path L, which stems from the positive charge e and terminates on the negative charge $-e$. The path is, of course, the Dirac string, an infinitely thin line which carries the flux of the electric field from the positive charge over to the negative charge.

Much has been written about Dirac strings, most of it by Dirac himself in his original paper. The string can be displaced at will, because a modification of the string can be shown to be equivalent to a gauge transformation and it does not modify the shape of the flux tubes which are formed between the quarks. One can also show that if an electron flies by a Dirac string, its wavefunction develops a different phase depending on whether it passes on the right or on the left of the string. The string would therefore become observable, unless the electric and magnetic charges e and g are related by the famous Dirac charge-quantization condition $eg = 2n\pi$. This is not the subject of our discussion.

2. THE LANDAU-GINZBURG MODEL OF A DUAL SUPERCONDUCTOR

The Landau-Ginzburg model of a dual superconductor is often referred to by particle physicists as the dual abelian Higgs model. The model lagrangian is:

$$L = -\frac{1}{4}\left(\partial_\mu B_\nu - \partial_\nu B_\mu + \bar{G}_{\mu\nu}\right)^2 + \frac{g^2 S^2}{2}\left(B_\mu + \partial_\mu\varphi\right)\left(B^\mu + \partial^\mu\varphi\right)$$
$$+ \frac{1}{2}(\partial_\mu S)(\partial^\mu S) - \frac{b}{2}\left(S^2 - v^2\right)^2 \qquad (7)$$

- B^μ is the gauge field associated to the dual field tensor $\bar{F}^{\mu\nu} = \partial^\mu B^\nu - \partial^\nu B^\mu + \bar{G}^{\mu\nu}$.
- S is a scalar field.
- $\frac{b}{2}\left(S^2 - v^2\right)^2$ is a potential term which forces the scalar field S to acquire a non vanishing expectation value $S = v$ in the ground state of the system.
- φ is a field which, in fact, represents the phase of the complex scalar field $\psi = e^{ig\varphi}S$ in terms of which the Landau-Ginzburg lagrangian is usually expressed.
- g is a magnetic charge.
- $\bar{G}^{\mu\nu} = \varepsilon^{\mu\nu\alpha\beta}G_{\alpha\beta}$ is the dual of the Dirac string term. It couples the system to the electric current j^μ because it is required to satisfy the equation $\partial_\alpha G^{\alpha\mu} = j^\mu$.

The scalar field represents some sort of condensation of color-magnetic charges, or monopoles as they are called, and the exact nature of the monopole condensation is a subject of current research [9]. It is possible to make a gauge transformation which eliminates the phase φ. The gauge in which $\varphi = 0$ is called the unitary gauge and in this gauge, the Dirac string runs along the center of the flux tube. This has not always been fully appreciated.

The scalar field S develops a mass $m_H = 2v\sqrt{b} \simeq 1.36\,GeV$ which particle physicists call a Higgs mass and the gauge field B^μ develops a mass $m_V = gv \simeq 1.31\,GeV$. The inverse masses are the penetrations lengths of the dual superconductor. The near equal values of the Higgs and vector masses are obtained [10] by fitting the model parameters to the lattice calculation shown on Fig.1. They suggest that the QCD vacuum is close to the boundary separating type I and type II superconductors.

Consider the system composed of two static color-electric charges discussed above. The charge current and density are given by (4) and the string term is given by the expression (6). After a little algebra, we can obtain the following expression for the energy of the system:

$$\mathcal{E}\left(\vec{B},S\right) = -\frac{e^2}{4\pi R} + \int d^3r \left[\frac{1}{2}\left(-\vec{\nabla}\times\vec{B} + \vec{\nabla}\times\vec{B}_0\right)^2 + \frac{g^2 S^2}{2}\vec{B}^2\right.$$

$$\left. + \frac{1}{2}\left(\vec{\nabla}S\right)^2 + \frac{b}{2}\left(S^2 - v^2\right)^2\right] \quad (8)$$

where $\vec{B}_0$ is the line integral:

$$\vec{B}_0(\vec{r}) = \frac{e}{4\pi}\vec{\nabla}_r \times \vec{e}_z \int_{-\frac{R}{2}}^{\frac{R}{2}} dz' \frac{1}{\sqrt{\rho^2 + (z-z')^2}} \quad (9)$$

which can be evaluated analytically. The energy (8) is a functional of two fields, namely the vector gauge field $\vec{B}(\vec{r})$ and the scalar field $S(\vec{r})$. It is expressed in the unitary gauge, in which the string term runs along the center of the flux tube. It is then a simple matter to calculate the fields $\vec{B}(\vec{r})$ and $S(\vec{r})$ which minimize the energy. The energy is minimized by fields which, in cylindrical coordinates, have the form $S(\vec{r}) = S(\rho,z)$ and $\vec{B}(\vec{r}) = \vec{e}_\theta B(\rho,z)$. When the distance R between the electric charges is much larger than the width of the flux tube, and far from the electric charges, the fields depend only on the distance ρ from the center of the flux tube:

$$S(\vec{r}) = S(\rho) \quad \vec{B}(\vec{r}) = \vec{e}_\theta B(\rho) \quad \left(-\frac{R}{2} \ll z \ll \frac{R}{2}\right) \quad (10)$$

In this region, the electric field runs parallel to the z-axis:

$$\vec{E}(\vec{r}) = -\vec{\nabla}\times\vec{B}(\vec{r}) = -\vec{e}_z \frac{1}{\rho}\frac{d}{d\rho}(\rho B) \quad (11)$$

and a magnetic current $\vec{k}_\theta$ circulates around the flux tube:

$$\vec{k}_\theta(\vec{r}) = -\vec{\nabla} \times \vec{E} = -\vec{e}_\theta \frac{d}{d\rho}\frac{1}{\rho}\frac{d}{d\rho}(\rho B) \tag{12}$$

This is precisely what is found in lattice calculations, as shown on Fig.1.

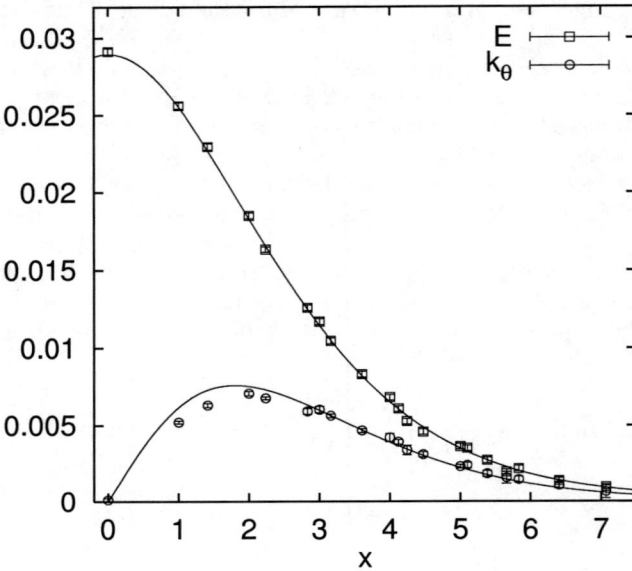

FIGURE 1. Profiles of the electric field E and of the magnetic current k_θ plotted against the distance x to the centre of the flux tube. The curves are obtained from the Landau-Ginzburg model and the data are from lattice calculations in the in the maximal abelian gauge [10].

We can try to include dynamical quark fields $q(x)$ by adding the term $\bar{q}(i\partial_\mu \gamma^\mu - m)q$ to the lagrangian (7). The Dirac string solution for the source term might be:

$$\vec{E}_{st}(x) = \int_{s_1}^{s_2} ds \left(\frac{d\vec{R}}{ds}\rho(x - Z(s)) - \frac{dT}{ds}\vec{j}(x - Z(s)) \right)$$

and it satisfies the equation $\vec{\nabla} \cdot \vec{E}_{st} = \rho$ with $\rho = q^\dagger q$. This leads to a lagrangian which is quadratic in the fermion fields, and which we could quantize assuming, for example, that the other fields remain classical. However, it remains to be proved that the energy functional obtained in this way is independent of the string shape. It is indeed tantalizing that, so far, nobody has been able to calculate the spectrum of bound light $q\bar{q}$ states in such a model.

REFERENCES

1. H. B. Nielsen and P. Olesen, "Vortex-line models for dual strings", *Nucl. Phys.* **B61**, 45, 1973.
2. Y. Nambu, "Strings, monopoles, and gauge fields", *Phys. Rev.* **D10**, 4262, 1974.

3. M. Creutz, *Phys. Rev. D10*, 2696, 1974.
4. G. t'Hooft, *High Energy Physics*, Editorice Compositori Bologna, 1975.
5. S. Mandelstam, *Phys. Rep. C23*, 245, 1976.
6. Georges Ripka, *Dual Superconductor Models of Color Confinement*, Lecture Notes in Physics LNP 639, Springer Verlag Berlin Heidelberg, 2004, hep-ph/0310102.
7. V.L. Ginzburg and L.D. Landau, *Zh. Eksp. Teor. Fiz. 20*, 1064, 1950. An English translation may be found in "Men of Physics: L.D. Landau", vol.I , p.138.
8. P.A.M. Dirac, *Phys. Rev. 74*, 817, 1948.
9. J.M. Carmona, M. D'Elia, A. Di Giacomo, B. Lucini and G. Paffuti, "Color confinement and dual superconductivity of the vacuum - III", *Phys. Rev. D64*, 114507, 2001, hep-lat/0103005.
10. Gunnar S. Bali, Christoph Schlichter and Klaus Schilling, "Probing the QCD vacuum with static sources in maximal abelian projection", *Progr. Theor. Phys. Suppl. 131*, 645, 1998, hep-lat/9802005.

LIST OF PARTICIPANTS

- **Arleo François**, LPTHE, 4 place Jussieu, F 75252 Paris cedex 05, France;
  arleo@lpthe.jussieu.fr
- **Beyer Michael**, Fachbereich Physik, Universität Rostock, 1 Universitätplatz, D 18051 Rostock, Germany;
  michael.beyer@uni-rostock.de
- **Blaschke David**, GSI Darmstadt, Planckstr. 1, D 64291 Darmstadt, Germany, and
  Bogoliubov Laboratory for Theoretical Physics, JINR Dubna, RU 141980 Dubna, Russia
  Blaschke@theory.gsi.de
- **Braun Jens**, Institut für Theoretische Physik, Universität Heidelberg, Philosophenweg 19, D 69120 Heidelberg, Germany;
  J.Braun@tphys.uni-heidelberg.de
- **Costa Pedro**, Centro de Física Teórica, Departamento de Física, Faculdade de Ciências e Tecnologia, Universidade de Coimbra, Rua Larga, P 3004 516, Portugal;
  pcosta@teor.fis.uc.pt
- **Courtoy Aurore**, Physique théorique fondamentale, Bât. B5, Université de Liège, 17 allée du 6 Août, B 4000 Liège 1, Belgium;
  Aurore.Courtoy@student.ulg.ac.be
- **Cudell Jean-René**, Physique théorique fondamentale, Bât. B5, Université de Liège, 17 allée du 6 Août, B 4000 Liège 1, Belgium;
  JR.Cudell@ulg.ac.be
- **Cugnon Joseph**, Physique théorique fondamentale, Bât. B5, Université de Liège, 17 allée du 6 Août, B 4000 Liège 1, Belgium;
  J.Cugnon@ulg.ac.be
- **Dechambre Alice**, Physique théorique fondamentale, Bât. B5, Université de Liège, 17 allée du 6 Août, B 4000 Liège 1, Belgium;
  Alice.Dechambre@student.ulg.ac.be
- **Gérard Jean-Marc**, FYMA, Université Catholique de Louvain, 2 chemin du Cyclotron, B 1348 Louvain-la-Neuve, Belgium;
  gerard@fyma.ucl.ac.be
- **Goeke Klaus**, Fakultät für Physik und Astronomie, Institut fur Theoretische Physik II, Ruhr-Universität Bochum, 44780 Bochum, Germany;
  klaus.goeke@ruhr-uni-bochum.de
- **Grigorian Hovik**, Fachbereich Physik, Universität Rostock, 1 Universitätplatz, D 18051 Rostock, Germany;
  hovik.grigorian@uni-rostock.de
- **Grünewald Daniel**, Institut für Theoretische Physik, Universität Heidelberg, Philosophenweg 19, D 69120 Heidelberg, Germany;
  daniel@tphys.uni-heidelberg.de

- **Guzey Vadim**, Fakultät für Physik und Astronomie, Institut fur Theoretische Physik II, Ruhr-Universität Bochum, 44780 Bochum, Germany; vadimg@tp2.rub.de
- **Hüfner Joerg**, Institut für Theoretische Physik, Universität Heidelberg, Philosophenweg 19, D 69120 Heidelberg, Germany; huefner@tphys.uni-heidelberg.de
- **Iancu Edmond**, Service de Physique Théorique, CEA/DSM/SPhT Unité de recherche associée au CNRS, Centre d'Etudes de Saclay, F-91191 Gif-sur-Yvette Cedex, France; iancu@spht.saclay.cea.fr
- **Itakura Kazunori**, Service de Physique Théorique, CEA/DSM/SPhT Unité de recherche associée au CNRS, Centre d'Etudes de Saclay, F-91191 Gif-sur-Yvette Cedex, France; itakura@spht.saclay.cea.fr
- **Kalinovsky Yura**, Physique théorique fondamentale, Bât. B5, Université de Liège, 17 allée du 6 Août, B 4000 Liège 1, Belgium; kalinov@qcd.theo.phys.ulg.ac.be
- **Kirschner Roland**, Institut für Theoretische Physik, Universität Leipzig, 10 Augustusplatz, D 04109 Leipzig, Germany roland.kirschner@itp.uni-leipzig.de
- **Kivel Nikolai**, Fakultät für Physik und Astronomie, Institut fur Theoretische Physik II, Ruhr-Universität Bochum, 44780 Bochum, Germany; nikolai.kivel@tp2.ruhr-uni-bochum.de
- **Klein Bertram**, Institut für Theoretische Physik, Universität Heidelberg, Philosophenweg 19, D 69120 Heidelberg, Germany; bklein@tphys.uni-heidelberg.de
- **Lansberg Jean-Philippe**, Physique théorique fondamentale, Bât. B5, Université de Liège, 17 allée du 6 Août, B 4000 Liège 1, Belgium; JPH.Lansberg@ulg.ac.be
- **Lorcé Cédric**, Physique théorique fondamentale, Bât. B5, Université de Liège, 17 allée du 6 Août, B 4000 Liège 1, Belgium; C.Lorce@ulg.ac.be
- **Matagne Nicolas**, Physique théorique fondamentale, Bât. B5, Université de Liège, 17 allée du 6 Août, B 4000 Liège 1, Belgium; nmatagne@ulg.ac.be
- **Pirner Hans**, Institut für Theoretische Physik, Universität Heidelberg, Philosophenweg 19, D 69120 Heidelberg, Germany; pir@tphys.uni-heidelberg.de
- **Polyakov Maxim V.**, Physique théorique fondamentale, Bât. B5, Université de Liège, 17 allée du 6 Août, B 4000 Liège 1, Belgium; Maxim.Polyakov@ulg.ac.be
- **Raufeisen Joerg**, Institut für Theoretische Physik, Universität Heidelberg, Philosophenweg 19, D 69120 Heidelberg, Germany; J.Raufeisen@tphys.uni-heidelberg.de

- **Ripka Georges**, Service de Physique Théorique, CEA/DSM/SPhT Unité de recherche associée au CNRS, Centre d'Etudes de Saclay, F-91191 Gif-sur-Yvette Cedex, France;
  ripka@cea.fr
- **Röpke Gerd**, Fachbereich Physik, Universität Rostock, 1 Universitätplatz, D 18051 Rostock, Germany;
  gerd.roepke@physik.uni-rostock.de
- **Schmidt Erik**, Fachbereich Physik, Universität Rostock, 1 Universitätplatz, D 18051 Rostock, Germany;
  schmidt@rubin.physik2.uni-rostock.de
- **Semenov Kirill**, Physique théorique fondamentale, Bât. B5, Université de Liège, 17 allée du 6 Août, B 4000 Liège 1, Belgium;
  cyr_stsh@mail.ru
- **Soyez Grégory**, Service de Physique Théorique, CEA/DSM/SPhT Unité de recherche associée au CNRS, Centre d'Etudes de Saclay, F-91191 Gif-sur-Yvette Cedex, France, and
  Physique théorique fondamentale, Bât. B5, Université de Liège, 17 allée du 6 Août, B 4000 Liège 1, Belgium;
  g.soyez@ulg.ac.be
- **Stancu Floarea**, Physique théorique fondamentale, Bât. B5, Université de Liège, 17 allée du 6 Août, B 4000 Liège 1, Belgium;
  fstancu@ulg.ac.be
- **Stassart Pierre**, Physique théorique fondamentale, Bât. B5, Université de Liège, 17 allée du 6 Août, B 4000 Liège 1, Belgium;
  Pierre.Stassart@ulg.ac.be
- **Strakovsky Igor I.**, Center for Nuclear Studies, Department of Physics, The George Washington University, Washington, D.C. 20052, USA
  igor@gwu.edu
- **Strauß Stefan**, Fachbereich Physik, Universität Rostock, 1 Universitätplatz, D 18051 Rostock, Germany;
  stefan.strauss@uni-rostock.de
- **Szymanowski Lech**, Physique théorique fondamentale, Bât. B5, Université de Liège, 17 allée du 6 Août, B 4000 Liège 1, Belgium, and
  Department of Nuclear Reactions, Andrzej Soltan Institute for Nuclear Studies, Nuclear Theory Department Hoza 69, PL-00-681 Warsaw, Poland;
  lechszym@fuw.edu.pl
- **Vanderheyden Benoît**, Institut Montefiore, Bât. B28 Université de Liège, 10 Grande Traverse, B 4000 Liège 1, Belgium;
  B.Vanderheyden@ulg.ac.be
- **Yudichev Valery L.**, Joint Institute for Nuclear Research (JINR), Joliot-Curie str. 6, BLTP, 141980 Dubna, Russia;
  yudichev@thsun1.jinr.ru

Author Index

A

Anikin, I. V., 51
Arleo, F., 133
Arndt, R. A., 41
Azimov, Y. I., 41

B

Beyer, M., 139, 212
Bissey, F., 61
Blaschke, D., 151
Braun, J., 162, 193

C

Costa, P., 173
Cudell, J. R., 61
Cugnon, J., 61, 235

D

de Sousa, C. A., 173

F

Frederico, T., 139

G

Gies, H., 162
Grigorian, H., 182
Grünewald, D., 71
Guzey, V., 3

H

Hüfner, J., 245

I

Iancu, E., 97
Itakura, K., 111

J

Jackson, A. D., 220
Jaminon, M., 61

K

Kalinovsky, Y. L., 173
Kirschner, R., 120
Kivel, N., 81
Klein, B., 193

L

Lansberg, J. P., 11, 61

M

Matagne, N., 22
Mattiello, S., 139, 212

P

Pire, B., 51
Pirner, H.-J., 162, 193, 252
Polyakov, M. V., 41

R

Raufeisen, J., 202
Ripka, G., 262
Ruivo, M. C., 173

S

Soyez, G., 88
Stancu, F., 22, 32
Stassart, P., 61
Strakovsky, I. I., 41
Strauß, S., 139, 212
Szymanowski, L., 51

T

Teryaev, O. V., 51

V

Vanderheyden, B., 220

W

Wallon, S., 51
Weber, H. J., 139
Workman, R. L., 41